EARTH COMMUNITY
EARTH ETHICS

ECOLOGY AND JUSTICE

An Orbis Series on Global Ecology

Advisory Board Members
Mary Evelyn Tucker
John A. Grim
Leonardo Boff
Sean McDonagh

The Orbis *Ecology and Justice* Series publishes books that seek to integrate an understanding of the Earth as an interconnected life-system with concerns for just and sustainable systems that benefit the entire Earth. Books in the Series concentrate on ways to:

- reexamine the human-Earth relationship in the light of contemporary cosmological thought
- develop visions of common life marked by ecological integrity and social justice
- expand on the work of those who are developing such fields as eco-social ecology, bioregionalism, and animal rights
- promote inclusive participative strategies that enhance the struggle of the Earth's voiceless poor for justice
- deepen appreciation for and expand dialogue among religious traditions on the issue of ecology
- encourage spiritual discipline, social engagement, and the reform of religion and society toward these ends.

Viewing the present moment as a time for responsible creativity, the Series seeks authors who speak to ecojustice concerns and who bring into dialogue perspectives from the Christian community, from the world's other religions, from secular and scientific circles, and from new paradigms of thought and action.

ECOLOGY AND JUSTICE SERIES

EARTH COMMUNITY
EARTH ETHICS

Larry L. Rasmussen

ORBIS BOOKS

Maryknoll, New York 10545

Fourth printing, October 2001

The Catholic Foreign Mission Society of America (Maryknoll) recruits and trains people for overseas missionary service. Through Orbis Books, Maryknoll aims to foster the international dialogue that is essential to mission. The books published, however, reflect the opinions of their authors and are not meant to represent the official position of the society.

Manufactured in the United States of America

Library of Congress Cataloging-in-Publication Data

Rasmussen, Larry L.
    Earth community earth ethics / Larry L. Rasmussen.
       p. cm. – (Ecology and justice)
    Includes bibliographical references and indexes.
    ISBN 1-57075-086-6
    1. Human ecology – Religious aspects – Christianity. 2. Human ecology – Moral and ethical aspects. 3. Christian ethics. 4. Social ethics. I. Title. II. Series.
    BT695.5.R37 1996
    261.8′362–dc20
                                        96-26923
                                          CIP

*With gratitude
to Unit III (Justice, Peace, Creation)
of the World Council of Churches,
and its ancestors*

Variety is the pledge that matter makes to living things.
Think of a niche and life will fill it,
think of a shape and life will explore it,
think of a drama and life will stage it.
— Diane Ackerman

The time of the finite world has come
in which we are under house arrest.
Nature no longer exists in the classic sense of the term.
— Boutros Boutros-Ghali

The heart, after all, is raised on a mess of stories.
Then it writes its own.
— Joe Wood

Contents

Preface

There is a true yearning to respond to
The singing River and the wise Rock.
So say the Asian, the Hispanic, the Jew
The African, the Native American, the Sioux,
The Catholic, the Muslim, the French, the Greek,
The Irish, the Rabbi, the Priest, the Sheik,
The Gay, the Straight, the Preacher,
The privileged, the homeless, the Teacher.
They hear. They all hear
The speaking of the Tree.
They hear the first and last of every Tree
Speak to humankind today.
Come to me,
Here beside the River.
Plant yourself beside the River.
. . . .
The horizon leans forward,
Offering you space
To place new steps of change.
Here, on the pulse of this fine day
You may have the courage
To look up and out and upon me,
The Rock, the River, the Tree, your country.
. . . .
Here, on the pulse of this new day
You may have the grace to look up and out
And into your sister's eyes,
And into your brother's face,
Your country,
And say simply
Very simply
With hope —
Good morning.[1]

---

1. Maya Angelou, "On the Pulse of Morning" (New York: Random House, 1993), n.p. The poem is copyright ©1993 by Maya Angelou. This and subsequent quotations from the poem are reprinted by permission of Random House, Inc.

These pages are a long echo of Maya Angelou's poem. The key of the music changes now and then. And the form, regrettably, is not poetry. But all is
in keeping with the poet's intention. "Our Country, the Planet,"[2] home to the
Rock, the River, the Tree, the Peoples, is the shared horizon.

This began differently, as a work in the busy corners of environmental ethics.
But before the first chapter had dried, the subject had changed. To earth and
its distress.

The difference is not small. "The environmental crisis" does not adequately
describe what ails us. "Environment" means that which surrounds us. It is a
world separate from ourselves, outside us. The true state of affairs, however, is
far more interesting and intimate. The world around us is also within. We are
an expression of it; it is an expression of us. We are made of it; we eat, drink,
and breathe it. And someday, when dying day comes, we will each return the
favor and begin our role as a long, slow meal for a million little critters. Earth is
bone of our bone and flesh of our flesh. This is not "environment" so much as
the holy mystery of creation, made for and by all earth's creatures together.

None of this intimacy is carried by the word "environment." Nor does our
responsibility ring as clear as it ought when we name our woe "the environmental
crisis" and offer "environmental ethics" as the antidote. By way of contrast, no
talk in these pages of "A Rock, A River, A Tree" excludes the fate of "the Asian,
the Hispanic, the Jew, the African, the Native American, the Sioux." Or vice
versa. The integrity of creation forbids division of human from otherkind or
human from one another.

So does consideration of earth's agonies. Thriving, even surviving, now lives
with a terrifying insight. Earth — all of it — is a community without an exit.
Our problems — people-to-people and humankind-to-otherkind — are genuinely ours all together, for worse and for better. The key terms in this text —
"sustainability," "earth faith," "earth ethics," "the integrity of creation" — all assume that the "bordered countr[ies]" of which Angelou speaks are no longer
truly bordered at all. Acid rain falls on the just and unjust alike. "Armed struggles
for profit" leave "collars of waste" upon earth's shores and "currents of debris"[3]
upon her breast, without worrying over passports and fences. The world, all of
it, has become game and booty and landfill.

But if this is a work about earth and its distress, rather than a volume in
environmental ethics, what kind of work is it? The thesis is clear enough: our
most basic impulses and activities must now be measured by one stringent criterion — their contribution to an earth ethic and their advocacy of sustainable
earth community. Fidelity to earth is the plea here.

In some curious sense that may render this a work in religious ethics. It argues
for a dedication to earth in the manner of the sacred and sacramental and cou-
ples this with a sense of wonder that is protective of all life. Too, the moral yield

2. The phrase is not Angelou's but that of Shridath Ramphal, from his book *Our Country, the Planet* (Washington, D.C.: Island Press, 1992).
3. Angelou, "On the Pulse of Morning," n.p.

drawn upon here is religiously rooted and watered. These pages move within the open circle of ethical monotheism. That means Judaism and Christianity principally. But other traditions sing as well.

So is this a book in religious ethics? Yes, if the way we understand reality and our place in it sits somewhere near the center of a religious ethic and marks it (as it does). Where we belong in the scheme of things and how we exercise our powers are certainly preoccupations of this book. At the same time, whole chapters here could appear just as easily in works of history, science, economics, or policy studies, all without bothering religion or ethics.

Why are they here then? Because something objective is happening to the planet that needs to be accounted for but is not well understood.

Explanations vary and compete. Some fasten on the hyperactive centuries trailing in the wake of the industrial revolution. These explanations argue that industrializing, for all its benefits, also holds the key to incremental destruction. Others name conquest and colonization and reach farther back than the steam engine and assembly line to five hundred years of Western-based globalization. It was adventurous sailors and soldiers and merchants who began globalizing the economy that now threatens the environment. Still others contend for yet deeper causes. Earth's agonies are sunk in soils as aged as the neolithic revolution and the rise of city-states. The "carrying capacity" of one locale was appropriated by folks in another. On a large planet with few people that may have been benign. On a small one peopled by billions it is fevered and fateful. In any event, historical sweep, the reports of science, policy debates, and proposals are gathered here as an analysis, an "Earth Scan" (part 1). Works in religious ethics often assume such analysis or put it to the side as something other than strict matters of faith, cult, and morality. This work does neither. A beloved world is steadily being destroyed by cumulative human activities. This is a compelling matter for religious ethics if ever there was one, in the same moment that it is a matter for other kinds of analysis.

A sustainable future — the very purpose for writing this volume — requires more than analysis, however. Constructive, inspired response is a life imperative now. Humanity's future is no longer guaranteed by (the rest of) nature. It must be kept open by humans themselves.[4] Much thus turns on the substance of "Earth Faith" and "Earth Action" (parts 2 and 3), those fundamental enterprises that orient human vocation. What makes for sustainability — social, environmental, spiritual, and moral sustainability — becomes the question. The answer will entail proposals both innovative and realistic. Such proposals themselves will, if they would see the light of day, require the imagination and tenacity characteristic of religious energy and devotion.

Yet where to shelve this book does not matter. What matters is whether these pages help untangle the skein of issues that allow us so little wiggle room these

---

4. Lukas Vischer, "The Theme of Humanity and Creation in the Ecumenical Movement," in *Sustainable Growth: A Contradiction in Terms?* report of the Visser't Hooft Memorial Consultation (Geneva: WCC Publications, 1993), 85.

days, issues that also frustrate every good leader's efforts to rouse and encourage our better angels.

Considering that we are very far from safe shores and uncertain how to get there, readers may find both the hopeful tone of these pages and its undisguised faith a little surprising, especially since many chapters are not the least reticent about registering bad news. Readers may find the specific commitment to Christian faith especially surprising, given Western Christianity's complicity in modernity's destructiveness. Surprising or not, the intent is to convey the presence of a healing and sustaining power that runs deep in life itself, a certain disposition and grace that mean to bring the reader, when she or he turns the last page, to one thing only. To look up and around and

> Say simply
> Very simply
> With hope —
> Good morning.

Perhaps this *is* a work in religious ethics.

Acknowledgments

Any book worth the tree it still carries is a community enterprise, even when only ten fingers work the keyboard. This volume is no different. The number complicit is large and their advice and counsel considerable. Not all were able to reply within the confines of a cramped schedule, but all agreed to read portions or all of what follows. So I thank Heinrich Bedford-Strohm, Elizabeth Bounds, Ellen Davis, Judith Diers, Mark Edwards, Wesley Granberg-Michaelson, Paul Gorman, George Tinker, Lisa Stoen Hazelwood, Beverly Wildung Harrison, Dieter Hessel, David Hallman, Carol Johnston, Shannon Jung, James Martin-Schramm, Gary Matthews, Vernice Miller, James Nash, Michael O'Connor, Janet Parker, Anna Mae Patterson, Richard Pemberton, Dudley Raynal, Andrew Rasmussen, Kusumita Pederson, Daniel Spencer, Stephen Viederman, Rob van Drimmelen, David Wellman, and Vincent Wimbush. I owe them all in the way friends do. Likewise I owe far-flung audiences for responses to this or that portion of what has become this book: at the Universities of Uppsala and Lund, Sweden, as well as the Stockholm School of Theology; the University of Michigan, Ann Arbor; Drake University, Des Moines; Ghost Ranch, Abiquiu, New Mexico; Holden Village, Chelan, Washington; the Unit III (Justice, Peace, Creation) meeting of the World Council of Churches in Larnaca, Cyprus; the Seventh International Bonhoeffer Congress in Cape Town, South Africa; the Society of Christian Ethics annual meeting in Washington, D.C.; at the conference entitled "Theological Education to Meet the Environmental Challenge" in Stony Point, New York; the Theology for Earth Community Conference at Union Theological Seminary, New York, Pittsburgh Theological Seminary for the 1996 Schaff Lectures, the National Religious Partnership for the Environment Retreat in Greensboro, North Carolina; and in forums in more church basements than I care to remember.

Some of the chapters here have been published elsewhere, though none appears precisely as it did in the earlier version. Permissions have thus been sought and secured for the following: "Earth Community," in Donald B. Conroy and Rodney Peterson, eds., *Creation as Beloved of God* (Atlantic Highlands, N.J.: Humanities Press, 1997); "Cross and *Oikos*," in Dale Irwin and Akin Akinade, eds., *The Agitated Mind of God* (Maryknoll, N.Y.: Orbis Books, 1996); "Next Journey: Sustainability for Six Billion and More," in Daniel Maguire and Larry Rasmussen, *Ethics for a Small Planet: New Horizons on Population, Consumption, and Ecology* (Albany: State University of New York Press, 1997); "Rocky Road from Rio," in a collection of papers on sustainable development copyrighted by the Presbyterian Church (U.S.A.), National Ministries Division, 1993; "Re-Beginnings," in *Human Values and the Environment*, a booklet of conference addresses distributed by the Institute for Environmental Studies, University of Wisconsin—Madison, 1993; "'The Integrity of Creation': What Can It Mean for Christian Ethics?" *Annual of the Society for Christian Ethics, 1995* (Georgetown University Press); "Epilogue," in Dieter Hessel, ed., *Theology for Earth Community* (Maryknoll, N.Y.: Orbis Books, 1996); "Cosmology and Ethics," in Mary Evelyn Tucker and John Grim, eds., *Worldviews and Ecology* (Maryknoll, N.Y.: Orbis Books, 1994); "Returning to Our Senses," in Dieter Hessel, ed., *After Nature's Revolt* (Minneapolis: Fortress, 1992); "Redemption: An Affair of the Earth," *Living Pulpit* 2, no. 2 (April-June 1993); "Toward an Earth Charter," *Christian Century* 108, no. 30 (23 October 1991); "The Integrity of Creation," *Christianity and Crisis* 51, nos. 16–17 (18 November 1991); "Conversion to Earthfaith," *The Cresset* 59, no. 5 (May/Pentecost 1966): 5–9; "A Community of the Cross," *Dialog* 30 (spring 1991): 150–62; "Theology of Life and Ecumenical Ethics," in David Hallman, ed., *Ecotheology: Voices from South and North* (Maryknoll, N.Y.: Orbis Books, 1994); "Sustainable Development and Sustainable Community: Divergent Paths," in *Development Assessed: Ecumenical Reflections and Actions on Development* (World Council of Churches, Unit III, 1995).

This volume is dedicated to Unit III (Justice, Peace, Creation) of the World Council of Churches for reasons that will become obvious in the text itself. What is not obvious is that these themes first came my way in the 1960s at Union Theological Seminary through the tutelage of Roger Lincoln Shinn and John Coleman Bennett (1903–95), both active ethicists in the ecumenical movement. My thankfulness for the WCC extends to them.

While I have, for theological reasons, used inclusive language in preparing this text, I have left most all quoted materials as originally published. Any changes made are noted.

Any book's production relies on the good work of many people. Personnel at Orbis Books and the World Council of Churches have provided that good work. I am grateful. But special thanks go to Bill Burrows at Orbis Books. The editor for this volume, he is also the one who approached me, years ago now, about writing it in the first place and nudged it along in his gentle, engaging manner. If I have tested his patience, it has not shown.

"The trouble with socialism," George Bernard Shaw once complained, "is that is takes all one's evenings." This book has taken far too many evenings. But the support of all the above has made it far more a pleasure than it otherwise would have been.

Introduction

EARTH AND ITS DISTRESS

Forty-second Street, Times Square, New York City, is sleazy and exciting. The New Amsterdam Theater, prematurely old and scheduled for a face-lift by its new owner, Disney, carries a figure over the entrance to the foyer. It goes largely unnoticed by the hundreds who pass through the doors and down worn aisles to uncomfortable seats. The figure is "Progress." Progress is an allegorical female figure cradling a sailing ship in her arms. The sun's rays and stars fill the space behind her.

Worn, fatigued Progress is emblematic. A kind of exhaustion is written in the grain of the grimy surface. It stands in for the mood at century's end, and the century itself.

Centennial

Commentators who must offer the mandatory reviews of the century wonder why spirits languish and falter now and a certain world-weariness creeps into each day. After all, the twentieth century not only promised far more than any previous century. It actually delivered! Production of goods and services increased fiftyfold, with more certain to come. Millions have been lifted from misery. The good life has been extended over lifetimes that doubled. Folks at century's end are taller and heavier and better fed than in 1900, even if by sheer number the poor and malnourished are also more numerous. The world's capacity to produce wealth is immeasurably greater than at the century's beginning, and most of its generations have lived better than their parents — provided they were lucky enough to live where the engines of wealth were firmly in place. Certainly more people were better educated in 1995 than in 1905. Even in the depths of the Depression, well before the greatest surge of all (after World War II), the Chicago World's Fair declared this "a century of progress" without batting an eye.

A faltering spirit, then, seems out of place. Nonetheless, when the secretary-general of the United Nations told the world at the Earth Summit of 1992 that "we are under house arrest," "the time of the finite world has come," and "progress is [no longer] necessarily compatible with life,"[1] not a soul shouted him down, called for his resignation, or tried to book him for heresy. Neither, we should add, did anyone venture to say how we secure a modicum of peace and

---

1. Cited by Wesley Granberg-Michaelson, *Redeeming the Creation, the Rio Earth Summit: Challenge to the Churches* (Geneva: WCC Publications, 1992), 7.

1

justice if progress and life *are* incompatible. But the point is that the secretary-general was heard in sober silence because the mood of this exuberant era is no longer exultant, despite unimagined triumphs in record time. Confidence has changed to foreboding.

Another look may explain the worry. Eric Hobsbawm argues that a golden age running from 1947 to 1973 gathered the forces that still shape our lives, but the consequences are no longer golden. Swimming with the current, we hardly noticed how powerful the stream has been. It may actually have ended seven or eight millennia of human history. Until the third quarter of this century, the overwhelming majority of the human race lived by growing food and herding animals.[2] No longer. The long stretch of the human tale that began with the invention of agriculture in the Stone Age may be history.

Yet Hobsbawm's chief attention is more immediate. Ours is a first era of global crises because for the first time in history the golden age has created a single, increasingly integrated world economy. This economy, which bears down on all of us in different ways, hops national frontiers with ease and generally ignores an international system that depended on territorial, sovereign, and independent nation-states for its stability.[3] Governance efforts themselves are pulled apart by these transnational economic forces on one level, while ethnic and regional secessionist conflicts pull from another direction. The problems we thought the golden age had solved thus reappear all over again. Yet now they surface with global reach: mass unemployment, severe cyclical slumps, the spreading distance between rich and poor in a confrontation of limousine plenty with homelessness, and limited state revenues for limitless expenditures. Thus does people's fear of breakdown heighten. The evening news is an enervating affair.[4]

Differently said, a certain helplessness sets in when we realize that domestic economies "are reduced to complications of transnational activities." And when crises go global, as increasingly they do, domestic political systems simply are not up to the task. That a revolution in transport and communications has also "virtually annihilated time and distance"[5] does nothing to still the uneasiness. On the contrary, the acceleration of globalization via the fast lane of the information superhighway only exacerbates the inability of public institutions and people's collective behavior to live with an information revolution grafted onto an industrial one.

More alarming, social bonds have disintegrated to an extraordinary degree. To take a common, if extreme, example — no other century comes close for sheer deadliness. Perhaps as many as 187,000,000 people have died as the result of warfare in this century, the equivalent of one-tenth of the total world population

2. Eric Hobsbawm, *The Age of Extremes: A History of the World, 1914–1991* (New York: Random House, 1994), 9.

3. Ibid., 9–10. Hobsbawm is not arguing that the nation-state system actually provided this stability but that this system was expected to do so. His book in fact documents the powerfully disintegrative forces of nationalism in this century, forces channeled as much by nation-states as checked by them.

4. Ibid., 10.

5. Ibid., 15, 12.

when the century began.[6] And as the century closes, even the relief of reduced superpower nuclear conflict fails to quell the suspicion that global warfare has changed rather than disappeared. Warfare has returned to vicious internecine struggles that are easier to sustain than peace itself.

Were this not enough, demographics add another shudder. The expectation is that somewhere around 2030, the population will stabilize at about ten billion people, five times the number in 1950![7] This explosion, accomplished in the lifetime of someone born in 1950, is a very heavy multiplying factor for the other forces of disequilibria.

In a word, the century ends with a global disorder that was not expected. We walk the edge of too many quagmires to ably concentrate on the programs that must be put in place. The soul can hardly be found who believes that our present institutions have significant control over the collective consequences of our cumulative actions. And the souls we do find testify to their own disorder.

The disorder is more than the fact that hope and history rarely rhyme and that our expectations normally outstrip our experience. Ours is the loss of the framework of expectations itself. The narrative that fueled our hopes, fails. The sheen and glitter of the golden age are gone. Progress, still an exuberant memory for some, is a defeated one for most. "Every flag that flies today is a cry of pain," Adrienne Rich writes. We've lost our moorings and are unsure of what "behooves us."[8]

The New Amsterdam Theater is thus an apt emblem. Progress as the humanly cradled ship set amidst shining sun and stars is more temple ruin than house of worship. Like the finish itself, the inspiration dissipates. Theatergoers who do notice may pause to acknowledge past glory. But mostly they just nod in quiet resignation and move on, wondering what will replace the faded damsel, other than annoying scaffolding and two hours of Disney entertainment.

The Omission

There is, of course, something perilous about writing history as the record of recent events. Circumspect students know that omissions from the list are as telling as what is included. Too, the list of big stories, like the list of the rich and famous, never holds. Each generation reshuffles the list, makes casual deletions, and adds critical developments hardly noticed by the very people who initiated them. Considering the degree to which history is a steady record of human folly, it is the better part of wisdom to forgo confident prophecy. Even if the twentieth century *is* the epochal break we suspect, that will, like the identity of true prophets, not be known for awhile.

Nonetheless, we must risk a judgment. The twenty-first century may well decide that the biggest twentieth-century story was not on most 1999 lists. That

---

6. Ibid., 12.
7. Ibid., 568.
8. Adrienne Rich, "One Night on Monterey Bay the Death-Freeze of the Century," in *An Atlas of the Difficult World: Poems 1988–1991* (New York: Norton, 1992), 23.

story is this: the relationship of the human world to the rest of earth changed
fundamentally and dramatically from the onset of the twentieth century to its
close. Techno-economic power sufficient to destroy the material conditions of
human and other life is the hallmark of that change, together with the explosion
of both human numbers and consumption.[9]

When the century began, neither human technology nor human numbers
were powerful enough to alter planetary life-systems, or at least had not done
so. "The world of the born" was not yet on a collision course with "the world
of the may" (e. e. cummings). Nor was it the case "that every natural system on the planet [was] disintegrating."[10] Or at least the rate and extent did
not register. Soil erosion was not exceeding soil formation at century's onset (or at least it wasn't noticed). Species extinction was not exceeding species
evolution. Carbon emissions were not exceeding carbon fixation. Fish catches
were not exceeding fish reproduction. Forest destruction was not exceeding forest regeneration. Fresh water use was not exceeding aquifer replenishment.[11]
Half the world's coastlines, the most densely populated human areas, were
not imperiled.[12] An ominous, if bland, word thus appears at century's close —
"unsustainability." It wasn't there at century's beginning, at least in people's consciousness. Neither were "carrying capacity," "the integrity of creation," and
"sustainable development."

Consider Sophocles' *Antigone* as one measure of change:

> Many the wonders but nothing more wondrous than man.
> This thing crosses the sea in the winter's storm,
> making his path through the roaring waves.
> And she, the greatest of gods, the Earth —
> deathless she is, and unwearied — he wears her away
> as the ploughs go up and down from year to year
> and his mules turn up the soil.[13]

"Deathless she is, and unwearied." Yes, earth is amazingly resilient. But no,
earth is not indefatigable. Nature, never spent, can be. She can be overwhelmed.
One particularly powerful and errant species is overwhelming her.

Still, most end-of-century commentators will not say, with Brian Swimme
and Thomas Berry, that "assault" rather than "communion" now names our re-

---

9. Hobsbawm, *Age of Extremes,* 584.

10. Paul Hawken, *The Ecology of Commerce: A Declaration of Sustainability* (San Francisco: Harper
Business, 1993), 22.

11. See Lester R. Brown, Hal Kane, and David Malin Roodman, *Vital Signs, 1994: The Trends
That Are Shaping Our Future* (New York: Norton, 1994), 15–21.

12. "Half of World's Coastlines Are Found to Be in Peril," *New York Times,* November 7,
1995, C4. The article is reporting on the study of the World Resources Council delivered at the
international meeting on biological diversity in Jakarta, Indonesia, November 1995.

13. Cited in Hans Jonas, *The Imperative of Responsibility: In Search of an Ethics for the Technological Age* (Chicago: University of Chicago Press, 1984), 2. This is from Jonas's initial chapter, "The
Altered Nature of Human Action."

lationship to earth.[14] Few will acknowledge the new, wild fact that "for the first time, our power to destroy outstrips the earth's power to restore."[15] And those who do acknowledge the epochal change will find the temptation to deny deeply changed reality as hard to resist as the temptation to grasp the remaining treasure we can. Nonetheless, just this altered relationship to earth will be the tattoo by which the next century identifies and laments this one.

But of course this centennial shift has not gone utterly unnoticed. Nor have the perplexing ethical issues been left unspoken. Thoughtful observers have always acknowledged, for example, that failure to distribute nature's bounty fairly has been the perplexity social justice consistently faces. Many will now add that increased natural constraints ratchet up the difficulty. Yet how many will explore what it means that we stand astride global threats to nature's capacity both to produce for its human members *and* to regenerate itself? Pursuing *this* complication on a saturated planet is little attended to, perhaps because earlier generations could ignore it. And did, since only one hundred years ago nature was still resilient and forgiving, and we could do largely as we wished. It seems our social genes prepare us poorly for our accomplishments.

If the great new fact of our time is that cumulative human activity has the power to affect all life in fundamental and unprecedented ways, then *what might and ought to be* is precisely what needs to be taken into account. This means the ascendancy of ethics for our era, as an utterly practical affair. How *ought* we to live, and what ought we *do* in view of a fundamentally changed human relationship to earth, a relationship we only partially comprehend?

Differently said, the radical increase in cumulative human power over this century means a new account of responsibility. It requires moralities and ways of life that extend responsibility to include everything that has life and is necessary to life. Sustainable ways of life, and living into them, await us as necessities that must come to pass.

So we are not wholly at sea. The overriding issue is clear enough. It's *power and sustainability* — human power, trying to eke out a way for all of us, on the one hand; and earth's sustainability on terms acceptable to us and millions of other creatures, on the other. This is the reality, some of it old as the hills, much of it new, that pushes these pages. The subject is earth — all of it — and its distress.

CONVERSION

The preface said the mood here is hopeful. That is so, despite the specter of eco-apocalypse some people feel. While understandable, eco-apocalypse is not the staple in these pages. Even when the very soul of responsibility on a planet

---

14. Brian Swimme and Thomas Berry, *The Universe Story: From the Primordial Flaring Forth to the Ecozoic Era* (San Francisco: HarperCollins, 1992), 226.

15. Daniel Maguire, *The Moral Core of Judaism and Christianity: Reclaiming the Revolution* (Minneapolis: Fortress, 1993), 5.

waving a "No Vacancy" sign requires clear-eyed realism about degraded conditions, these pages do more than entertain the dire warnings of the prophet. Gospel is intended here. And whatever "gospel" may mean, it is not more bad news, or even no more bad news. Something drawing us to life and extending life from us toward others must be uncovered and embraced. It may be as morally earnest and demanding as any prophetic oracle. But it cannot be without joy. Absent the joie de vivre, our fate is surely sealed. That is the one apocalyptic thought this volume entertains.

Earthbound Loyalty

A chapter in *The Brothers Karamazov* titled "Cana of Galilee" captures the kind of spirit and earthbound loyalty that begins the gospel search. The closing scene centers on Alyosha, "the little man of God" who goes late at night to the monastery cell where his spiritual mentor, the saintly old Zossima, lies in his coffin. The scene comes, of course, only after the tumultuous tears and tortured relationships of love, betrayal, and loyalty that inhabit all thick Russian novels. In the dark candlelight Alyosha hears the account of Jesus' first miracle of water to wine at the wedding in Galilee. Exhausted, he drifts from the words into prayer and his own running commentary on the familiar verses, all in the presence of old Zossima's body and spirit. In this dreamlike state, Alyosha suddenly hears the voice of Zossima, who extends his hand and raises him from his knees. "Let us make merry," "the dried-up old man" says, "let's drink new wine, the wine of new gladness, of great gladness." And a little later: "Begin your work, my dear one, my gentle one." It is in this moment of tender invitation and new vocation that Alyosha, who had in truth fallen asleep on his knees in prayer, suddenly awakens. He goes to the coffin briefly, turns abruptly, and exits the cell. Dostoyevsky writes:

> Alyosha did not step on the steps, but went down rapidly. His soul, overflowing with rapture, was craving for freedom and unlimited space. The vault of heaven, studded with softly shining stars, stretched wide and vast over him. From the zenith to the horizon the Milky Way stretched its two arms dimly across the sky. The fresh, motionless, still night enfolded the earth. The white towers and golden domes of the cathedral gleamed against the sapphire sky. . . . The silence of the earth seemed to merge into the silence of the heavens, the mystery of the earth came in contact with the mystery of the stars. . . . Alyosha stood, gazed, and suddenly he threw himself flat upon the earth. He did not know why he was embracing it. He could not have explained to himself why he longed so irresistibly to kiss it, to kiss it all, but he kissed it, weeping, sobbing and drenching it with his tears, and vowed frenziedly to love it, to love it for ever and ever. Water the earth with the tears of your gladness and love those tears, it rang in his soul. What was he weeping over? Oh, he was weeping in his rapture even over those stars which were shining for him from the abyss

of space and he was not ashamed of that ecstasy. It was as though the threads from all those innumerable worlds of God met all at once in his soul, and it was trembling all over as it came in contact with other worlds. He wanted to forgive everyone and for everything, and to beg forgiveness — oh! not for himself, but for all and for everything, and others are begging for me, it echoed in his soul again. But with every moment he felt clearly and almost palpably that something firm and immovable, like the firmament itself, was entering his soul. A sort of idea was gaining an ascendancy over his mind — and that for the rest of his life, for ever and ever. He had fallen upon the earth a weak youth, but he rose from it a resolute fighter for the rest of his life, and he realized and felt it suddenly, at the very moment of his rapture. And never, never for the rest of his life could Alyosha forget that moment.

"Someone visited my soul at that hour!" he used to say afterwards with firm faith in his words.... Three days later he left the monastery in accordance with the words of his late elder [Zossima] who had bidden him [now new-born to earth] to "sojourn in the world."[16]

There the chapter ends.

Alyosha threw himself upon the *earth* and arose from it a resolute fighter. He did not throw himself upon "the environment"! As noted earlier, the difference is not slight and is the reason this is not a treatise on the environment and its distress but a meditation on earth and its distress. If the subject were the environment, earth's distress would be a crisis of nature in which nature is the sumptuous stage for the human drama and its stock of steady resources. In this environmentalist view, nature is threatened by human use and encroachment. The crisis is thus the degradation of natural habitat and capital. So considered, "the environment" gets slotted somewhere on the same miserable list with racism, poverty, domestic violence, crime, and homelessness, and joins the unhappy competition for attention and resources. People who speak of earth's distress as a crisis of the environment are certainly correct about one thing: what is happening to earth *is* the great drama of our time. The crisis is not one of nature, however, as nature is commonly conceived. Nature, however degraded and diminished, will survive in a million different forms and for a very long time. Earth's distress is a crisis of culture. More precisely, the crisis is that a now-globalizing culture *in* nature and wholly *of* nature runs full grain *against* it. A virile, comprehensive, and attractive way of life is destructive of nature and human community together — this is the crisis. Soils, peoples, air, and water are being depleted and degraded together. (Or, on our better days, are being sustained together.) It is not "the environment" that is unsustainable. It is a much more inclusive reality, something like life-as-we-have-come-to-know-it. What we call "the environmental crisis" is a sign of cultural failure, then. It is the

---

16. Fyodor Dostoyevsky, *The Brothers Karamazov*, translated and introduced by David Magarshack (Baltimore: Penguin, 1963), 2:426–27.

failure to submit human power to grace and humility, and to work "toward the habitation of the places in which we live"[17] on terms that respect both human limits and the rest of nature's. Life-as-we-have-come-to-know-it is eating itself alive. Modernity devours its own children.

Here the whiff of apocalypse *is* in the wind and blows strong. The foreboding is captured in two modern renditions of the most famous of Chinese poems, "Spring View." Composed by Du Fu, "Spring View" was written at the time of the An Lushan rebellion in 755 C.E. Lamenting the rebellion and in search of consolation, the poet opens with this line:

> The State is destroyed, but the mountains and rivers survive.

A twentieth-century Japanese poet, Nanao Sakaki, has seen fit to reverse Du Fu:

> The mountains and rivers are destroyed, but the State survives.[18]

As though this reversal were not enough, Thomas Fraser Homer-Dixon's study of causes of acute conflict in effect extends Sakaki:

> The mountains and rivers are destroyed, and the State survives — but society unravels.[19]

Most North American readers have to travel to appreciate this. To China itself, to Papua New Guinea, to the Philippines and communities in Latin America, to West Africa and India, to Haiti, to tracts in Siberia and Eastern Europe, to Egypt, to the homelands of native and indigenous peoples most anywhere, or to most any of their own central cities. Social and environmental disintegration is rampant.

Yet even when degradation is apparent, many do not truly comprehend what is happening to earth. Peoples of dominant cultures and their extensions around the world lack the right categories. "For too long," Daniel Deudney says, "we've been prisoners of 'social-social' theory, which assumes there are only social causes for social and political changes, rather than natural causes, too."[20] This habit of mind (tagged a version of "apartheid thinking" in a later chapter in this book) has been regnant since the industrial revolution, a revolution that lifted us wholesale from the rest of nature as a species apart. In both working habits and morality, industrialism assumed the universe that counts is the human universe, plus props. That remains the assumption of most daily practices. So it is

---

17. David C. Toole, "Farming, Fly-Fishing, and Grace: How to Inhabit a Postnatural World without Going Mad," *Soundings* 76, no. 1 (spring 1993): 87.

18. Gary Snyder, "Survival and Sacrament," in *Practice of the Wild* (New York: Farrar, Straus and Giroux, 1990), 175.

19. Thomas Fraser Homer-Dixon, with Jeffrey H. Boutwell and George W. Rathjens, "Environmental Change and Violent Conflict," *Scientific American* (February 1993), 38–45.

20. Cited in Robert D. Kaplan, "The Coming Anarchy," *Atlantic Monthly* (February 1994): 60.

only now, when nature shows both vengeance and vulnerability and suffers the grim interplay of demographic, environmental, and social stress, that children of industrialized civilizations finally discover what most other civilizations have long known: the fate of mountains, rivers, and societies is a single fate; nature is not what is around us or where we live, but the reason we are alive at all; nature is the reason each and every society and culture that ever existed did so.

This *internal,* not external, linkage of society and environment, and of nature, history, and culture, all of a piece at century's end, is what these pages call "earth and its distress."[21] We will speak only occasionally of "the environmental crisis" or "the ecocrisis," since the proper name is some variation on the threat and promise to *all* things as they are bundled together from the inside out. Society and nature together — that is, earth — is a community, without an exit.

In this light, Alyosha's moving moment serves only as prelude to a prolonged conversion to earth. Reconciled of soul with earth, falling to earth a weak being and rising a resolute fighter with a new vocation — this serves us well, as does Alyosha's weeping and watering the earth with the tears of gladness, and loving those tears. ("The capacity to weep and then do something is worth everything," Greta Gaard writes in *Ecofeminism.*)[22] Yet the figure of Alyosha does not convey the complexity of earth and its distress. Alas, there is no *one* best analysis for our reality. Nor does *one* figure, picture, or model fit it, unless perhaps it be earth itself as *oikos,* as Habitat Earth, the only life-form in the universe we know of to date. Unless it be earth *as* comprehensive community.

We do not, however, have a common language, even for something as basic as earth community. We are temporarily stuck in the awkward space between worlds we trusted and ones strangely new to us. This is a puzzling, sometimes bizarre place to be. And because it is difficult to comprehend, it is difficult to articulate.[23] No doubt a multiplication of strategies and actions as well as symbols, perspectives, and voices is essential. Dostoyevsky is only one of these stirring conveyors of life's adventure and passion.

Yet certainly the right starting point is the Hebrew prophets: it's mountains and rivers and society all together, or not at all. The "vines languish," and "the merry-hearted sigh" together or, alternatively, thrive together (Isaiah). Everything is caught up in a complicated fate laced with both the "dearest freshness"

---

21. The phrase is Dietrich Bonhoeffer's, from a 1929 address we shall discuss in the chapter entitled "Song of Songs."

22. Greta Gaard, "Living Connections with Animals and Nature," in *Ecofeminism,* ed. Greta Gaard (Philadelphia: Temple University Press, 1993), 3.

23. The truth itself rings strange to many ears in such times. As Albert Schweitzer wrote at midcentury about ethics in European and, in general, Western society: "It is the fate of every truth to be an object of ridicule when it is first acclaimed. It was once considered foolish to suppose that black men were really human beings and ought to be treated as such. What was once foolish has now become a recognized truth. Today it is considered an exaggeration to proclaim constant respect for every form of life as being the serious demand of a rational ethic. But the time is coming when people will be amazed that the human race existed so long before it recognized that thoughtless injury to life is incompatible with real ethics. Ethics is in its unqualified form extended responsibility with regard to everything that has life" (Albert Schweitzer, *Memories of Childhood and Youth* [New York: Macmillan, 1950], 40).

of "deep down things" (Gerard Manley Hopkins) and new levels of vulnerability. Creation has its integrity. But it's for worse as well as for better. Earth is full of life but under house arrest.

Religion and Fidelity to Earth

Daniel Maguire puts things crisply. "If current trends continue, we will not. And that is qualitatively and epochally true. If religion does not speak to [this], it is an obsolete distraction."[24]

Religion speaking to earth and its distress in nondistracting ways *is* a challenge! It is the challenge taken up here. And the thesis is that religion not only *should* help effect our conversion to earth but *can* help it. At the same time, it cannot do so without reformation. The reformation is that *all* religious and moral impulses of whatever sort must now be matters of unqualified earthbound loyalty and care. Faith is fidelity to earth and full participation in its ecstasy and agony. There is no room, for example, for the earth avoidance carried in the teaching of *contemptus mundi* (contempt of the world). Or any other forms of otherworldliness that work to "make [persons] think lightly of their responsibilities."[25] Nor is there room for perhaps the most popular religious metaphor of all, the metaphor of ascetic ascent, throwing off the corruptible things of earth for the precious booty of heaven. For these time-honored spiritualities, the entanglements of mundane existence are left behind by ascending the great chain of being, step-by-step. Serious devotees climb from the lower regions of the material and base to higher realms of pure spirit and free communion with the divine. Early Christianity, to bring forward one prominent case, often cited Socrates' statement in Plato's *Theaetetus* so as to bolster this strain of common asceticism: "We ought to try to escape from earth to the dwelling of the gods as quickly as we can; and to escape is to become like God, as far as this is possible; and to become like God is to become righteous and holy and wise."[26] Imitating God meant leaving earth behind.

We must, of course, listen with the heart to all human yearnings. This includes those anguished cries that find no hope or consolation anywhere on earth, much less joy and quiet satisfaction. Yet precisely in the face of such anguish, "felt in the days when hope unborn died,"[27] this volume judges as ethically valid only those human and religious energies that unapologetically serve earth's care and redemption. Traditions will be revisited in this view, both in search of trea-

---

24. Maguire, *Moral Core*, 13.

25. F. D. Maurice, *The Kingdom of Christ* (London: SCM, 1958 [1842]), 2:284–85, as cited by J. Philip Wogaman, *Christian Ethics: A Historical Introduction* (Louisville: Westminster/John Knox, 1993), 172.

26. Cited from Plato's *Theaetetus* 176B in the Loeb Classical Library translation and reported by Wayne Meeks in *The Origins of Christian Morality: The First Two Centuries* (New Haven, Conn.: Yale University Press, 1993), n. 236.

27. From the hymn text of James Weldon Johnson's "Lift Every Voice and Sing," no. 562 in the *Lutheran Book of Worship* (Minneapolis: Augsburg, 1978).

sure still there and in quest of religious creativity that has not yet found its full voice as part of earth's cause.

Among those traditions, Christianity is an example of an ambivalent faith that might yet become an earth faith. Christianity is a radically incarnational faith, chiefly because its elder and birth home is Judaism, and its founder and foundation was a Jew understood as the earthly incarnation of God. Thus Christianity carried from its origins Judaism's age-old insistence that both creation and redemption are matters of earth, affairs of history and nature together in time and space as we know them. Yes, there are classic religious motifs of sojourn, pilgrimage, and world-alienation in Judaism, Christianity, and Islam. These are the familiar motifs of God's faithful as "strangers and foreigners on the earth" (Hebrews 11); of the life of nomadic wandering; and of apocalyptic overthrow of all that presently oppresses. But in the end even these earth-alien and earth-restless themes are decisively qualified by the foundational conviction that creation is good and the work of a God committed to its fullest possible flourishing, nothing less than its redemption in toto. So the themes of pilgrim, sojourner, and stranger all have variants, only one of which is ascetic ascent and world contempt. There is pilgrimage to *the good land*, for example, or pilgrimage to *the holy mountain*. Both are rich, earthbound images of the very soil from which all things are created, on which they all depend, and to which they all return. Pilgrimage is finally about homecoming.

Even the theme of nomadism indicates less rootlessness than at first might appear. "Nomad" is from the Greek, signifying "law" (*nomos*). The word conveyed the notion of dividing justly and splitting up shares properly. Nomads could survive only if they knew how to share pastures with other nomads. Their way of life could survive only if they treated earth as a genuine commons. Without such law, there could not be "nomads." So it is not coincidental that *the* great legislator, Moses, is the leader of a nomadic, sojourning people and the mediator of the covenant between God, this wandering band, and the land.[28]

The *contemptus mundi* of some religious traditions can itself be turned to ways that serve rather than abandon earth.[29] In fact, contempt for the world is the first and most basic heresy only when it means contempt for the good clay of God's earth, human clay included. Its nonheretical meaning is relevant and powerful. The purpose of spiritual practices is, after all, to make us more fully alive by heightening the capacity of the senses to experience grace and the giftedness of life itself. Feasting and fasting are thus a rhythm, as are self-denial

---

28. Almost equally intriguing is that Jacques Attali, former president of the European Bank for Reconstruction and Development and one of the architects of the European Union, finishes his book (*Millenium: Winners and Losers in the Coming Order* [New York: Random House, 1991], trans. Leila Conners and Nathan Gardel) with this image. It comes after his survey of the great tasks of coming years, a survey that ends with this: "Above all, a new sacred covenant must be struck between man and nature so that the earth endures, so that the ephemeral gives way to the eternal, so that diversity resists homogeneity" (129). It is in this context that he invokes the image of nomad.

29. See Vincent L. Wimbush and Richard Valantasis, eds., *Asceticism* (New York: Oxford University Press, 1995). The *contemptus mundi* tradition is varied, complex, and more nuanced than the brief treatment here allows.

and self-fulfillment, withdrawal to the wilderness and return to the marketplace, solitary spiritual discipline and vigorous public engagement. *Contemptus mundi* is not contempt for *earth*, then, but contempt for the world when "world" is shorthand for the "principalities and powers" of inner and outer variety that oppress, enslave, and distract. These demons of our lives — structural, cultural, social, and psychosocial — merit *contemptus mundi* and call for practices that exorcise the powers that plague and enslave. An asceticism that loves the earth fiercely as God's own and understands spiritual disciplines as ways to make body, self, and society even more alive to earth and life is only to be affirmed.

Even the literature of apocalypse, perhaps the supreme example of world- and earth-hating motifs, can be turned another way to serve the cause of earth. It is a way analogous to the language of "heaven" in African-American spirituals. Here a radical withdrawal from the world as the place of madness and oppression is matched by a judgment upon that world for its grave injustice. This in turn is matched by a persistent attempt to lay claim to a more equitable world envisioned as "a new earth."[30] "Heaven" is in fact a way of speaking about earth. It is the dream of a new world and redeemed earth, with the throne of God in the midst of the city and rivers of crystal waters flowing in four directions from the throne, the riverbanks themselves lined with trees of life bearing abundant fruit.[31] Like Maya Angelou's Rock, River, Tree, and Peoples "on the pulse of morning," apocalyptic yearning may in fact be the ultimate earth patriotism, since all hope, energy, and dedication are here invested in a new earth, despite superhuman odds. The rage against wrong and the dream of revolt and overthrow that fires an imagination will not let earth be the enemy's forever. The language of heaven thus works to name the vision *of* and *for* earth redeemed. It is creation made new and made whole.

And what is "faith" here but the capacity to see earth as it might be transformed by grace and to affirm it in spite of everything in us that fears and rejects it? Faith "resists the actual in the name of the possible," fighting "the three great instabilities" — injustice, unpeace, creation's disintegration.[32] Faith is hope in things yet unseen (Paul). It is "a bird at dawn, already singing in the dark" (Indian saying).

If we were to extend this analysis of apparently earth-denying themes in Christianity, we would discover that there is a warning within the apocalyptic and *contemptus* traditions themselves. The warning is not to let world-weariness

---

30. I have taken this insight from the interpretation by Raphael Gamaliel Warnock in his master of divinity thesis, "Churchmen, Churchmartyrs: The Activist Ecclesiologies of Dietrich Bonhoeffer and Martin Luther King, Jr." (Union Theological Seminary, New York, 1994), 28–29. See also James H. Cone, *The Spirituals and the Blues: An Interpretation* (New York: Seabury, 1972), especially the chapter, "The Meaning of Heaven in the Black Spirituals." As I trust is clear from the text above, I distinguish such apocalyptic visions as this sharply from those sectarian movements, often isolated and in the woods, whose rites and rituals and collective existence are meant to jump-start the apocalypse itself.

31. See the last chapter of the most apocalyptic of all the New Testament books, Revelation.

32. Douglas John Hall, *Professing the Faith: Christian Theology in a North American Context* (Minneapolis: Fortress, 1993), 313.

cut the chord of hope for this world, drain us of action, weaken faith's resolve, or substitute heaven for earth's redemption. Martin Luther King's last and most apocalyptic sermon, preached the eve of his assassination and entitled "I See the Promised Land," includes the following:

> It's alright to talk about "long white robes over yonder," in all of its symbolism. But ultimately people want some suits and dresses and shoes to wear down here. It's alright to talk about "streets flowing with milk and honey," but God has commanded us to be concerned about the slums down here, and his children who can't eat three square meals a day. It's alright to talk about the new Jerusalem, but one day, God's preacher must talk about the new New York, the new Atlanta, the new Philadelphia, the new Los Angeles, the new Memphis, Tennessee. This is what we have to do.[33]

This rereading of apparently earth-distracting traditions acknowledged, the note must be registered again that earth and its present distress call for nothing less than religious and social re-formation, a conversion far from fully effected. It is fundamentally dishonest to argue that believers need only bring to earth's agonies their own favored Vincentian canon ("what the church has always and everywhere taught and believed"). The need for religious and social imagination and reformation is more basic, exacting, and far-reaching than that. Christianity, to stay with the example, will have to address far more profoundly than it has the "residual...ambivalence" it holds "about sex, the body, women's bodies in particular and the natural world in general."[34] The right direction is "the healthier Hebrew earthiness"[35] still at home in Judaism. But even here it bumps up against a residual ambivalence almost as deep and strong — the association in Judaism of pagan taboo with nature religions. So no matter which turn is taken, at some juncture the turn to earth rubs up against an internal problematic resident within most every faith or philosophy presently contending for our allegiance. Viable earth faiths, it seems, require rerooting virtually all religious and moral traditions, even when some meet our trauma better than others. Ours is not a time for the religiously and socially timid or the intellectually fainthearted.

33. Martin Luther King Jr., "I See the Promised Land," in *A Testament of Hope: The Essential Writings of Martin Luther King, Jr.*, ed. James M. Washington (San Francisco: Harper and Row, 1986), 282.

34. Catherine Keller, "Chosen Persons and the Green Ecumenacy," in *Ecotheology: Voices from South and North*, ed. David Hallman (Geneva: WCC Publications; and Maryknoll, N.Y.: Orbis Books, 1994), 302. Though from a different direction, the French priest-scientist Teilhard de Chardin makes a similar point: "The world will not be converted to the heavenly promise of Christianity unless Christianity has previously been converted to the promise of the earth" (cited in Jackson Lee Ice, "The Ecological Crisis: Radical Monotheism vs. Ethical Pantheism," in *Moral Issues and Christian Response*, ed. Paul T. Jersild and Dale A. Johnson, 5th ed. [Fort Worth, Tex.: Harcourt Brace Jovanovich, 1993], 198).

35. Keller, "Chosen Persons," 306.

Can these faiths become earth faiths? Is there a conversion to earth in the making or the offing? Yes, but it is important to remember that established religions are *never* ready to meet new events and developments. As religious creatures we are never well prepared for emancipation movements of all kinds, the intensity of religious pluralism on our home turf, new technologies, or the ecosocial crisis. It isn't only that religions are bridled by the standard weaknesses of mortals tending matters of divinity (a wonderful but plainly foolish vocation). It's that religions are inherently conservative of past meaning, authority, rites, rituals, and moralities. Thus the religiously committed are often poorly poised by their traditions to discern the new, the deviant, and the non-normative, much less to judge these with clarity, insight, and foresight. At the same time, it is precisely a deep cultural crisis that initially gives birth to religions and later to their reform. Religions typically come to be and take their distinctive shape in the breakup of worlds. They accompany the long and painful birth of epochs in the making and breaking. Such religions and their ways of life do not, of course, start from scratch. In a world long underway, there is no scratch to start from. Rather, new and reforming faiths emerge as major transformations of lively or newly enlivened traditions and practices. What we should ask, then, is not whether something is orthodox or not. Like ancient and familiar scriptures, the oldest teachings may be generative all over again. What we should ask is whether orthodoxies have grown humorless and constricted. For what is reprehensible about religions (and ways of life generally) is not their initial lack of readiness to meet new crises or their bungling on morally vexed and volatile issues. That can be expected. What is reprehensible is the refusal of such traditions to be transformed, in the Spirit, *by* their crises and the culture's.[36] What is reprehensible is the unwillingness to wash all things anew "in the purifying waters of contrition and gratitude."[37]

So the answer to the question of whether religious faiths and other life philosophies can become genuine earth faiths and potent sources of earth patriotism and earth ethics is an emphatic yes. But only on the condition they are open to being transformed *by* earth and its distress. Truth be told, humans don't really long for another world, far beyond the ordinariness of this one, so much as they long for their own world in all its hidden beauty and possibility. The insistence of the biblical religions is correct: the universe in its essence has justice and love as its foundation, and it is our alienation from a just and loving totality that feeds our deep frustration with the order of things.[38] Ridding earth of its distress, in order to trot out its glory for all to see, is the soul's deepest desire.

---

36. John F. Haught, *The Promise of Nature: Ecology and Cosmic Purpose* (New York: Paulist, 1993), 2–4.

37. Dietrich Bonhoeffer and Maria von Wedemeyer, *Brautbriefe: Zelle 92* (Munich: Beck, 1992), 176; trans. mine.

38. Michael Lerner, *Surplus Powerlessness: The Psychodynamics of Everyday Life* (Atlantic Highlands, N.J.: Humanities Press International, 1991), 310.

COMMUNITY AND RESPONSIBILITY

The effort to understand what is happening and how we are responding (part 1, "Earth Scan") and the effort to revisit creatively the world of religion, ethics, and human symbols in the interest of necessary conversion to earth (part 2, "Earth Faith") could be races run for their own sake. They are worthy. But not here. Here they are woven as an ethic organized around a theme. The theme is responsibility for the sake of sustainable community (part 3, "Earth Action").

That nature is a community is *the* scientific discovery of the twentieth century. That earth, human society included, is *also* a community has not yet registered with us. At least how to sustain it as a community has not.

The issue is more than knowing how to track long-playing consequences of cumulative human activity. A sense of the whole is absent, a sense that understands the interrelationship of the varied elements and knows what to do with that. Or, if a sense of the whole is *beyond* our ken — a true human possibility — then the issue is what to do with *that* in a world where our actions largely outstrip our capacities to track and direct them. How do we exercise responsibility in a world where consequences no longer bother with borders or care much about the quarterly report and the fiscal year? Are we even capable of hands-on responsibility if the cumulative scale of impact is global, swift, and long-term?

Differently said, what habits of heart and mind, and what kind of policies and institutions, are required when nature and earth are factually a complex community, yet effective human responsibility has not tripped its way into that reality?

The good news is that faced with a compelling need for change and some new habits, people don't simply sit back, sigh, and pour another drink. They try to answer. The current lead answer is the international proposal of "sustainable development." It was the centerpiece at the UN Conference on Environment and Development of 1992 and near the center of subsequent UN conferences in Vienna, Cairo, Copenhagen, Beijing, and Istanbul. Moreover, it is not only a lead notion. It has become the subject of treaties and comprehensive international action plans that have emerged from these global gatherings. A Presbyterian report is correct: "'Sustainable Development' is now the most widely recommended antidote to the global social, economic, and ecological crises. It is also the most widely varied."[39]

Yet in the investigation of power and sustainability in these pages, sustainable development is not the preferred direction. Sustainable community is. Sustainable development and sustainable community are contrasting paths into the future. They carry different notions of the structure of human society and responsibility. The differences turn most of all on the issues that most impact nature and earth as communities: namely, the globalizing economy and

---

39. Task Force on Sustainable Development, the Reformed Faith, and U.S. International Economic Policy, "Hope for a Global Future: Toward Just and Sustainable Human Development" (Presbyterian Church [U.S.A.]), 67 of draft dated August 8, 1995. Cited with permission of the Advisory Commission on Social Witness.

the distribution of power among the peoples of the world. These pages favor a downward distribution of economic and social power and heightened status for all forms of life, human and other.

Both "sustainable development" and "sustainable community" speak the same popular language of community, empowerment, and responsibility. Analysis of UN documents will show this. Thereby hangs an important point: when everyone from Exxon and Mobil Oil to Greenpeace and the Rainbow Warrior is an environmentalist, when everyone talks of community and no one wants to be identified as antidevelopment, and when all sides offer their course of action as the responsible and sustainable one, then all disagreements and battles are internal. Consequently, the rhetoric and action have to be decoded on every side. For this reason, the next chapters here take time to uncover, layer-by-layer, the key ideas used on all sides. Notions such as sustainability itself, sustainable development, sustainable community, the integrity of creation, even nature and the environment, require upfront scrutiny. On an even grander scale, current human self-understandings require scrutiny, as do rival notions about our vocation as the presently dominating species. (For some curious reason, we rarely think of ourselves as a species. Rather, we consider ourselves peoples, nations, races, clans, or competing customers. When we do think species-thoughts, we usually conceive ourselves a species apart.)

Yet this is to rush ahead. The initial point was our lack of a sense of the whole. On some level, we may understand nature and earth as *oikos* (the ancient vision of earth as a single public household). Yet at century's end ours is "a story without a center, a formless flurry of incidents and events."[40] Information triumphs rather than understanding, and we live by a strained hope that knowledge will, by some strange alchemy, turn to wisdom if we just keep piling it on and sending it around. Meanwhile the endless accounts of disaster and doom numb us, and we either drift off into confusion and resignation or stare at the separate pieces of the puzzle in the hope that the whole will reveal itself.[41] Failing that, we pray that the whole will somehow care for itself despite our confusion.

Yet a sense for the whole is insistent, even when it seems out of reach. We require a common story to live by, some narrative or set of narratives to render events intelligible. Something moving in that direction was offered in a Fourth of July speech in Philadelphia by Vaclav Havel. It serves nicely to combine an analysis of where we are with where we might go. To our advantage, Havel both assumes earth as a community and tries to make sense of our present failure to live out that community. He does this with an eye to human responsibility and self-understanding. His seems an analysis worthy of our time.

The Czech president begins by describing transitional periods of history. Their distinguishing features "are a mixing and blending of cultures, and a plurality or parallelism of intellectual and spiritual worlds." Furthermore these are

---

40. Theodore Roszak, *The Voice of the Earth* (New York: Simon and Schuster, 1992), 308, as cited by Stephen Viederman, "The Economics of Sustainability: Challenges," 6 (the paper is available from the Jessie Smith Noyes Foundation, New York City). Used with permission.

41. Roszak, *Voice of the Earth*, 308, as cited by Viederman, "Economics," 6.

periods "when all consistent value systems collapse" and "cultures distant in time and space are discovered or rediscovered." New meaning is gradually born amid the turmoil, from the sometimes harsh encounter of many different elements.[42]

Havel's analysis next moves to modernity as an age coming to a close. He faults modern sciences for failing "to connect with the most intrinsic nature of reality, and with natural human experience." We know immeasurably more about the universe than our ancestors, yet "it increasingly seems they knew something more essential about it than we do, something that escapes us." The same is true of nature and of ourselves: "The more thoroughly all our organs and their functions, their internal structure and the biochemical relations that take place within them are described, the more we seem to fail to grasp the spirit, purpose and meaning of the system that they create together and that we experience as our unique 'self.'" Havel concludes as many now do: "The world of our experiences seems chaotic, disconnected, confusing. There appear to be no integrating forces, no unified meaning, no true inner understanding of phenomena in our experience of the world. Experts can explain anything in the objective world to us, yet we understand our own lives less and less."[43]

Next Havel speaks as the titular head of state and responsible world citizen: "The central political task of the final years of this century . . . is the creation of a new model of coexistence among the various cultures, peoples, races and religious spheres within a single interconnected civilization." New organizational, political, and diplomatic instruments will necessarily play a role, but "such efforts are doomed to failure if they do not grow out of something deeper." Then, in what must have come as a surprise to the Fourth of July crowd, he tactfully rejects "the fundamental ideas of modern democracy" as the "something deeper" upon which to draw. The solution democratic ideas offer is still, as it were, modern, "derived from the climate of the Enlightenment and from a view of man and his relation to the world . . . characteristic of the Euro-American sphere for the last two centuries." While the Enlightenment idea of human rights and freedoms must be integral to any meaningful world order, even moral treasures such as these need to be "anchored in a different place, and in a different way, than has been the case thus far." In the end, the problem of modernity is a "lost integrity," and it simply will not suffice to ground "inalienable human rights" or other moral obligation in the notion of *Homo sapiens* "as a being capable of knowing nature and the world" who regards itself "the pinnacle of creation and lord of the world."[44]

To put the matter as bluntly as Havel does: the world of "modern anthropocentrism" is deeply, even fatally, flawed. The notions and institutions that issue from its ethics and spirituality, and depend upon them, must be set aside. A moral universe limited to the human universe will not, under present circum-

---

42. Vaclav Havel, "Address of the President of the Czech Republic, His Excellency Vaclav Havel, on the Occasion of the Liberty Medal Ceremony, Philadelphia, July 4, 1994," p. 2 of photocopied manuscript made available by the Czech Republic Mission, New York City. Used with permission.

43. Ibid., 3, 3–4, 4.

44. Ibid., 5–6, 7.

stances, even understand life, much less serve it. Earth community requires a biocentric or a geocentric knowledge, ethic, and faith.

Havel then turns from his analysis of a "lost integrity" and an imploding world to a possible newfound integrity in the "anthropic cosmological principle." Since the usual Fourth of July gathering normally does not deal in anthropic cosmological principles, Havel takes on a teaching role. With a touch of the scientist-turned-poet, he explains:

> We are not at all just an accidental anomaly, the microscopic caprice of a tiny particle whirling in the endless depths of the universe. Instead, we are mysteriously connected to the entire universe, we are mirrored in it, just as the entire evolution of the universe is mirrored in us. Until recently it might have seemed that we were an unhappy bit of mildew on a heavenly body whirling in space among many that have no mildew on them at all. This was something that classical science could explain. Yet the moment it begins to appear that we are deeply connected to the entire universe, science reaches the outer limits of its powers [and finds itself] between formula and story, science and myth. In that, however, science has para-doxically returned, in a roundabout way, to man, and offers him — in new clothing — his lost integrity. It does so by anchoring him once more in the cosmos.[45]

To this anthropic cosmological principle Havel adds his rendition of the "Gaia hypothesis." We belong to a larger whole, this hypothesis insists, and our destiny is dependent not so much on what we do for ourselves but on what we do for earth as a whole. Havel writes: "If we endanger her, she will dispense with us in the interests of a higher value — that is, life itself." Together the cosmological principle and the Gaia hypothesis tap a forgotten awareness encoded in all religions and in most philosophies and cultures, an awareness perhaps even inscribed in the unconscious of all of us, resident there in the form of primordial archetypes. This is the awareness of our being "anchored in the Earth and the universe, the awareness that we are not here alone nor for ourselves alone." Rather, we are an integral part of higher, mysterious entities "against whom it is not advisable to blaspheme."[46]

Havel nears the end of his oration by citing an anonymous modern philosopher: "Only a God can save us now." The Czech president exegetes this by arguing that the basis for any new world order, including a universal respect for human rights and democratic processes, is not laid unless the imperatives derive from respect for the miracle of Being, the miracle of the universe, the miracle of nature, and the miracle of our own existence. Morally this means that only those who "submit to the authority of the universal order of creation," and value

---

45. Ibid., 7–8.
46. Ibid., 8–9.

participation in it, can genuinely value themselves and their neighbors, thus honoring their neighbors' rights as well as their own. It follows that in today's world, the truly reliable path to peaceful coexistence and creative cooperation must start from what is at the root of all cultures and what lies "infinitely deeper in human hearts and minds than political opinion, convictions, antipathies or sympathies." The name for this is "transcendence." Here is Havel's final gloss:

> Transcendence as a hand reached out to those close to us, to foreigners, to the human community, to all living creatures, to nature, to the universe; transcendence as a deeply and joyously experienced need to be in harmony even with what we ourselves are not, what we do not understand, what seems distant from us in time and space, but with which we are nevertheless mysteriously linked because, together with us, all this constitutes a single world. Transcendence as the only real alternative to extinction.[47]

Readers may take issue with Havel on points large or small. The present point is not so much agreement, disagreement, or critique, however. It is Havel's articulation of what by another name is called "the integrity of creation." With that he joins others, this volume included, who carry on a search that moves between a description of the presently convulsed world and an effort to articulate an integral sense of the whole that carries us beyond present division and destruction. In the end, this is the search for an earth cosmology and earth ethic, carried out in the recognition that nature and earth comprise a single community. Whether we like it or not, it's life together now or not at all. Earth faith and earth community — this is humanity's next journey.

---

47. Ibid., 9–10.

PART I

EARTH SCAN

The triptych is a worn, distinguished feature of Christianity and a presence in many a sanctuary. Framing the altar from behind, it has three hinged panels. Each different, together they portray a single drama.

Part 1 is a triptych. Each panel portrays what is happening to earth. Together, the whole is an Earth Scan.

The Earth Summit in Rio de Janeiro (1992) was the world gathered on one beach. But there were two meetings. At RioCentro, a huge convention center out past the favellas and safely away from the city center, the largest meeting of heads of state ever held on environment and development hammered out treaties line by line. Rows of national flags, each flag waving a more or less bloody and glorious history, symbolized the world's separate sovereignties and announced the meeting and its stakes. Downtown in Flamengo Park the Global Forum met, a hothouse of the world's peoples and their innumerable social-environmental causes as represented by nongovernmental organizations (NGOs). No flags here, just one poster, everywhere. The image? Earth from space, a marbled sphere caught for a second in the iris of an astronaut. This is the quiet, uncluttered image on the first panel, a cosmic oasis carefully adrift in the universe, held in place by mysterious forces of gravity. Lewis Thomas's description of it is this:

> Viewed from the distance of the moon, the astonishing thing about the earth, catching the breath, is that it is alive. The photographs show the dry, pounded surface of the moon in the foreground, dead as an old bone. Aloft, floating free beneath the moist, gleaming membrane of bright blue sky, is the rising earth, the only exuberant thing in this part of the cosmos. If you could look long enough, you would see the swirling of the great drifts of white cloud, covering and uncovering the half-hidden masses of land. If you had been looking a very long, geologic time, you would have seen the continents themselves in motion, drifting apart on their crustal plates, held aloft by the fire beneath. It has the organized, self-contained look of a live creature, full of information, marvelously skilled in handling the sun.[1]

But what information is this "live creature" full of? What does the distant gaze reveal? That the earth is curved and well wrapped, a closed living sphere. Seven land masses bent over a great round water. Revealed is creation's integrity.

The second panel, unlike the first, is neither quiet nor distant. It is a "grunt-ing, lowing, neighing, crowing, chirping, snarling, buzzing, self-replicating"

---

1. Lewis Thomas, *The Lives of a Cell* (New York: Bantam, 1974), 170.

entourage on the move.[2] It is, in fact, Sweet Betsy from Pike County, Missouri, and her lover, Ike, making their way. They were not alone. Hundreds of wagon trains and thousands of roosters, hogs, bulls, cows, hens, bees, seeds, and plants joined their entourage. "Sweet Betsy from Pike," in addition to being a good American folk song, is a snapshot; a snapshot of the movement of peoples and plants and animals so globally successful that we now stand face-to-face with one another on a planet that is, as Thomas noted, the most "exuberant" place around. It is also a place in plain jeopardy. Earth up close is a fractious, deadly, saturated locale, hardly the picture of serene beauty circling in space. We need to visit some history. How did we get from there to here in such a way that sustainability is no longer assured?

The third panel bears an ark. It could be the little boat of the Children's Defense Fund, with the child's plea: "Dear Lord, be good to me. The sea is so wide, and my boat is so small." Or it might be the Good Ship Oikoumenē itself, the "whole inhabited earth" (*oikoumenē*) rocking on the great waters. (The *oikoumenē* ship is the logo of the World Council of Churches [WCC].) It could also be earth as Noah's Ark, the ark of life alone in the waters of deep space. Or it might be the ship in the foyer of the New Amsterdam Theater, New York City, cradled in the arms of a female figure named Progress. In any case, the subject is the same: earth's carrying capacity, its powers for sustaining present and future generations, the ability of natural and social systems to live together indefinitely. What now taxes those powers is the subject. So is a constructive response for the sake of the present and future. Getting from here (unsustainability) to there (sustainability) is the preoccupation.

The triptych thus tells a story, an earth story, a story of beauty and terror, our own and otherkind's story, the story of earth as a single vast household (*oikos*).

As this story, the triptych frames what we shall later ponder as we face the altar itself (part 2, "Earth Faith," and part 3, "Earth Action").

Any one triptych can only tell fragments of one story. It is not the only story. And it is not the future story. But it is where we are now. Unless we prefer rank illusion, we should begin nowhere else.

---

2. Alfred W. Crosby, *Ecological Imperialism: The Biological Expansion of Europe, 900–1900* (Cambridge: Cambridge University Press, 1986), 194.

A Slow Womb

EARTH: A WORK IN PROGRESS

Earth's is a slow womb.[1] Life came late, with trauma. It continues by jerks and starts, dyings and rebeginnings.

The beginning of beginnings was probably 15 or so billion years ago.[2] Everything began all at once as stupendous energy. The galaxies — all of them — were seeded in a single flash.

Still, it took another 5 billion years for the primal stars to form and smaller galaxies to coalesce into larger ones, and yet another 5 billion years for a disc-shaped cloud in one arm of one galaxy, the Milky Way,[3] to birth a sun and solar system. On one of the newborn planets, earth, an atmosphere, oceans, and continents — all wildly different from ours at present — were formed as molten rock cooled over several tens of thousands of years. Roughly 4 billion years ago, after the meteor bombardment subsided, the first cell emerged from the primordial soup. At 3.9 billion the first act of photosynthesis took place. Blue-green algae similar to pond scum came on the scene. It was apparently another 2 billion years before complex cells adept at using a newer gas — oxygen — evolved. About 1.7 billion years ago multicellular plants living on the sea floor evolved. Yet from 2 billion years ago to a mere 570 million, relatively little happened.

---

1. The metaphor is taken from Vikram Seth's *The Golden Gate:*

> Think how the crown of earth's creation
> Will murder that which gave him birth,
> Ripping out the slow womb of earth.

Seth is describing a nuclear holocaust, not another onslaught on earth by humans. The image nonetheless pertains. At the Vancouver assembly of the WCC in 1983, "the integrity of creation" was introduced in response to *both* threats, the nuclear and the environmental. Seth is cited from Shridath Ramphal, *Our Country, the Planet: Forging a Partnership for Survival* (Washington, D.C.: Island Press, 1992), 28.

2. Dates are uncertain. They presently range from 8 to 20 billion years ago. Fifteen billion is commonly cited as being as near a consensus as exists among scientists at present. See "'Age of Universe Is Now Settled,' Astronomer Says," *New York Times*, March 5, 1996, C1, C8.

3. The shortcoming of this sketch is its failure to stay long enough in any one place to notice details. Consider just this one paragraph about the Milky Way: "The Milky Way galaxy is a celebration of diversity, abounding with hundreds of billions of stars, each different from every other. The Milky Way's brightest stars emit more light in a single day than the sun will generate for the next two thousand years, while the faintest stars glow so feebly that if one of them replaced the sun, noon would be darker than a moonlit night.... The Milky Way's oldest stars date back to the galaxy's formation, 10 to 15 billion years ago; its youngest are younger than you or I" (cited by Owen Gingrich, "Starstruck," *New York Times Book Review*, May 7, 1995, in his review of Ken Croswell, *The Alchemy of the Heavens: Searching for Meaning in the Milky Way* [New York: Anchor/Doubleday, 1995]).

Then came an eruption. In an astonishingly short time, forms of life proliferated. Insects, earthworms, corals, sponges, mollusks, and animals with rudimentary backbones — all the major body plans still with us — appeared. Just as suddenly most of these were eliminated. Roughly 80–90 percent of life collapsed forever (the great Cambrian extinctions). Beginning largely anew, shellfish, finned fish, vertebrates, insects, trees, and reptiles strode on the scene over the next 300 million years, only to experience the Permian extinctions of 245 million years ago (75 to 95 percent of all species simply vanished). Life began yet again, and dinosaurs and flowers, the first mammals and birds all appeared. So did the "Atlantic" Ocean. The mass continent of Pangaea had begun to break up and drift on earth's crustal plates, "held aloft by the fire beneath."[4] Then 67 million years ago, in the Cretaceous era, more extinctions occurred. Nonetheless, life began (or continued) again. Various orders of mammals extended their family trees (40 million years ago); cats and dogs and apes first appeared (30 million years ago), as did camels, bears, and pigs (only 4.5 million years ago). Grasslands spread. But so did glaciers. (The current cycle of ice ages began 3.3 million years ago.) About 2.6 million years ago *homo habilis* made its mark, using stone tools. Around 1.5 million years ago *homo erectus*, the hunter, emerged, as did bison, sheep, and a plethora of mammals. (Mammals as a whole saw their best days 1 million years ago.) *Homo sapiens* first walked tall only very recently — some 200,000 to 400,000 years ago. Some, the argument is made, walked to the "Americas" 35,000 years ago. Ritual burials were conducted 100,000 years ago, and caves were the scene of early interior decorating 18,000 years ago. Agriculture began 10,000 to 12,000 years ago; cattle were tamed 8,800 years ago; farmers in China grew millet 7,500 years ago; Andean pottery was used 5,300 years ago. Mesopotamian civilizations, and the wheel and writing, were known in 3500 B.C.E., Nile civilization in 3000 B.C.E., and Greek philosophy in 600 B.C.E. (The world human population was 5 to 10 million in 3500 B.C.E.) Modern nation-states began to emerge 400 years ago. (The population was 500 million by then.) The industrial revolution started about 200 years ago. (In 1900, halfway through, the population was 1.6 billion; by 2000 it will be the inverse, 6.1 billion.)[5]

Last Words

Time lines give hurried information and some sense of a great work in progress, the universe story itself.[6] Yet lists of dates capture life's drama poorly. We feel

---

4. The vivid, though not scientifically precise, phrase of Lewis Thomas (*The Lives of a Cell* [New York: Bantam, 1974]).

5. This time line is adapted from the more elaborate one of Brian Swimme and Thomas Berry in *The Universe Story: From the Primordial Flaring Forth to the Ecozoic Era* (San Francisco: HarperSanFrancisco, 1992), 269–78. I have updated Berry and Swimme with "Three Finds Clarify Life's Murky Origins," *New York Times*, October 31, 1995, C1, and "New Fossils Take Science Close to Dawn of Humans," *New York Times*, September 22, 1994, A1, A11.

6. The reference is to the title of the book by Brian Swimme and Thomas Berry, cited above. The time line here was the best available at the time of Swimme and Berry's writing. These dates

neither the struggle nor the gift of earth, and we barely grasp the human factor in this greatest of all sagas. We fathom neither humankind's minuscule role nor its gigantic one. Let's try again.

Robert Overman uses a nice device to bring the whole story within range of our humble capacity to imagine earth's events.

Suppose, restless creatures that we are, that we simply skip the first two-thirds of the universe (perhaps 10 billion years) and fast-forward to earth's appearing. And suppose we conceive the 5 billion years of earth as a 10-volume set of books, 500 pages per volume. Each page thus tells a tale 1 million years in the making. Life itself is a late episode, it seems, at least cellular life. Its first mention is in volume 8. Most of that volume is devoted to plants, plants going terrestrial together with amphibians. Then on page 440, reptiles have their most prodigious hour, only to be superseded by birds and warm-blooded animals 5 pages later (i.e., 5 million years).

On page 499 of volume 10 humankind appears, probably making claims to "dominion" and supposing that the species was the reason the drama of the previous 4,599,600,000 years was choreographed in the first place! The last two words on the very last page recount the human story from the beginnings of civilization (6,000 years ago) to the present.[7] In geological terms, that is, earth terms, all of human history is virtually instantaneous, a late case of spontaneous combustion.

What might all this mean? Or, shrinking from such a question at the hands of a species so new to the story, what might the last two words say? At least this. Humans entered a world billions of years in the making, a world that for all its fits, starts, and rebeginnings is a biologically rich one with a fine-tuned capacity for life. But not simply "life." Rather, life befitting *Homo sapiens*. With every breath and chew and innovation, we are part and parcel of nature as womb and lifegiver for nature's most recent of innumerable forms, most of them stranger than fiction.

The last two words also give double notice. While humans contributed very little to life, on land, in the water, in the air, we are utterly dependent upon the rest for ongoing life. Such is the wont of creatures at the top of the food chain, busy beyond comparison, and numerous in the extreme.

A *New York Times* article, "Bugs Keep Planet Livable Yet Get No Respect," makes the point well:

---

are in constant flux, however, as new astronomical, paleontological, and archaeological discoveries are made. For a limited sampling of the debates, see "Astronomers Debate Conflicting Answers for the Age of the Universe," *New York Times*, December 27, 1994, C1; "New Fossils Take Science Close to Dawn of Humans," *New York Times*, September 22, 1994, A1; "A Research Team Detects Many More Galaxies," *New York Times*, April 7, 1995, A24. Intriguing as these debates are, for our purposes they do not alter the main point of earth as a slow work in progress in which life emerges late on a rough journey (more like "the reckless fury of a rock concert" than "the grandeur of a classical symphony," to adapt Ken Croswell's comment about the origins of the Milky Way), and human life arrives very late.

7. Robert Overman's device as reported by John Cobb in *Sustainability: Economics, Ecology, and Justice* (Maryknoll, N.Y.: Orbis Books, 1992), 119–20.

Humans may think they are evolution's finest product, but the creepies, crawlies, and squishies rule the world. Remove people from the face of the earth and the biosphere would perk along just fine, ecologists say. Remove the invertebrates — creatures like insects, spiders, worms, snails and protozoans — and the global ecosystem would collapse, humans and other vertebrates would probably last only a few months, and the planet would belong mostly to algae and bacteria. Invertebrates, it turns out, are the biological foundation of ecosystems and crucial to every one of the ecosystem processes.[8]

The article goes on to say that scientists are concerned about the diminution, even extinction, of numerous invertebrate species. The consequences are far graver than the loss of those creatures to which the public gives high attention — a whale, an orchid, a snail darter, an owl. The issue is more basic; it is ecosystem life and health, indispensable to all those "higher" forms.

Still, the startling reality is not our embeddedness in earth as the only home fitted to our nature. Were that not so, we wouldn't be here! The astonishing thing is the last syllable of the last word of the last volume. Here humans turned the great tide against life itself. Here the process of slowly closing down major life-systems began at human hands. (In earth time it's happening at breakneck speed.) Slow or fast, it is astonishing "that all that has been produced over a billion years is so vulnerable to destruction by this latecomer."[9] It is astonishing that an organic world aeons in the making could be so easily jeopardized by the species that claims to master its secrets and care for it as a watchful steward. It is astonishing that *Homo sapiens*, by nature painfully aware of life's comings and goings in death and birth, should have considered earth indestructible and inexhaustible. It is astonishing that we treat nature as zoo, stage, sink, treasure chest, reservoir, and dumpster. Alongside our minuscule tenure the thing most astonishing is our gargantuan imprint.

These are the ten volumes of earth to date. We begin volume 11 as the twentieth century comes to an end.

Universe Principles

We are eager to get volume 11 right. The first task is understanding the nature of nature and our failure to account for it rightly in our present way of life. So we need to keep circling the subject and spy it out, first from one angle, then another. Perhaps most important is to understand the universe as a dynamic evolutionary one, earth included.

Take contingency, for example. Everything in the natural order — even galaxies, even dinosaurs, even humans — comes and then goes. None is necessary, all

---

8. "Bugs Keep Planet Livable Yet Get No Respect," *New York Times*, December 21, 1993, C1.

9. Cobb, *Sustainability*, 120. Cobb's reckoning of "a billion years" is conservative. Much we depend upon and affect, such as the atmosphere, is longer in the making.

die. In fact, only through the coming and going, the processes of living and dying, does life, especially new life, come to be.[10]

Humans, through clever technologies that enhance native powers, can reconfigure nature so as to stave off some of contingency's assaults. But not all of them and not forever. Life can be prolonged. It cannot, however, be exempted from finitude and mortality.

Yet the contingency is not anarchic. Evolution, too, has its principles, its "form-producing powers."[11] Differentiation, autopoiesis, and communion are three crucial ones.

Differentiation means that "in the universe to be is to be different." Nature has "an outrageous bias for the novel" and uses it for survival and development.[12] Diversity, complexity, variation, disparity, wild articulation — all these, at the very heart of cosmic evolution, guarantee a constant flowering of the unique. The universe fifteen billion years ago was nothing on the order of what it is now. Nor was life on this planet even one million years ago. Some millions of years hence it will be another world as well, wildly different.

Autopoiesis is the capacity of nature to self-organize. Whether we call it self-manifestation, subjectivity, or interiority, it is the power of self-articulation, and it runs from an atom's self-organization to a cell's to an ecosystem's to a star's and galaxy's. There is an integral, endemic creativity here. The articulation never ceases. And it happens in connection with all else.

Communion is the third universal principle, a synonym for the internal relatedness and interdependence of creation, the reciprocity and affiliation manifest everywhere "in earth and sky and sea" (to recall a hymn line). There is a deep, aboriginal kinship, since all is stardust. All the "createds" are "relateds." Yet the forms all interrelated with one another are myriad. "Nothing is itself without everything else." The "togetherness of [changing] things" is a systemic feature of evolution itself.[13] In some instances the scale may be small; a cell is a communion of organelles; a tissue is a communion of cells; a body is a communion of tissues. Or it may be huge; a galaxy is a communion of stars and their worlds. Whatever the scale, it is invariably complex, from the cell to the galaxy.

Swimme and Berry offer an example of this third principle that nicely illustrates the other two as well:

> An unborn grizzly bear sleeps in her mother's womb. Even there in the dark with her eyes closed, this bear is already related to the outside world. She will not have to develop a taste for blackberries or for Chinook salmon. When her tongue first mashes the juice of the blackberry its delight will be immediate. No prolonged period of learning will be needed for the difficult task of snaring a spawning salmon. In the very shape of

---

10. Langdon Gilkey, *Nature, Reality, and the Sacred: The Nexus of Science and Religion* (Minneapolis: Fortress, 1993), 162.
11. Swimme and Berry, *Universe Story*, 70.
12. Ibid., 74.
13. Ibid., 72, 77.

her claws is the musculature, anatomy, and leap of the Chinook. The face of the bear, the size of her arm, the structure of her eyes, the thickness of her fur — these are dimensions of her temperate forest community. The bear herself is meaningless outside this enveloping web of relations.[14]

Swimme and Berry's description of nature as evolving in keeping with these principles leaves us with the empirical meaning of "the integrity of creation." While Swimme and Berry hesitate to assign the word "organism" to the earth itself — a worthy caution — they do allow "organic." Organic describes the "inner coherence and integral functioning of the planet." "Indeed the Earth is so integral in the unity of its functioning that every aspect of the Earth is affected by what happens to any component member of the community."[15] If this is so, the implication is utterly clear: the well-being of the planet is the condition of the well-being of its member communities. Jeopardy for the one is jeopardy for the other. The implications are striking:

> To preserve the economic viability of the planet must be the first law of economics. To preserve the health of the planet must be the first commitment of the medical profession. To preserve the natural world as the primary revelation of the divine must be the basic concern of religion. To think that the human can benefit by a deleterious exploitation of any phase of the structure or functioning of the Earth is an absurdity. The well-being of Earth is primary. Human well-being is derivative.[16]

Lest all this drift from our main concern — our failure to account for the nature of nature in our destructive way of life — we return to our own time and place. At the end of volume 10, what do we find?

Forced changes in natural systems at human hands are one reason for earth and its distress. *That* such changes happen is not a surprise to nature, since everything participates in the rest. But *what* occurs can be a surprise, not least to us.

Take the changing atmosphere. This venerable atmosphere has seen dramatic changes. Some millions of years ago early life-forms consumed the carbon dioxide then dominant and generated oxygen as waste. The surfeit of oxygen eventually killed off most of those life-forms. But new species evolved that utilized oxygen "pollution" rather well. We presently live in an age when the balance of atmospheric gases and temperatures is part of subtle physical arrangements highly conducive to plants and animals, together with their elaborate food chains. You are reading these words because millions of changes and trillions of exchanges issued in a certain complex and relatively stable mix of gases friendly

---

14. Ibid., 77–78. Examples can be multiplied. In the desert Southwest of the United States each species of yucca requires a separate and single species of moth to pollinate it. Each moth and each yucca depends on the other for life.

15. Ibid., 243.

16. Ibid.

to mammalian life. Had much higher levels of oxygen evolved, fires would have been a continuous plague. Had the levels been much lower, respiration would have required other systems than those we have. We are as fine-tuned to our atmospheric environment as the bear and the Chinook salmon. At the same time we, and not they, threaten to affect those balances in ways we never have, on a timescale too rapid to maintain atmospheric quality, and with elements too alien. This is the significance of excessive greenhouse gases, the destruction of the ozone shield, and acid rain. Worse, what goes up must come down, and it may be that humans have started writing the history of weather in ways we never have. We do not yet fathom the dimensions of this. We *do* know that we are not in control. The biosphere, including the atmosphere, is too enormous and complex for that and lives by another schedule. Thus we do not see the consequences of forced atmospheric change until long after we set the forces in motion that create them. (Current ozone destruction is from carbons released years ago. Those of recent years are still casually wending their way to the stratosphere like the slow waters of meandering streams.) Government and business policy notwithstanding, this time lag means we cannot wait to "see what happens" before deciding what to do. Using the planet as a laboratory for massive, uncontrolled trial-and-error experiments — this is our present case — is hazardous. "We are safe, well ventilated, and incubated," Lewis Thomas says, "provided we can avoid technologies that might fiddle with the ozone, or shift the levels of carbon dioxide." It may be hard "to feel affection for something as totally impersonal as the atmosphere, and yet there it is, as much a part and product of life as wine and bread."[17]

In conclusion, Charles Marsh apparently got it right in a best-seller published a century and a quarter ago. Marsh, an oddball naturalist because he wrote of relationships instead of organisms, said this in 1874:

> Man is dealing with dangerous weapons whenever he interferes with arrangements pre-established by a power higher than his own. The equation of animal and vegetable life is too complicated a problem for human intelligence to solve, and we can never know how wide a circle of disturbance we produce in the harmonies of nature when we throw the smallest pebble into the ocean of organic being.[18]

APARTHEID HABITS

Understanding the concrete meaning of differentiation, autopoiesis, and communion as universal principles of nature, that "ocean of organic being," is a

---

17. Thomas, *Lives of a Cell*, 173–74.
18. From Charles Marsh, *Man and Nature* (New York: Charles Scribner, 1874), as cited in William Ashworth, *The Economy of Nature: Rethinking the Connections between Ecology and Economics* (New York: Houghton Mifflin, 1995), 11. Ashworth doesn't cite the page number of the quotation.

start. But where would that lead us? What other notions are needed to get volume 11 right?

Learning language is learning culture. Language plays a shaping role in our perception of the world, how we think about it and respond to it. For this reason, learning to perceive, think, and talk differently often leads to acting differently.

"The integrity of creation" requires a kind of return to early childhood in that we need to learn anew the categories we think *with* when we think *about* things. Learning to think with different categories happens in part by learning new words or putting familiar words together in new ways.

We could learn to speak, for example, not of humanity *and* nature, but of humans *in* and *as* nature. Not of culture *and* nature, or history and nature, but of culture and history *in* and *as* nature. Not of society *and* nature, but of society *in* and *as* nature humanly construed. After all, what we call society and culture are dramatic episodes of what earth does through us as part of earth itself, in the form of nature called us.

Or we could acknowledge that humans never rise above nature, never transcend it. Rather, our considerable spiritual capacities are nature's own flowering in the form of nature that is us, just as our considerable freedom is the gift nature in the form that is us has fashioned as one of its own freedoms. Our rational and moral powers are likewise not alien additives but nature's own home-grown powers. From within such a world, the human is that creature of earth in whom is found earth community's own capacity to reflect upon and celebrate itself in conscious self-awareness.[19] In this creature, the universe learned to pray.

With attention to the worlds that language and notions build inside our heads, we might even find a way to banish the apartheid habit of distinguishing "human" from "nonhuman" as categories that divide in order to identify. Just as *blankes* (whites) and *nie-blankes* (nonwhites) carved the entire human universe along racial lines in South Africa and elsewhere, so "human" and "nonhuman" still cleave creation in our present mind-set and habits. Such apartheid thinking leaves us imagining we are an ecologically segregated species. Such thinking violates the integrity of creation and puts it at risk.

How so? Apartheid mind-sets identify all else by declaring what it is *not*. "Nonhuman" is not us. "Not us" is its significance, moral and otherwise. This fashions a negative identity. That is apartheid's wont. In human affairs such apartheid habits are disastrous over the long haul, whether "us" and "not us" are racially construed or construed on lines of ethnicity, class, gender, or sexual orientation. "Not us" means "other than" as "less than." "Us" means "more than" as "better than." "We" are not "they," and "they" are "the other" as objects rather than subjects.

It is no less so for humans and (the rest of) nature. Indeed, it is potentially even more dangerous, since all humans everywhere through all generations are

---

19. Thomas Berry, *Technology and the Healing of the Earth,* Teilhard Studies 14 (Chambersburg, Pa.: Anima, 1985), 15.

totally dependent upon the rest of nature, breath for breath, chew for chew, comfort for comfort, livelihood for livelihood.

Yet "nature" to an apartheid mind-set is the great aggregate, a faceless mass without personality. It may merit a name — even *nie-blankes* is a name of sorts. But the name is often little more than "resources," "natural capital," or some other term that refers to the grand stage for the human drama, even the divine-human drama. ("Creation is the external — and only the external — basis of the covenant," wrote the twentieth century's most famous Protestant theologian.)[20]

An utterly utilitarian treatment commonly results. Since World War II especially, a kind of planetary bulldozing has taken place. Chemical, automotive, agricultural, construction, electronic, nuclear, aeronautical, military, and space industries have fashioned transnational networks that are aggressive in the extreme. (This has been labeled "development aggression" by some Two-Thirds World peoples.) Whole "townships" of life have been moved and removed — another analogy to apartheid. The entire planet is inventoried and transformed for the global economy. Nature has become "the other" of choice.

To break this habit of human and nonhuman expressed in industrial-technological ascendancy, and to begin mirroring the integrity of a more fragile creation than it allows, we might modulate our language. We might say "human-kind" and "otherkind," for example. It may seem precious little progress, and it is, since humans are still abstracted from the rest of nature, which is still an aggregate. Yet at least difference is recognized and marked in a way that identifies nature as more than "not us."

Fortunately, truth once released often worms its way inch by inch and demands more. What begins as "otherkind" soon blooms and darts and crows and turns golden in autumn. It dies and rots and grows and suckles and climbs green up stone walls. It insists upon attention as unique lives lived with and for us, as well as apart from us and long before us. After us as well. Once we escape apartheid thinking, "otherkind" takes on life and articulation.

We do not live with "nature," then. We live with trees, animals, birds, and insects of nearly infinite variety; with winds, clouds, and the spirited gases of the atmosphere; with mountains, lakes, streams, oceans, beaches, forests, grasslands, and deserts; with bacteria and amoebae and viruses; with the sun, moon, and fifty billion galaxies. We do not live on a stage that serves as the grand prop for history. We live deep inside critical daily, historical exchanges large and small with all things at once, exchanges near and far and much too complex and wondrous to understand in either their minutiae or their totality. We do not live "on" earth. We live as part *of* earth's articulation — in subtle bioregional communities internally tuned to tropics and arctics; seacoasts and mountains; plains, woodlands, and deserts; and urban settlements amidst these.

Maybe, with practice, we will even learn to see. Then we might match Annie Dillard's communion with Ellery:

---

20. Karl Barth, *Church Dogmatics* (Edinburgh: T. and T. Clark, 1958), 3/1:97.

This Ellery cost me twenty-five cents. He is a deep red-orange, darker than most goldfish. He steers short distances mainly with his slender red lateral fins; they seem to provide impetus for going backward, up, or down. It took me a few days to discover his ventral fins; they are completely transparent and all but invisible — dream fins. He also has a short anal fin, and a tail that is deeply notched and perfectly transparent at the tapered tips. He can extend his mouth, so it looks like a length of pipe; he can shift the angle of his eyes in his head so he can look before and behind himself, instead of simply out to his side. His belly, what there is of it, is white ventrally, and a patch of this white extends up his sides — the variegated Ellery. When he opens his gill slits he shows a thin crescent of silver where the flap overlapped as though all his brightness were sunburn.

For this creature, as I said, I paid twenty-five cents. I had never bought an animal before. It was very simple; I went to a store in Roanoke called "Wet Pets"; I handed the man a quarter, and he handed me a knotted plastic bag bouncing with water in which a green plant floated and the goldfish swam. This fish, two bits' worth, has a coiled gut, a spine radiating fine bones, and a brain. Just before I sprinkle his food flakes into his bowl, I rap three times on the bowl's edge; now he is conditioned, and swims to the surface when I tap. And, he has a heart.[21]

TIME IN, RATHER THAN ON, OUR HANDS

Apartheid's categorical errors help explain why we violate creation and tear at earth's slow womb. But there are other dysfunctional habits of mind.

Take our sense of time and history. We — some we's more than others — continue to live as children of the great divorce of nature from history. Modernity's time belongs to a notion of history that snubs the realization we can only have history *in* and *as* nature. Historical time is human time itself by our reckoning, time abstracted, like us, from the rest of nature. This is time of our own making, much as we then come to think the world is! Utter dependence upon biological and geological time for every breath we take and morsel we taste is obscured. Like one masterpiece painted over another, we superimpose our cycles on the rest of nature's and lose the capacity to see the original work. It would be one small step for humankind if for a season we spoke only of "nature-history," "culture-nature," or "society-nature." Then we might habituate ourselves to see indissoluble connections and pause to ask whether our ways are outstripping the rest of nature's. Not least we might begin to recognize how historical nature is. Nature is not unchanging cycles, shorn of memory and bearing a future identical with its past. The butt log of a red oak, for example, is a compressed treasure of time and memory. The original sapling alive

---

21. Annie Dillard, *Pilgrim at Tinker Creek: A Mystical Excursion into the Natural World* (New York: Bantam, 1975), 126.

perhaps a hundred years ago is still distinct as dark wood at the center, and
knots here and there across narrow and wider layers of wood record branches
since outgrown as well as the vagaries of the weather from one season to another.[22] Of course there is a difference here from human time. But it is not,
Erazim Kohák explains, "the putative difference between human historicity and
the . . . atemporality of nature." The difference runs instead "between the natural
temporality of all living beings, *including humans,* and the illusory mechanical
temporality of the [modern, human-made] world."[23] The latter is the "o'clock"
world, the increasingly artifactual world, the world of highly particular cultural
rhythms imposed upon the rest of nature's rhythms. These artifacts and constructs, including our sense of time itself, *can* serve human and other life well
and often do. We do not want a world without them, anymore than we want
nature without improvement. But the constructs serve well only when they are
the human way of articulating a more fundamental truth: namely, "the rhythm,
the seasons of life."[24] They serve well only when they do not superimpose mechanical temporality upon natural temporality and demand that the latter bend
to the former come what may.

How deep has apartheid thinking sunk into our brains? Western religion,
philosophy, and science offer a good test, since each in its own way seeks to describe essential reality. Noteworthy, then, is that virtually all the philosophies and
theologies prominent in the West during the decades the causes of earth's distress were gathering their finest head of steam — the years after World War II —
share the deadly illusion of historical time as human time. They blessed the divorce of nature from history and forgot that the womb of earth is a slow womb,
and alive. Existentialism, neoorthodoxy, liberalism, political theology, most liberation theologies in their early stages, and virtually all of fundamentalism and
evangelicalism assumed nature to be little more than stage, external habitat,
and resource. At best nature was inspiration and the stuff of poetry, song, and
meditation. An aesthetic use supplemented a basic utilitarian one.

Curiously, such theology took a page from the very sciences that had little
time for theology and, like those sciences, treated nature as "outside" us, ruled
by a causality we could know and manipulate as spectators to reality. Both theology and science were blind to nature as the rest of *us.* Nature was chiefly
useful matter, even when theologians were quick to add that it was also the
good creation of God for which we bore responsibility as stewards. Despite such
high stewardly calling, or perhaps because of it, the ethical universe was shrunk
to our size. Only other humans counted as true companions. The rest of nature was "not us," homogenized as the mute other. This is surely a diminished
Community of Life. And surprisingly lonely, even with six billion of us.

Thomas Berry's striking analogy for this is autism. The modern West's has
been an autistic theology and worldview. Autistic children are often intelligent,

22. Erazim Kohák, *The Embers and the Stars: A Philosophical Inquiry into the Moral Sense of Nature* (Chicago: University of Chicago Press, 1984), 76.
23. Ibid., 77; emphasis added.
24. Ibid., 78.

but they are enveloped by their own tight world. They relate hardly at all to others emotionally and are so isolated that others penetrate their austere sanctuary with the greatest difficulty. We moderns have been autistic in our relation to the rest of nature. We cannot listen to the earth or "think like a mountain." We talk only to ourselves. "We have broken the great conversation."[25] No one is let in, and no one is home.

It will not do, in understandable self-defense, to tally all the achievements of this five-hundred-year epoch, in order to provide a "balanced" account or show the accomplishments of even autistic generations. The achievements are plainly numerous and real. Many are enormous and genuinely lasting. Average human life span has been extended a generation and more; diseases that caused misery for millions have been eliminated or controlled; wealth has been produced en masse and distributed to millions, even billions, so that life could be more than the compulsions of survival. Yet claiming these as the sweat of our salvation will not do, *if* this way of living is ultimately ruinous. And ruinous it is, not finally because it is violent, spiritually comatose, and expensive, though it is all three. It is ruinous for a reason far simpler. Nature, ourselves included, cannot afford it on terms hospitable to life as we know and cherish it. In short, our relationship to the rest of nature has grown so wildly asymmetrical that we violate the basic law of life itself — exchange and reciprocity, giving and receiving in some relatively fair measure, living and dying and dying in order to live, with sacrifice a requirement of all parties. Creation doesn't work otherwise. We cannot live as an ecologically segregated species (apartheid and autism). Nor can human historical time run on another track from (the rest of) nature's rhythms, rhymes, and length of days. To right ourselves we will have to learn to think against the grain of much of our ingrained thought.

But let us put the matter positively, and as a summary.

All lives are integral to earth. All share in the integrity of creation. All belong to the Community of Life. All deserve names and the recognition of differences that bestow character. And — here is the payoff — all share certain realities essential to an earth ethic. For brevity's sake, we use Paul Taylor's conclusions. (1) We and all other creatures "face certain biological and physical requirements for our survival and well-being." (2) All of us have a respective good of our own, the realization of which "depends on contingencies that are not always under either our or [other creatures'] control." (3) Although free will, autonomy, and social freedom are distinctly human freedoms, they rest in an even more basic freedom shared in common. Namely, that every living thing struggles to live the kind of life of which it is capable. To these commonalities are added the differences we noted earlier. (4) As a species, we're the newcomers in an order of life established hundreds of millions of years before we stood erect somewhere in East Africa. (5) And while we cannot do without otherkind's membership in

---

25. Thomas Berry with Thomas Clarke, S.J., *Befriending the Earth* (Mystic, Conn.: Twenty-Third, 1991), 97.

earth's Community of Life, otherkind can do without ours. "Our dependence on the general integrity of the whole realm of life is absolute."[26]

Perhaps we should all modify our humble sense of time and place in the manner of Austin Dobson: "Time goes, you say. Ah no! Alas, time stays, we go."[27] At least we should recognize that earth's is the only womb we have, and slow.

---

26. Paul Taylor, *Respect for Nature: A Theory of Environmental Ethics* (Princeton, N.J.: Princeton University Press, 1986), 101–2, 114.

27. As cited as a sidebar quotation for an editorial, "As Busy as We Wanna Be," *Utne Reader* (January/February 1994): 52.

Sweet Betsy and Her Avalanche

At least since the creation of agriculture and settled human communities, humans have stressed their surroundings. We cannot exist without modifying and exploiting ecosystems. So Clive Ponting is probably right: "The most important task in all human history has been to find a way of extracting from the different ecosystems in which people have lived enough resources for maintaining life — food, clothing, shelter, energy and other material goods."[1] Local environmental damage thus began somewhere just east of Eden and has spotted the globe ever since. Chinese felling of forests beginning three thousand years ago created an erosion that colored the water fifty miles out at sea. Hence the name "the Yellow Sea." Greek isles pictured as paradisiacal in ancient literature heard complaints of deforestation and ruin already from Plato. What is now the northern Sahara of North Africa was the Roman Empire's granary.

Nonetheless, the last two words of volume 10 are wildly different from the foregoing paragraphs and pages of earth's tale. The tide itself, rather than an occasional destructive wave, has turned against life itself.

Why? The answer cannot simply be that humans are the only species capable of destroying the ecosystems upon which they depend, even though that appears to be the case, at least on a mass scale. This potential cannot of itself be the reason. If it were, our present circumstances would have appeared much earlier in the story. A likelier explanation is that humans alone of all species have now invaded every terrestrial ecosystem and, with powers amplified through technology, have dominated them. (We are presently extending this to marine ecosystems.)

But to many people the most obvious reason is sheer human numbers. The case is a strong one. The forty-five years between 1930 and 1975 (one letter in the last word of volume 10) *added* 2 billion of us, a steep climb even from the *total* population only three decades earlier (1.6 billion). Now another Mexico City is added every sixty days, another Brazil each year. Delegates to the UN International Conference on Population and Development in September 1994 noted that 2.2 million people were added during the nine days they met to solve population problems!

Unfortunately, catching the meaning of such addition taxes human imagination beyond its powers. We have little capacity to see, even in the mind's good eye, the difference human presence makes for the planet when 2000 B.C.E. is compared with 2000 C.E., or even 996 C.E. with 1996.

This eludes us in part because body count and century jumps are but a slim slice of the whole. Adding another Mexico City or Bombay is certainly of grave

---

1. Clive Ponting, *A Green History of the World: The Environment and the Collapse of Great Civilizations* (New York: Penguin, 1991), 17.

significance in a cramped world. But it is far different from adding, say, another New York or Los Angeles. New York and Los Angeles, or Berlin and Tokyo, are hubs in nations that presently consume 75 percent of world resources and account for 75 percent of all solid and toxic waste. Thus we have the jarring fact that Chicago (at 2.9 million citizens hardly a teeming urban monster today) consumes as much as one South Asian nation of 97 million people (Bangladesh).[2] The estimated 50 million people added to the U.S. population over the next forty years will have roughly the same negative global consumption impact as 2 billion people in India.[3] Prime Minister Gro Harlem Brundtland reported to the Oslo Symposium on Sustainable Consumption that "if 7 billion people were to consume as much energy and resources as we do in the West today we would need 10 worlds, not one, to satisfy all our needs."[4]

To say, then, that "we humans" are presently degrading "nature" in unprecedented degree is not unimportant, just as it is not unimportant that humans now inhabit and exploit every terrestrial ecosystem. But such truth obscures more than it reveals. Different peoples in different locales at different times living very different kinds of lives with different consequences for the planet — this is more clearly the subject than sheer aggregates. In like manner, the subject is not what is happening to "nonhuman nature" as some nebulous mass entity. The subject is what is happening to particular life-systems and their citizens in particular locales, to a certain layer of gas in a dynamic atmosphere at a particular location above the poles, or to the endowment of specific nonrenewable resources on the ocean floor or beneath forests, grasslands, and sand dunes. Here Liebig's law of the minimum is critical: it is not general conditions that ultimately determine "carrying capacity," but that single factor that is in short supply in a given ecosystem. All of which is to say that undifferentiated thinking of apartheid minds will never resolve earth's distress. It will only exacerbate it. We must dive into the details and bring a comprehensive analysis along.

Whatever the details and however complex the analysis, the issues and goal are quite simple. (At least *stating* them is simple.) The issues are two: (1) how to live in a sustainable relationship with the rest of earth; and (2) how to live with the play of human power, both in human-human configurations (the nature of society) and in human-biosphere configurations (society in nature). And the

---

2. Information from the Pew Global Stewardship Initiative, in "Who's Consuming the Planet?" *New York Times*, April 18, 1994, A15.

3. Paul Hawken, *The Ecology of Commerce: A Declaration of Sustainability* (San Francisco: Harper Business, 1993), 161.

4. Gro Harlem Brundtland, "The Challenge of Sustainable Production and Consumption Patterns" (keynote address, Symposium on Sustainable Consumption, mimeographed), 1. Paradoxically she goes on to say: "Some view decline in the standard of living for industrialized countries as necessary steps in bringing total global consumption to a sustainable level. The [UN] Commission [on Environment and Development] did not, however, conclude that such measures were needed, nor that they were desirable. Global change requires thriving economies, large-scale investment, and technological change. Neither change nor full employment will come about as a result of reduced economic activity" (2). Here the tensions, even contradictions, of economy and environment are blatant.

goal is sustainability as the capacity of natural and social systems to survive and thrive together indefinitely.

These issues and this goal in mind, we turn to an analysis that lets us understand what has transpired as the last word of volume 10. It may also poise us to write the first word of volume 11. In any case, such long-term investigative reporting should show us why earth's distress is less a crisis of nature than the crises of certain cultures *in* nature living full grain against it. Such is the subject of this and the next two chapters, against the backdrop of earth as a slow womb.

THINGS OLD, THINGS NEW

In human historical perspective, living sustainably and the play of human power are relatively recent as issues humans needed to worry much about. *Homo habilis* appeared about two million years ago. But agriculture, a big word in the human vocabulary, is among the very last words of volume 10 of cosmic history, invented only about ten thousand years ago. By then hunter-gatherer societies had already sauntered through 99.5 percent of the human odyssey itself! The rest — all .5 percent — is us in more recent ensembles.

From what we know with any confidence, hunter-gatherer societies were quite remarkable on the perennial matters of living sustainably with the environment and configuring power for sustainable society. Apparently rather egalitarian among themselves, they also lived on sustaining terms with their home terrain for very long stretches of time. Nonetheless, there were constraints and collapses, and instances of these issued in what eventually became settled agriculture.

Yet that 99.5 percent can mean little to us now. Six billion of us on a planet waving a "No Vacancy" sign can hardly take up hunting and gathering as sustainable livelihoods. The .5 percent is thus everything. In fact, the whole object of sustainability is to improve upon that unimpressive fraction.

Differently said, sustainability/unsustainability as an issue did not first appear with the 1972 publication of the MIT report, *Limits to Growth*. It has been a human preoccupation since the dawn of human time and a perennial issue since the coming of agriculture. There are genuinely new and critical dimensions of late, to be sure. But it is important to understand that sustainability is an issue as old as human records themselves. As *the* human pastime it has real standing.

Any and every civilization lives from the economy of nature. All civilizations depend finally on the ecological viability of their agricultural base and other natural resources, or the carrying capacity they have appropriated from others through commerce, conquest, or both. With this as backdrop, John Gowdy offers a short history of sustainability. More precisely, a history of unsustainability.[5]

---

5. John Gowdy, "Place, Equity, and Environmental Impact: Lessons from the Past," *Ecojustice Quarterly* 14, no. 2 (spring 1994).

Gowdy's studies took him to the three areas where intensive agriculture apparently arose almost simultaneously: Mesoamerica, China, and Southwest Asia. The rise of intensive agriculture seems to have gone hand in hand with three developments: (1) Some human populations in limited spaces grew relatively rapidly. Faced with too many people for one patch of earth, *Homo sapiens* needed to become either celibate or clever. Predictably, says Alfred Crosby, they took the latter course. But it did not always work.[6] (2) This meant that the taxing of local resources was added to population issues. Terracing, crop selection, animal husbandry, and irrigation were all eventually developed to increase yield and thereby sustain increased numbers on roughly the same land. (3) Nonetheless, in some instances local carrying capacity was exceeded, and a third development occurred, the collapse of relatively prosperous, or at least relatively "steady," human living in these locales.[7]

To his study of development in these regions, Gowdy added Middle Eastern and Mediterranean societies as well as the Easter Island calamity. Looking back on all of them, he noted an important *social* process accompanying societies' collapse. It seems that stratified society accompanied intensified agriculture, a social characteristic largely missing in hunter-gatherer societies. Furthermore, the social stratification had important ecoconsequences. Problems of overgrazing or deforestation, soil erosion or salinization were apparently driven as much by social dynamics as they were by, say, ignorance of nature's ways or technical ineptitude. After Gowdy checks in with studies of other civilizations, he concludes that "evidently there has never been an agrarian socioeconomic system, with its need for specialization and social stratification, that was sustainable for more than a few hundred years, with the possible exception of a succession of societies along the Nile Delta."[8]

This datum is disturbing enough for any whose goal is sustainability as the capacity of natural and social systems to survive and thrive together indefinitely! We will stay with it only long enough, however, to put in place four robust facts that affect subsequent analysis. First, the cohabitation of human beings with one another and the rest of nature has been problematic for a long while in human terms — ten thousand years. Unsustainability has real seniority as a human issue. It simply is *not* the case that societies normally lived harmoniously with the rest of nature for long stretches until modern interlopers leapt from history as a great exception. Ecologist Peter Farb may in fact be right about a certain nasty trait: "It appears to be a characteristic of the human [evolutionary] line — perhaps the one that accounts for its domination of the earth — that from the very beginning *Homo* [*sapiens*] has exploited the environment up to his

---

6. Alfred W. Crosby, *Ecological Imperialism: The Biological Expansion of Europe, 900–1900* (Cambridge: Cambridge University Press, 1986), 20.

7. For a history of the rise and fall of civilizations as related to environment, see Ponting, *Green History*.

8. Gowdy, "Place, Equity, and Environmental Impact," 13.

technological limits to do so."[9] Farb adds a note, probably to keep us moderns sober: "But until recently the harm this exploitation could cause was limited, for ancient man's populations were low and his technology primitive."[10]

Gowdy's second datum is equally unsettling: there is no guarantee that intimate human ties to restricted and well-known places translates as sustainability. A strong case can be made for community self-reliance and bioregionalism — this book argues vigorously for both. Yet human societies with these characteristics have also been among those that have literally bitten the dust, often dust of their own making. Plantation societies living close to the land, feudal fiefdoms, or village life generally are not paragons of either virtue or sustainability.

This leads Gowdy, third, to conclude that "the lesson to be learned for our own quest for sustainable development" is that "justice and equality are important factors in assuring sustainability."[11] The backdrop to this is Gowdy's finding that highly stratified societies controlled by a relatively small elite generate institutions that aim to keep social arrangements intact even when the cost is obvious and there is progressive environmental deterioration. Differently said: where maldistributed power and inequality reign, powerful forces usually continue to benefit from the depletion of both peoples and the land — until it is too late. Thus Gowdy's judgment that sustainability and unsustainability correlate roughly with just and unjust social arrangements.[12] Evidently, sustainability owes as much to its socio-ethical character as it does to its technical prowess and knowledge base. Gowdy is here echoing C. S. Lewis's comment of decades ago that what we call human power over nature is actually the power exercised by some people over others, using nature as a tool. For both Lewis and Gowdy, the outcome is the same: injustice is a great instability.

Fourth, and closely related, the arrangements of social power in the rise and fall of civilizations are usually linked to environmental components. The linkages may be complex, but they are normally present. So far as human beings affect conditions, social and environmental well-being thus tend to walk the same direction, if not always hand in hand. Social and environmental degradations also tend to be linked.

This awareness of sustainability and power as perennial issues since the rise of agriculture is without doubt important. Yet it may not divulge what sustainability and unsustainability are for us. That depends upon whether our circumstances parallel earlier ones. Some parallels exist, including our tendency (noted by Farb)

9. Peter Farb, *Ecology* (New York: Time-Life Books, 1970), 164, as cited by James Nash, *Loving Nature* (Nashville: Abingdon, 1991), 89.

10. Farb, *Ecology,* as cited by Nash, *Loving Nature,* 89.

11. Gowdy, "Place, Equity, and Environmental Impact," 13.

12. Ibid. Gowdy's anthropological evidence is confirmed from another angle in recent studies in ecological economics. See, for example, James K. Boyce, "Inequality as a Cause of Environmental Degradation," *Ecological Economics* 11 (1994): 169–78. Using economic and social data, Boyce tests two hypotheses: "First, the extent of an environmentally degrading economic activity is a function of the balance of power between the winners, who derive net benefits from the activity, and the losers, who bear net costs. Second, greater inequalities of power and wealth lead, all else equal, to more environmental degradation" (169).

to exploit to the limits of our technical capacities. Sustainability and power also show new forms and faces, however. Basically these are that cumulative human power now acts both *on* nature and *into* nature in ways that move sustainability from a local issue and requirement to a planetary one with both global and local impact. Thus while unsustainability is a ten-thousand-year-old problem, and local ecosystems have often been jeopardized at human hands, only now is the biosphere itself under siege and sustainability a *global* public worry. Planetary life-systems are in jeopardy, on land, in the air, under the sea, and the relationship of one species to the rest has been fundamentally altered in a short time, with consequences for all. Scale and extent of impact are wildly different at the conclusion of this century and millennium.

But how did we get here? And what might we learn about sustainability itself by telling the story one more time, from the front end forward and shorn of apartheid-infested categories? That brings us to Sweet Betsy and her lover, Ike.

SWEET BETSY

The *New York Times Book Review* ran a piece entitled "Developing Ourselves to Death." It briefly investigates Bruce Rich's *Mortgaging the Earth: The World Bank, Environmental Impoverishment, and the Crisis of Development.* The reviewer opens with this: "Economic development may seem so much a part of modern times as to render self-evident Bruce Rich's premise that it 'is now the organizing principle for almost every society and nation on the planet.'"[13] Yet Rich notes that it is "a relatively new idea in history, spreading from western Europe in the 17th century to conquer the world over the next three centuries."[14] Then the reviewer says: "These are the same three centuries in which people have most extensively altered the global environment."[15] *Mortgaging the Earth* goes on from here. It does not pursue the relatively recent domination of economic development or the linkages suggested by the juxtaposition of economic development and environmental transformation. Rather, it investigates policies of the World Bank as an agent of destructive modern development. But what about those linkages and the currents that economic development ("a relatively new idea in history") set in motion? What connections are there when, for example, the following three citations by four famous authors in three famous books are laid side by side? Or are the ties ephemeral, wisps of smoke accidentally crossing the same sky?

The first is from Adam Smith in *Wealth of Nations.* Writing in the precocious year of 1776, Smith said, with apparent confidence: "The discovery of America, and that of a passage to the East Indies by the Cape of Good Hope, are the

13. William Stevens, "Developing Ourselves to Death," *New York Times Book Review,* March 6, 1994, 29.

14. Stevens, "Developing Ourselves," citing Bruce Rich, *Mortgaging the Earth: The World Bank, Environmental Impoverishment, and the Crisis of Development* (Boston: Beacon Press, 1995).

15. Stevens, "Developing Ourselves," 29.

two greatest and most important events recorded in the history of mankind."[16] Little more than a half-century later, Charles Darwin, in *The Voyage of the Beagle* (1839), noted that "[w]herever the European had trod, death seems to pursue the aboriginal. We may look to the wide extent of the Americas, Polynesia, the Cape of Good Hope, and Australia, and we find the same result." And only nine years after Darwin, in 1848, Karl Marx and Friedrich Engels included the following in the *Manifesto of the Communist Party:* "The discovery of America, [and] the rounding of the Cape opened up fresh ground for the rising bourgeoisie. The East Indian and Chinese markets, the colonization of America, trade with the colonies, the increase in the means of exchange and in commodities generally, gave to commerce, to navigation, to industry an impulse never before known, and thereby, to the revolutionary element in the tottering feudal society, a rapid development."[17]

Intriguing, is it not, that a moral philosopher tracking early capitalist economies, a biologist describing natural selection and developing evolution as a theory, and social historians and critics of industrial capitalism, none of whom is reading the other, should all mention the discovery of America and the rounding of the Cape of Good Hope as epoch-making events? Even more intriguing are the possible historical connections of the "most important events recorded in the history of mankind" (Smith) to the revolutionary change and development of European society (Marx and Engels) and then both of these to the death of aboriginal peoples around the world (Darwin). And all this is recorded, not in 1992 with a gaze back over five hundred tumultuous years on the occasion of the rediscovery of Columbus by America, but independently of one another within three-quarters of a century beginning with 1776!

These three citations (the fourth is set aside for a moment) are epigraphs to Alfred Crosby's *Ecological Imperialism: The Biological Expansion of Europe, 900– 1900.* "Biological" expansion of Europe? Imperialism as an "ecological" event? Yes, and therein lie the quiet clues to the connections we seek. In due course we will have to slip even further back, to neolithic cultures, in order to understand what happened to nature and culture in their common trek *to* the present and why they now live under a common threat *in* the present. But for now we begin exactly where Crosby does, on page 1. The book opens thus:

> European emigrants and their descendants are all over the place, which requires explanation.
>
> It is more difficult to account for the distribution of this subdivision of the human species than that of any other. The locations of the others make an obvious kind of sense. All but a relatively few of the members of the many varieties of Asians live in Asia. Black Africans live on three continents, but most of them are concentrated in their original latitudes, the tropics, facing each other across one ocean. Amerindians, with few ex-

---

16. Cited from the frontispiece of Crosby, *Ecological Imperialism.*
17. Ibid., 1.

ceptions, live in the Americas, and nearly every last Australian Aborigine dwells in Australia. Eskimos live in the circumpolar lands, and Melanesians, Polynesians, and Micronesians are scattered through the islands of only one ocean, albeit a large one. All these peoples have expanded geographically — have committed acts of imperialism, if you will — but they have expanded into lands adjacent to or at least near to those in which they had already been living, or, in the case of the Pacific peoples, to the next island and then to the next after that, however many kilometers of water might lie between. Europeans, in contrast, seem to have leapfrogged around the globe.[18]

Or, in an image that Crosby takes from bee-keeping, "Europeans...have swarmed again and again and have selected their new homes as if each swarm were physically repulsed by the others."[19]

Europeans have indeed swarmed, in two senses. Both are important to the biological expansion of Europe and ecological imperialism.

Europeans first swarmed as population. The highest population growth rates in recorded history are not those of Asians, Latin Americans, or Africans, but Europeans in the period 1750–1930, that is, roughly the three centuries Stevens noted and that Smith, Marx and Engels, and Darwin at different times and in different ways documented. The number of Caucasians in this period increased over 5 times, as compared with a 2.3 increase for Asians and less than 2 for Africans (and African-Americans). In what Crosby calls the "neo-Europes" (the places Europeans settled in leapfrogging around the world), the population between 1750 and 1930 increased phenomenally — more than 14 times over! The jump for the rest of the world was by comparison a paltry fraction — 2.5 times. In the more recent slice of this period, from 1840 to 1930, the European population at home grew from 194 million to 463 million, still double the rate of increase of the rest of the world.[20]

During this same century the second "swarming" of Europeans occurred. Not population growth this time, but emigration. "The deluge" occurred between 1820 and 1930. Over fifty million Europeans migrated to establish or solidify neo-Europes abroad. Recorded emigration has never happened on that scale, before or since. Fifty million people were in fact one-fifth of the entire European population at the beginning of this period.[21]

What is happening here, and what counts as the significance of these swarmings? Certainly the conquest of "new worlds" is happening and the establishment of Western sources for what today is misleadingly termed "globalization." Moreover, this state of affairs seems utterly taken for granted by Caucasians even as it turns transnational and multicultural. Crosby offers the aside that whites' securing of thirty million square kilometers of land around the globe is one of

---

18. Ibid., 2.
19. Ibid., 3.
20. Ibid., 302–3.
21. Ibid., 5.

the "very greatest aberrations in the demographic history of the species." But aberrant or not, it is evidently "a situation this minority considers permanent."[22] Considered permanent as well, to recall Bruce Rich, is the "self-evident" organizing principle for every society and nation this side of the dual swarming — "economic development."

Yet to read the five hundred years since Columbus in the manner of raw political and economic imperialism *only,* as led by demographics, prevents us from seeing what we most need to see — history and nature, culture and environment impacted together and interlaced in ways both noticed and unnoticed. This certainly is the biological expansion of Europe in neo-Europes around the world. But what is the history of nature as intertwined with peoples in this still-unfinished saga?

Understand, first, two things: how strange the New World was to these alien swarms, and how confident they were about their mission of transformation, a transformation that included nature.

As those who live on this side of the great nature/culture upending, we can only partially sense the strangeness of the new worlds for European emigrants. Some perspective is gained by standing on the Manhattan side of the Hudson River, anywhere from Riverside Park north to Inwood Park, and gazing across the waters to the New Jersey Palisades. These cliffs, falling straightaway to the river, are a geological breach of millennial proportion (to understate it badly). This ancient rock exposed itself to the rest of creation at the sundering of Pangaea, that enormous land mass from which all present continents are the leftovers. Continents once huddled as Pangaea slowly rode tectonic plates hither and yon, held afloat by plastic rock and, deep down, the molten fire beneath. So what was "eastern" New England is now western Europe, as what was "eastern" Brazil and the Caribbean are now western and central Africa.

Fascinating to latecomers as this may be, why is it important? Because nature evolved differently after the breakup of Pangaea and the Big Drift. The flora and fauna of one portion of the planet, and their diseases as well, were not a photocopy of other portions at all, even at the same latitude and clime. Nero, after all, threw Christians to the lions, not the cougars.[23] (Both Nero and his particular lions are now extinct, of course, but that is another matter.) Or, in another amusing example, a Mr. Martin from Great Britain complained as he visited Australia in the 1830s:

> The trees retained their leaves and shed their bark instead, the swans were black, the eagles white, the bees were stingless, some mammals had pockets, others laid eggs, it was warmest on the hills and coolest in the valleys, [and] even the blackberries were red.[24]

---

22. Ibid., 303.
23. Crosby, *Ecological Imperialism,* 11.
24. Ibid., 6–7.

Noted from another angle, the very regions that today produce the majority of the foodstuffs of European provenance — grains and meats — had "no wheat, barley, rye, cattle, pigs, sheep, or goats whatsoever" five hundred years ago.[25] More than that, these particular grain- and meat-based systems have marginalized or exterminated almost all other forms of agricultural production worldwide, even though they themselves borrowed plenty from the New World and kept it as their own, as we shall see.[26] Since about 1450, most of the agricultural systems of what we until recently called the Third World have in fact experienced gradual genetic collapse.[27] We may think it quite "natural" that (immigrant) Argentineans, Brazilians, Costa Ricans, Hawaiians, Canadians, and Texans raise cattle to export to Europe, the United States, and Japan. But it is only "natural" in a European sense, just as to this day it has "naturally" never occurred to any European theologian or philosopher to contemplate what Third World genetic collapse means for creation and its peoples. Only now, as transnational pharmaceutical, petro-agricultural, and biotechnology firms are scrambling to locate and patent the remaining biogenetic material for the purposes of research, commerce, and profit, is this impacted and vastly altered nature suddenly an "issue" meriting international treaties and guaranteed access for "global" firms anchored in the neo-Europes of the world.[28]

The point cannot be labored, only stated. Europeans are not a different species and are genetically no more or less expansionist than other peoples infected with material desire, the glories of heroism, the adventure of conquest, and the heady rush of killing unwanted neighbors. But Europeans were the first to learn to master the great ocean currents and winds and were thus the first to cross the seams of Pangaea en masse and in such a way as to stitch them shut. Furthermore, they stitched them shut by acts of not only sociocultural imperialism but ecological imperialism as well. Thus Crosby:

> The seams of Pangaea were closing, drawn together by the sailmaker's needle. Chickens met kiwis, cattle met kangaroos, Irish met potatoes, Comanches met horses, Incas met smallpox — all for the first time. The countdown to the extinction of the passenger pigeon and the native peoples of the Greater Antilles and of Tasmania had begun. A vast expansion in the numbers of certain other species on this planet began, led off by pigs and cattle, by certain weeds and pathogens, and by the Old World peoples who first benefited from contact with lands on the other side of the seams of Pangaea.[29]

---

25. Ibid., 7.

26. See Timothy C. Weiskel, "In Dust and Ashes: The Environmental Crisis in Religious Perspective," *Harvard Divinity Bulletin* 21, no. 3 (1992): 8.

27. Ibid., 11.

28. Ibid. Much of the controversy at the Rio Earth Summit was over principles and treaties that create or retain access by firms and nations chiefly of the North to tropical forests and their lode of treasures, including those essential to biodiversity.

29. Crosby, *Ecological Imperialism*, 131.

"Drawn together by the sailmaker's needle" indeed! In 1492, the sailors first crossed the Atlantic in a way that "stuck." A few years earlier, in 1488, that other marker of Smith, Darwin, and Marx and Engels, the Cape of Good Hope, was rounded as ship after ship headed the opposite direction. Only three decades thereafter the entire globe was circumnavigated, and by 1600 private citizens, not just sailors, were onboard.[30] It would only be a matter of time until Broadway would symbolize Progress as a cradled ship in the foyer of a theater named New Amsterdam. New Amsterdam, New York, the fate of the Manhattan Indians, and the Palisades all testify to Pangaea's breakup and restitching. A very long geological and biological history was telescoped and climaxed by some few thousand sailors over a few hundred years.

HOW THE WEST WON

Yet understanding this as a transformation *of nature and culture together*, of the history of nature itself on a global scale, requires another kind of voyage as well. It requires a voyage into the consciousness of these adventurers, into their habits of heart, into their convictions, religion, and way of life, into their particular earth faith and earth ethic. And not just into the heads and hearts of adventurers at the helm or in the trading company offices of Marx's rising revolutionary bourgeoisie, but of private citizens as well, both those who joined the mass emigrations and those who stayed home. After all, people do not live in raw nature so much as in their pictures of nature, nature as humanly imaged and "cognized." Nor do people live in "cognized" nature only. To use equally dreadful English, they live in nature as "operationalized," nature as it lives and dies in peoples' transformations of it. "Culture," Donna Haraway says, "designates the political realization of natural materials."[31] And so it is.

Samuel Purchas was one of those who unwittingly bore testimony to the intimacies of culture and nature in the human consciousness of the age. A seventeenth-century Englishman of the cloth, this was his ode:

> [W]ho ever tooke possession of the huge Ocean, and made procession about the vast Earth? Who ever discovered new Constellations, saluted the Frozen Poles, subjected the Burning Zones? And who else by the Art

---

30. On a personal note and for purposes of illustration, I report a moment during a visit in Larnaca, Cyprus, for a WCC meeting. Cyprus, with nine thousand years of continuous human habitation, has a sixteenth-century English cemetery. One tombstone reads: "Mary the wife of Samuel Palmer Died the 15th of July 1720 and Here Lies Buried With Her Infant Daughter." What were the English doing in Cyprus in 1720? What were they doing there when times were tough and dangers abundant, like Mary Palmer dying with her child? Why this mass movement to foreign parts, often deadly ones for aliens as well as native inhabitants? Mary Palmer of Larnaca is but a glimpse into Crosby's discussion of the biological and ecological expansion of Europe.

31. Donna Haraway, *Primate Visions: Gender, Race, and Nature in the World of Modern Science* (New York: Routledge, 1989), 289.

of Navigation have seemed to imitate Him, which laies the beames of his chambers in the Waters, and walketh on the wings of the Wind?[32]

The answer, of course, is sailors, English sailors above all. The alert exegete uncovers something else here, however, something so "natural" to Purchas that the images of praise almost offer themselves of their own accord. In their possession, procession, and subjection, these sailors imitate none less than God. But which God, whose God? The God of the Book of Job and of the great creation psalm, Psalm 104! "Him, which laies the beames of his chambers in the Waters, and walketh on the wings of the Wind." No other passages of the Bible and few in any literature anywhere so majestically describe the creator of the cosmos as these do. But they do so in a way carefully designed by the author of Job and the Psalmist to set the greatest possible distance between the majesty of the creator and the creation, on the one hand, and the puniness, transience, and vulnerability of humanity on the other. Precisely this God, the God of unsurpassing transcendence, of possession, procession, and subjection in *cosmic* degree, is the God whose work, to Purchas's mind, compares favorably with little English sailors in small boats on high seas on dangerous voyages! Were we not all now and again familiar with such "innocent" arrogance and an ability to turn argument and fact on their heads, it would be hard to imagine how such conceits could so easily be conceived and held.

Consider other texts. One is the fourth epigraph to Crosby's book. It, too, belongs to a famous text by a famous scholar, Charles Lyell. His *Principles of Geology* of 1832 includes this: "Yet, if we wield the sword of extermination as we advance, we have no reason to repine the havoc committed."[33] Crosby might also have used a line from a great conservationist and hunter, Theodore Roosevelt. "The settler and the pioneer," he wrote in *The Winning of the West*, "have at bottom had justice on their side. This great continent could not have been kept as nothing but a game preserve for squalid savages."[34]

The last text is lengthier and in two parts. The first is from Kiowa nation history, the second from the English governor in the neo-Europe of Massachusetts. Both are about smallpox in a part of Pangaea that had never known it.

The Kiowa are a people of the Great Plains of North America. In one account the mythic hero of the nation, Saynday, comes across a stranger. The

---

32. Cited by Crosby, *Ecological Imperialism*, 131. An extraordinary discussion a couple centuries later, too lengthy to include here, is Walter Rauschenbusch's in his *Christianity and the Social Crisis* (New York: Macmillan, 1907), 211–12. When the nineteenth century came to an end, Rauschenbusch follows its spirit to "the vaulted chamber of the past" where the spirits of all the dead centuries reside. There the Spirit of the Wonderful Century (the nineteenth) boasts of the accomplishments of Progress, only to be met "with troubled eyes" by the other spirits. One among them, the Spirit of the First Century, then asks a series of unsettling questions, leaving the Spirit of the Nineteenth Century aware of "the wreckage" of Progress. "Give me a seat among you," the Spirit of the Nineteenth Century says, "and let me think why it has been so" (212).

33. Ibid., epigraph.

34. As cited by Maria Mies and Vandana Shiva in *Ecofeminism* (London: Zed Books, 1993), 32.

stranger is dressed in a black suit and wears a tall hat. It is the dress of Christian missionaries known to the Kiowa. The stranger speaks first:

"Who are you?"
"I'm Saynday. I'm the Kiowa's Old Uncle Saynday. I'm the one who's always coming along. Who are you?"
"I'm smallpox."
"Where do you come from and what do you do and why are you here?"
"I come from far away, across the Eastern Ocean. I am one with the white men — they are my people as the Kiowas are yours. Sometimes I travel ahead of them, and sometimes I lurk behind. But I am always their companion and you will find me in their camps and in their houses."
"What do you do?"
"I bring death. My breath causes children to wither like young plants in the spring snow. I bring destruction. No matter how beautiful a woman is, once she has looked at me she becomes as ugly as death. And to men I bring not death alone but the destruction of their children and the blighting of their wives. The strongest warriors go down before me. No people who have looked at me will ever be the same."[35]

Thus the old story of the Kiowa, told in the language of dreamy myth but raw with the experience of history and ecological imperialism.

Crosby, in next citing the fabled first governor of the Massachusetts Bay Colony, John Winthrop, comments that "the whites took a sunnier view of imported diseases." He then quotes lawyer Winthrop from May 1634: "For the natives, they are neere all dead of small Poxe, so as the Lord hathe cleared our title to what we possess."[36]

Taking title by such means, with such sanction, occupies the same universe and way of life as the possession and procession of European sailors in their faithful imitation of the great God Jehovah. The same consciousness of all things transformed, land and peoples alike and together, runs in the cultural bloodstream of these peoples as they press the sacred cause of "civilization," a cause laid on them by "the Laws of Nature and Nature's God."[37] The same consciousness was also part of a later episode with native peoples and smallpox,

---

35. Cited by Crosby, *Ecological Imperialism*, 208–9.
36. Cited in ibid., 208.
37. The Declaration of Independence is without doubt one of the great documents of ethics and liberty in human history. Its effort to express the cosmopolitan ethic of Enlightenment universalism and consider "all men" as "created equal" and "endowed by their Creator with certain inalienable Rights" is nonetheless part of a larger mind-set and cosmology that betray their universalism in favor of the missionary cause of European civilization. One of the declaration's complaints against King George himself is that he "has endeavored to bring on the inhabitants of our frontiers, the merciless Indian Savages, whose known rule of warfare is an undistinguished destruction of all ages, sexes and conditions" (cited from Henry Steele Commager and Richard B. Morris's volume of documentary readings, *The Spirit of 'Seventy-Six* [New York: Harper and Row, 1958], 319). The irony of which peoples most suffered "an undistinguished destruction of all ages, sexes and conditions" should not escape us. See Crosby's citation above from Charles Darwin about the fate of indigenous peoples.

when U.S. cavalry traded blankets they first deliberately infected with smallpox. Germ warfare did not first appear in World War I enemy trenches.

To most of those long at home in this place, this did not look like "civilization." To all but the Europeans and their embedded myth of cultural superiority, it was "predatory expansive agriculture and parasitic resource use"[38] cast in the mythology of expanding frontiers and manifest destiny. In whatever form, it put in place the grand illusion the West has lived from since and now globalizes. In Timothy Weiskel's words, "Having expanded upon the things in nature, the West came to believe that expansion was in the nature of things. Perpetual growth was considered natural, good, and inevitable." To this day, Weiskel adds, "[W]e are trying to sustain a 'frontier culture' in a post-frontier world."[39] The expansionist vision lives on even though earth is no longer flat. Differently said, a life without limits in a world of our own making (globalized neo-European) was the dream and drive of modernity from its earliest stirrings, long before Ronald Reagan and the invention of the first neoconservative. As noted earlier, even in the West's most calamitous moments, after the Great War and during the Great Depression, the World's Fair in Chicago could still confidently proclaim "a century of progress." Promethean power for the transformation of nature and culture together via economic messianism is modernity's very soul and pulse. The difference between 1492, 1776, and the present is that the transformation is now fully institutionalized, globalized, and "self-evident[ly]" organized as "economic development." It will not be easily dissuaded, much less rooted out. The bulk of Adam Smith's library, after all, is not in his native Scotland, or in England, or even at the University of Texas. It's in Tokyo.

Yet the principal subject is not imaged, cognized, and operationalized nature/ culture in the West now globalized. At least this is not the larger point. Nor is the main point the important conclusion we can safely draw: the sure outcome of economic messianism in a finite and postfrontier world is *un*sustainability. The larger point is the approach itself. We will not understand where we are, how we got here, where we might go, and how we might get there so as to live sustainably, until our analysis reads history and nature *as one piece*, until we read eco- and social location *together*. Clarity about sustainability and sustainable development means clarity about nature-in-culture and culture-in-nature on their long journey *together*. Other interpretations of this half millennium can be offered. But none should be offered that severs land and people, all the people and all the land and all the changes.

For his part Crosby uses the delightful image of Sweet Betsy to draw the tale together. Sweet Betsy and her entourage capture in one image the biological expansion and ecological imperialism of Europe together with its cultural expansion and imperialism. Crosby's source is, of course, the favorite frontier folk song, "Sweet Betsy from Pike." Betsy herself was from Pike County, Missouri, and she crossed the high mountains, the Rockies, with her lover, Ike. Not only

---

38. Weiskel, "Dust and Ashes," 10.
39. Ibid.

Ike, however, but "two yoke of oxen, a large yellow dog, a tall Shanghai rooster, and one spotted hog." As Crosby notes:

> Betsy was heir to a very old tradition of mixed farming, and whereas it must be pointed out that her oxen were castrated and the other animals without mates, Betsy's party was not the only one to cross the mountains; wagon trains had bulls and cows, plus hens and dogs and pigs of genders opposite to those of her animals. (Betsy herself had the foresight to bring Ike.) Rapid propagation of the colonizing species would be the rule on the far side of the mountains. Betsy came not as an individual immigrant but as part of a grunting, lowing, neighing, crowing, chirping, snarling, buzzing, self-replicating and world-altering avalanche.[40]

This, then, is the outcome of the past five centuries: a "world-altering avalanche" that slid outward from Europe and has upended culture and nature together. Such is the proper context for asking about sustainability and sustainable development.

Yet even this bead on what has brought us from "there" to "here" is too squinty-eyed. Sustainability requires a deeper, broader gaze. We return to the agricultural revolution and the nature-culture ensembles of the last .5 percent of human history (an infinitismal fraction of earth's history). That journey will bring us to the perplexities of the present.

---

40. Crosby, *Ecological Imperialism*, 193–94.

Three Revolutions or Four?

Ernst Troeltsch, in *The Social Teaching of the Christian Churches* (1911), set out in search of a usable past for the sake of the future. Among his conclusions was this: bourgeois modernity "will probably soon be an interlude between an old and a new civilization of constraint."[1] A few years earlier his good friend Max Weber had pondered whether capitalist culture was not building "an iron cage" trapping all of us. Technical rationality and mass society were denuding life of all value except instrumental value. This "disenchantment" of the world, Weber wrote in 1904, leaves us spiritually anorexic. We are "specialists without spirit" and "sensualists without heart" who imagine that "this nullity" is "a level of civilization never before achieved."[2] More recently E. F. Schumacher, in *Small Is Beautiful* (1973), contended that the world had never seen a quarter-century like that following World War II and would likely never again. This golden age was productivity without historical precedent. But when expanded globally and extended in time, it is utterly unsustainable. Earth cannot afford modernity. Even more recently, in the 1980s and 1990s, Christopher Lasch and Paul Kennedy have argued that Troeltsch's "new civilization of constraint," demanding new perspectives, institutions, and values, is hard upon us.[3]

Yet these prescient perspectives are not sufficiently deep nor comprehensive. Modernity — encompassing roughly the last five hundred years — is only the latest in a series of revolutions, *all* of which are qualitatively different from what is now required for sustainability. Thomas Berry describes these epochal waves in the language of geological and cultural evolution and offers notions of a new "Ecozoic Era." We are not simply passing through another human historical and cultural transformation, Berry contends. The real order of magnitude is both religious-civilizational and geobiological. We are terminating "the Cenozoic Era," a sixty-five-million-year stretch of stupendous creativity across the Community of Life. And we are trying, with blessed little success to date, to enter "the Ecozoic Era," defined by mutually enhancing human/earth relationships.[4] The Canadian National Round Table on the Environment and the Economy takes another route. It uses the language of social revolutions. That is

---

1. Ernst Troeltsch, *The Social Teaching of the Christian Churches,* trans. Olive Wyon (Chicago: University of Chicago Press, 1981), 2:992.
2. See Max Weber, *The Protestant Ethic and the Spirit of Capitalism,* trans. Talcott Parsons (New York: Scribner's, 1958), 180–82. The quoted phrases are from p. 182.
3. Christopher Lasch, *The True and Only Heaven: Progress and Its Critics* (New York: Norton, 1991); Paul Kennedy, *Preparing for the Twenty-First Century* (London: HarperCollins, 1993).
4. See the remarkable volume by Brian Swimme and Thomas Berry, *The Universe Story: From the Primordial Flaring Forth to the Ecozoic Era* (San Francisco: HarperSanFrancisco, 1992), esp. chaps. 7 and 13.

53

the path this chapter takes, in keeping with the conviction that "ecocrisis" is a poor label for a combined socio-environmental and civilizational-cultural peril. In any case, we need to explain how, in a mere .00000044 percent of earth time (the last few hundred years), our species has wrought more change on the planet than had taken place in the past billion years.[5]

FOUR REVOLUTIONS OR THREE?

The *Annual Review* of the Canadian Round Table hopes that we are now entering the world's fourth great revolution. The first three were the agricultural, the industrial, and the informational. They cover human history from approximately 10,000/8000 B.C.E. to the present and into the foreseeable future, a story arching from life in neolithic villages to present megalopolises. These revolutions all shared one crucial characteristic: they all *reorganized society* so as *to produce more effectively.*[6] To do so, and quite unlike most hunter-gatherers, these revolutions *self-consciously reconfigured nature for the sake of society.* Society became a set-apart, humanly designed, humanly ordered rendition of nature.

The fourth great revolution has not yet come to pass, but must, for survival's sake. It is the ecological revolution, and its socioeconomic characteristic is qualitatively different, the Canadians contend. Namely, to "reorganize society to produce without destructiveness."[7] "To *reproduce* without destructiveness" should be added, since a rising human population, especially at high levels of consumption, is a heavy multiplier of all socio-environmental problems.

Yet neither the Canadians nor the rest of us have much idea about the shape of mandatory, nonviolent systems that yield sustainable societies, that is, societies without destructiveness. We have little notion of what such an imperative means for our daily habits and consciousness or for organizing a world that can sustain six to ten billion people — and we are further overwhelmed when we realize that all these momentous transformations will happen in the normal lifetime of a child born today in the Northern Hemisphere.[8] We are stymied before the question of what this social revolution implies for current and alternative moral, philosophical, and religious visions and for our institutions, polities, policies, and laws. And we are brought up short before the pastoral issues this challenge raises: What kind of life together buoys hope, clears the

---

5. Lester W. Milbrath, *Envisioning a Sustainable Society: Learning Our Way Out* (Albany, N.Y.: State University of New York Press, 1989), 1–2. Milbrath's version of earth as a "slow womb" uses a movie as metaphor. The movie takes a full year to run, from the origin of earth to the present. Each day of the movie represents 12,602,240 years. On this scale, the era of the industrial revolution has lasted two seconds thus far (p. 2).

6. National Round Table on the Environment and the Economy, "The Challenge," *1991–92 Annual Review* (Ottawa, 1992): 4.

7. Ibid.

8. In addition to the Canadian materials, see Gerald O. Barney with Jane Belwett and Kristen R. Barney, *Global 2000 Revisited: What Shall We Do?* (Arlington, Va.: Public Interest Publications, 1993). This important volume of the challenges for the next generations uses graphs with an overlay of the lifetime of a child born in the 1990s.

eyes, and draws us out of our tribalisms just when greater constraint on all sides (Troeltsch's expectation) is the common experience? What structures of daily life and devotion let grace and the mystery of life wash our grimy souls, nudge courage, enable sacrifice, and fire the imagination, just when millions grieve over diminished dreams and clutch at what little (or much) they have? And how, in the end, do we learn, decide, and act prudently in the face of uncertainty and risk, within more crowded parameters?

Among our responses, one must be to understand the revolutions that have brought us thus far on the way. We take each in turn.

NEOLITHIC REVOLUTIONARIES

As mentioned, the shift to agricultural societies apparently happened in three separate core areas simultaneously: Southwest Asia, China, and Mesoamerica. On some counts, this must be reckoned the most important transition in human history. Dubbed the neolithic revolution, it was several revolutions dotting four or five thousand years. The growth of settled societies; the emergence of cities and of craft specializations; the rise of powerful religions and philosophies with equally powerful accompanying social elites; the development of writing, horticulture, pottery, weaving, and many of the arts; the domestication of animals and plants — all this belongs to the neolithic revolution. (By about 2000 B.C.E. all the major crops and animals that belong to present agricultural systems around the world had been domesticated, even though agricultural systems themselves were drastically changed after the neolithic revolution.)[9]

Not the least of neolithic innovations was population growth. After about 5000 B.C.E., the upward trend in human population, modest though it was, began and has continued ever since. This all happened in societies that established culture as agriculture and that likewise established village life as the normal context for human development, even though some impressive urban centers also emerged.[10] (The legacy of village life, too, extended well into the modern era. It is the lot of millions, even billions, still.)

Yet lines drawn too sharply between hunter-gatherer and agricultural society falsify the record. Hunter-gatherers were not passive toward their environment. They also intervened in ways affecting their locales. They followed animal populations and practiced controlled predation, sometimes virtually "herding" these wild populations, just as they sometimes seeded plants in the "wild" (our term, not theirs) and engaged in limited small-scale irrigation. It now appears that

9. Clive Ponting, *A Green History of the World: The Environment and the Collapse of Great Civilizations* (New York: Penguin, 1991), 52. On change and relocation, recall the "biological expansion" and "ecological imperialism" of the neo-Europes.

10. A fuller account is available in many places. Among those already cited earlier the reader can consult Swimme and Berry, *Universe Story*, esp. 173–81; Ponting, *Green History*, esp. 37–67; and Alfred W. Crosby, *Ecological Imperialism: The Biological Expansion of Europe, 900–1900* (Cambridge: Cambridge University Press, 1986), esp. 8–40. My summary is a brief composite of these sources.

not only dogs but pigs were domesticated before crops and settled agriculture.[11] Apparently basketry and some weaving predate the neolithic revolution as well.[12]

All this said, the agricultural transition was nonetheless a transformation that radically changed society and nature together. Ponting's summary extends ours and at the same time recalls Gowdy's earlier discussion:

> The development of agriculture, bringing with it intensive forms of food production and settled societies, had essentially the same effect all over the world. Surplus food was used to feed a growing religious and political elite and a class of craftsmen whose main role was to supply that elite. The redistribution of surplus food required extensive control mechanisms for transport, storage and reissue leading to powerful central institutions within society. These processes became self-reinforcing as the elites with political and social power took an ever greater degree of control and imposed greater discipline through enforced labour and service, first in labour gangs for major social projects such as temples or irrigation works and then in the rapidly growing armies. Societies that were broadly egalitarian were replaced by ones with distinct classes and huge differences in wealth.[13]

Ponting goes on to note two further consequences, one positive, one negative.

Virtually all major human cultural and intellectual achievements would have been impossible without the development of agriculture and a food surplus. The development of whole worlds of art, science, and culture followed from the fact that a growing number of people were no longer engaged in the direct production of food. Builders, architects, artists, priests, philosophers, and scientists, together with their creations, were invented.[14]

But increased coercion within society and warfare between societies were the parallel development. The great projects of these societies were only possible with huge amounts of human labor. Lacking sufficient volunteers, some wielded power and authority over others.

Too, settled societies meant defined territories and the ownership and control of resources. This increased the reasons and potential for serious jostling with the neighbors. (By about 7500 B.C.E. Jericho had a wall almost half a mile long around it, ten feet thick and thirteen feet high, with at least one tower thirty feet in diameter and twenty-eight feet high.[15] Megiddo, on the Plain of Armageddon, had been destroyed and rebuilt twenty times before the Common Era even dawned.)[16]

Ponting's conclusion overall, directed to our issues of sustainability and the play of human power, is that societies this side of the agricultural revolution

---

11. "First Settlers Domesticated Pigs before Crops," *New York Times*, May 31, 1994, C6.

12. See "Find Suggests Weaving Preceded Settled Life," *New York Times*, May 9, 1995, C1,10.

13. Ponting, *Green History*, 64–65. For a review of Gowdy's findings, see the opening pages of the previous chapter.

14. Ponting, *Green History*, 65.

15. Ibid., 66.

16. See Shemaya Ben-David, *Megiddo Armageddon* (n.p., n.p., 1979), 1.

provide the first examples of two phenomena we meet regularly from neolithic times forward: intensive human alteration of the environment, sometimes with major destructive impact; and the first societies that so damage the environment that they bring social collapse upon themselves.[17]

All we need reiterate is Crosby's point that the neolithic revolution happened after the long, hard breakup of Pangaea (it took millions of years). Thus the evolution of plants and animals, both with and without human intervention, took place with different results in different locales, the greatest differences often across the greatest expanses of water. This includes evolved diseases and immunities, as well as new crops, "weeds" and "vermin" (both inventions of agriculture and settled communities), and animal species. When the seams of Pangaea are then stitched shut by sailors' needles, and their cargo of culture, crops, animals, and home-grown pathogens is offloaded, the new explorers, extractors, and settlers not only will have different ways, means, and intentions for ordering the world they find but will also carry "the advantage of their infections."[18]

Yet pestilence is not the story. The story is the dynamic unsettling and resettling of nature and culture together since the agricultural revolution and by means of it. The story is sustainability and unsustainability, their whys and wherefores. And one of its questions is this: What kind of agriculture can sustain a world of six billion and more who belong to a species at the very top of the food chain? Since no one can live in a postagricultural world, the answer is important for all.

INDUSTRY

The industrial revolution is the second transformation enumerated by the Canadians. Its success and power are manifest. In the words of the Catholic theologian of capitalism, Michael Novak: "No system has so revolutionized ordinary expectations of human life — lengthened the life span, made the elimination of poverty and famine thinkable, enlarged the range of human choice — as democratic capitalism."[19] Of course, the industrial revolution is more than democratic capitalism or any other kind of capitalism. But Novak's unchecked applause for capitalism holds for industrialization generally. The revolutionized expectations he praises have come in a way that now generates unsustainability, however. (A subject that, we add, is missing in Novak's impressive output. Like

---

17. Ponting, *Green History*, 67.

18. Crosby, *Ecological Imperialism*, 32. Before Crosby, William McNeill argued that one people's conquering of another has, in the grand sweep of things, had less to do with the technical and scientific advantages of "more advanced" or "developed" peoples over "less advanced" or "less developed" ones than the pestilence brought by one population to the particular vulnerabilities of another. Pestilence most often took the form of diseases. But it could, as Crosby documents extensively, take the form of feral plants and animals as well. See William H. McNeill, *Plagues and Peoples* (Garden City, N.Y.: Anchor Doubleday, 1976).

19. As cited in Paul Hawken, *The Ecology of Commerce: A Declaration of Sustainability* (San Francisco: Harper Business, 1993), 7.

the people, systems, and economic culture he praises, Novak's thought, as most neoconservative thought, is ecologically empty.)

By and large, industrial ecology lives by economic principles and practices that run against the grain of nature's economy and literally undermine it. Too, industrial ecology, with its tendencies to urbanize human settlements and appropriate carrying capacity from elsewhere through commerce, puts the consequences of its drastic reconfigurations of nature out of sight and so out of mind as well, at least for urban dwellers. With that, the sense of responsibility is quietly put to sleep.

What other dynamics prevail? The agricultural revolution prepared the way for the industrial in that it created settled human communities in which not all persons needed to be engaged in securing food and shelter. At the same time, bringing more land under cultivation and intensifying the means of cultivation went hand in hand with a dramatic increase in human population. The estimate is that after two million years of hunter-gatherer societies, the population at the onset of agriculture ten thousand years ago was only four million.[20] The industrial revolution quickened this double trend dramatically. With the use of other energy sources, chiefly fossil fuels, and the spread of industrial production, far more persons could be, and were, engaged in nonagricultural endeavor at the same time that the basic needs of far more persons could be, and were, met (not least through the industrialization of agriculture itself). The resulting population explosion is strictly a phenomenon of the past two hundred years. Improvements in health and life span; higher agricultural production; a growing, now massive, place for creative commerce, industry, and services; and the production and distribution of unprecedented wealth brought a soaring birthrate and longer lives. The sharp drop in infant mortality rates and the explosive growth of population in fact correlate with only one major development in human history: industrialization.[21] (From 1800 to the present, human population increased fivefold. In the 1980s earth had to support about ninety million more people *each year;* ninety million was the *total* approximate population of earth twenty-five hundred years ago.)[22]

Described differently, commerce and culture in agricultural societies and before the industrial revolution were regulated almost entirely by natural energy flows, chiefly solar energy as captured by food, wood, and wind. The industrial revolution, by contrast, came about through the use of *stored* energy (fossil fuels).[23] As Paul Hawken notes, this permitted a huge transition. Commerce and culture could shift from necessarily working *with* natural forces to *overcoming them* for human ends. Production processes and people could both be separated from intrinsic ties to land. An artificial or artifactual world could be created, with few constraints from the very nature on which it depended utterly. It buried that

---

20. Ponting, *Green History,* 396.
21. Peter Drucker, "The Age of Social Transformations," *Atlantic Monthly* (November 1994): 59.
22. Ponting, *Green History,* 394.
23. Hawken, *Ecology of Commerce,* 130.

nature *in* the artifacts themselves or in the processes of their production, distribution, and consumption. The economy of nature will no doubt have the last word in all this. But for the time being industrial ecology had found the way, through stored energy and processes of industrialization, commerce, and trade, to massively reconfigure nature for the sake of *and as* human society. Industrialization found a way to live, at least for a season, against the grain of (the rest of) nature. And since it allowed humans to do better and better what they had always sought to do — grow more food, survive infancy and childhood, live longer, conquer disease, build better and more comfortable shelter, domesticate the rest of nature for human ends, enjoy a little leisure, and bequeath the children a better world than the ancestors knew — it looked for all the world like Progress at full sail.

Because industrialism reconfigured nature as society in a certain way, it became a culture as well. For human beings, culture is "second nature." And industrial culture is so much the skin we live in that we hardly notice its uniqueness in earth's story. We do not have community and society, Wendell Berry says, so much as "industry": "the power industry, the defense industry, the communications industry, the transportation industry, the agriculture industry, the food industry, the health industry, the entertainment industry, the mining industry, the education industry, the law industry, the government industry, and the religion industry."[24]

So accustomed are we to this massively reconfigured world of nature and culture together as "industry" that we barely recognize its difference from all that went before. In the economy of nature and in the longest stretches of human history (hunter-gatherer society), all creatures consume only renewable resources — fruit, nuts, plants, fish and animals, seeds, grass, berries, bark, and so on. Industrial society, however, draws from stored energy to use both renewable and nonrenewable resources (and renewables at a nonrenewable rate). But as Hawken notes, this newfound capacity for what we came to call progress means that "every day the worldwide economy burns an amount of energy the planet required 10,000 days to create. Or, put another way, 27 years worth of stored solar energy is burned and released by utilities, cars, houses, factories, and farms every 24 hours."[25] It hardly takes rocket scientist intelligence to read "unsustainability" writ large here, if such a way of life goes on for very long, in earth terms, or if very large numbers of persons pursue it even a short while. "No deposit, no return" as a way of life cannot work.

Still, from a human point of view abstracted from the rest of nature, industrialization looked like genuine progress. After all, "industriousness" modified and controlled environments in ways that met the needs of more and more people. It offered greater choices, more adventure, new freedoms, expanded opportuni-

---

24. Wendell Berry, "Does Community Have a Value?" in *Home Economics* (San Francisco: North Point Press, 1987), 179.
25. Hawken, *Ecology of Commerce,* 21–22.

ties. Whole new worlds were created and inhabited, with increased well-being for millions. Life was on a roll.

From an ecological point of view, however, what was happening was very different. To earth, industrialization looks more and more like a succession of more complex and environmentally disruptive and damaging ways to meet the needs and wants of one particular, inordinately aggressive species (more precisely, for some particular cultures of that busy species). From the bottom up, that is, from earth, industrialization is more and speedier deforestation; soil erosion and salinization; desertification; loss of life, even entire species; decrease of the genetic pool; higher levels of and new kinds of pollution and other waste; the destruction of cultures and even peoples alien to modernity; and highly unequal distribution of wealth and resources, including food itself.[26]

So it can hardly come as a surprise that the chief relationship of "industrious" humans to the rest of nature is domination and alienation. Industrialization's mighty victories and extensive benefits have been won precisely *through alienation:* humans are the subjects, and the rest of nature is fundamentally "other" objects. *And through domination:* humans overcome natural constraints in order to exploit nature for human ends. Apartheid and autism have their purposes.[27]

Nor can it come as a surprise that this revolution overturns the deeply ingrained ways of those whose living is "closer to the land." Culture is never severed from nature or nature from culture, so cultures whose ways of life were intimate with nature on *pre*industrial terms were as much the object of domination by industrialism as the rest of nature. (Darwin's "wherever the European has trod, death seems to pursue the aboriginal.")

But what kind of culture is industrialism? It is first of all a dream and a promise — to supplant poverty, disease, and toil with an abundance that permits the good life as enriching, expanded choice. That dream and promise (and partial success) have been irresistible and remain the lure. Industrialism is also a way of life, one of economic materialism and market rationality (Berry's "industry") with a certain set of cultural assumptions. We simply list the assumptions underpinning this way of life. Society and culture, after all, are a set of mental constructs as much as they are institutions and mores.

---

26. Ponting, *Green History*, 396–407.

27. G. Whit Hutchison's discussion of this way of understanding by way of investigating key words adds light. "Theory" is from the Greek *theoros,* meaning "spectator." We regard what we know, then, as "out there," exterior to us, perhaps on stage. If we are drawn into it unduly, we would lose our perspective as spectator and no longer be "objective." And claims that are not "objective" are a species of prejudice or passion. Furthermore, the Latin and German of "objective" mean to "stand over against" and "to put against" or "oppose," with the connotation that knowledge is there to order reality outside us. "Reality" itself is from the Latin, *res,* meaning property, possession, a thing (thus "real estate"). Reality is something we lay claim on, own, perhaps even possess and control (see G. Whit Hutchison, "The Bible and Slavery: A Test of Ethical Method" [Ph.D. diss., Union Theological Seminary, New York, 1995], 48). Of course, meanings do not reside in words in abstraction. The point of the discussion above is the meanings of the terms as expressions of industrial culture. Industrial culture is congruent with the argument of those prototypical modern philosophers, Francis Bacon and René Descartes, that knowledge as "objective" representation of nature as "other" is the most reliable form of truth and the key to human power and control over (the rest of) nature.

- Nature has a virtually limitless storehouse of resources for human use.

- Humanity has the commission to use and control nature.

- Nature is malleable and can be reconfigured for human ends.

- Humanity has the right to use nature's resources for an ongoing improvement in the material standard of living.

- The most effective way to promote the continuing elevation of material standards of living is through ongoing economic growth.

- The quality of life itself is furthered by an economic system directed to ever-expanding material abundance.

- The future is open; systematic material progress for the whole human race is possible; and through the careful use of human powers humanity can make history turn out right.

- Human failures can be overcome through effective problem-solving.

- Problem-solving will be effective if reason and goodwill are present, and science and technology are developed and applied in a free environment.

- Science and technology are neutral means for serving human ends.

- Modern science and technology have helped achieve a superior civilization in the West.

- What can be scientifically known and technologically done should be known and done.

- The things we create are under our control.

- The good life is one of productive labor and material well-being.

- The successful person is the one who achieves.

- Social progress and individual interests are best served by achievement-oriented behavior in a competitive and entrepreneurial environment.

- A work ethic is essential to human satisfaction and social progress.

- The diligent, hardworking, risk-taking, and educated will attain their goals.

- There is freedom in material abundance.

- When people have more, their freedom of choice is expanded and they can and will *be* more.[28]

---

28. This is a slight adaptation and change from the list in Bruce C. Birch and Larry L. Rasmussen, *The Predicament of the Prosperous* (Philadelphia: Westminster, 1978), 44–45.

Such culture is a distant cousin of that suggested in an amusing entry in a seventeenth-century European diary: "I'd rather face a thousand attacking Turks than one Calvinist bent on doing the will of God." Not that this culture is parochially Calvinist; it is decidedly more ecumenical. Frederic Morton's *Crosstown Sabbath*, narrating the experience of riding the Forty-second Street cross-town bus in New York, describes the fatigue on the faces of the passengers as "classically Judeo-Christian." We are imitating the God of our civilization, Morton says, the "Workaholic Supernal" who "assembled the world in Factory terms." "Before the Hebrews, no other people had a sabbath," he writes. "No other people needed one."[29]

Yet this culture did not emerge from the pages of Genesis or even Calvin's Reformation, whatever contributions were theirs. Nor did the industrial revolution, quite unlike the agricultural, develop separately and independently in different parts of the globe simultaneously. It was a process and culture dominated at the start by one part of the world, Europe, then exported and extended through the network of the neo-Europes, and chiefly North America. Of late some Asian countries have come to play a large role as well. Industrialization took a couple hundred years to develop in Europe and the neo-Europes, a mere wink in earth time. But it took less than a wink elsewhere because its base was in place; experience had developed tested and transportable processes; and the earlier episode of conquest and colonization had put in place a ready global infrastructure with essential links from centers to margins. Understanding this genealogy is vital. It means that both nature and culture have come under increasing control of the West over the past five centuries and especially the past two (industrialization's youth and best working years). Industrialized countries have thus established access to the resources of virtually the whole world through economic processes they continue to dominate, now quite apart from direct political control. (Once expected to be the big story of the twentieth century, decolonization dwindles as a story of autonomy and independence.)

All this was a dramatic change from previous human arrangements, when, despite considerable and even distant trade, societies were dependent for most all their significant ends upon the resources of their own area and use of *renewable resources only.*

We must return to the civilizational change wrought by industrialism as led by the West. The process by which preindustrial societies become industrial ones has come to be called "development." A synonym for economic progress, it was, until the ecocrisis, taken for granted by most as the proper course for societies wishing to enter the modern world. Such development is now so extensive and widely accepted that it is no longer considered parochial, but global. It is Bruce Rich's "self-evident" organizing principle of modern societies.[30]

Yet the terms "globalization," "the global economy," and "development" are

---

29. Frederic Morton, *Crosstown Sabbath: A Street Journey through History* (New York: Grove, 1987), 31.

30. See the discussion early in the previous chapter.

misleading. They mask the ways in which development and progress belong to a highly particular culture with specific assumptions and values (the list above) and a very short tenure in earth history. It is a culture that uses and dominates nature (and people it views as "close to nature") and alienates humanity from the rest of nature (and people from one another on the basis of a certain socialization). It is also a culture whose processes "disenchant" the world (Weber). Chiefly, this is the incapacity to receive the world as holy mystery and gift; and to stand in utter awe and celebration of life as holy mystery and gift. Industrial culture's assumptions have convinced us that we are the world's creators.

But what about this revolution's institutions? The key new institution, it turns out, was not the one usually cited, the nation-state. And this despite the fact the nation-state arose near the beginning of the industrial revolution and became an active and critical player. (Nation-statehood as a modern form of sovereignty is usually dated to the Peace of Westphalia of 1648.) The key institution was hardly even noted when it was born. It was the corporation as the little engine that could. And, like that engine, it led the rest of the train up the mountain, over it, and down the other side, an avalanche in tow.

The corporation was a brilliant invention that unleashed a new spirit of enterprise in a particularly potent form. Its history is fascinating. Corporations developed in the fifteenth and sixteenth centuries, about the time sailors were stitching shut the seams of Pangaea. At that time debts were transgenerational. Children and other relatives bore liability for their ancestors' recklessness or bad luck. They could land in debtors' prisons for the "sins of the fathers unto the third and fourth generation," and often did. But in order to sponsor exploration of the New World, England chartered corporations with a distinctive and economically powerful new provision: limited liability. If ships with their precious cargoes were lost to bad weather or piracy, or simply lost, stockholders' liabilities were limited to the investment they had made and no more. With such state-sponsored corporations, commerce found a way to do two vital things at once: absorb risk-taking and promote exploration and settlement (the neo-Europes). The charter of limited liability, unique among forms of enterprise, was a dazzling innovation unwittingly tailor-made for the transition from agricultural societies to globalizing conquest, commerce, and industrialism.[31]

Initially the corporation was subordinate to its sponsor, the state. This meant very strict, stipulated limits: limits on profits, indebtedness, and overall capitalization, together with the amount of land that could be acquired.[32] As late as the nineteenth century, corporations were considered, in the language of U.S. statutes, a "creature of the law and may be molded to any shape or for any purpose that the Legislature may deem most conducive for the general good."[33]

The U.S. Supreme Court's interpretation of the Fourteenth Amendment, meant to protect the rights of freed slaves, changed the status and power of

---

31. Hawken, *Ecology of Commerce*, 105–6.

32. Ibid., 106–7.

33. Ibid., 107, citing A. A. Berle and G. C. Means, *The Modern Corporation and Private Property* (New York: Macmillan, 1933), n.p.

corporations dramatically, however. Corporations now enjoyed the same status before the law as individual persons. On this basis hundreds, perhaps thousands, of state laws and regulations were overturned: on wages, working conditions, ownership, and capitalization. Since then corporations have found even more latitude to take risks far beyond their original borders. Before long they not only joined nation-states in jockeying for global economic benefit but also largely created the global competition for resources and profits itself. In a spectacular turnabout, nation-states now vie to be included in the wealth generated by corporations in the global economy. "When it comes to global markets these days," a *New York Times* article reports, "the motto of governments is: 'There they go, I must catch up, for I am their leader.'"[34]

The issue is not corporations per se, however. It is the role they play in the industrial revolution's exploration and extraction of world resources, the settlement and flourishing of the neo-Europes, the growth of trade, the globalization of production, the flow of technology and finance, and the promotion of living styles, ideologies, and values promotive of these processes. As of the 1990s the whole world has become a kind of "non-union hiring hall"[35] for these giants as they create the effective working matrix for what they casually combine as "human" and "natural" "capital" and "resources." And their economic clout is enormous. In the 1990s the one hundred largest corporations in the world had more economic power than 80 percent of the world's people.[36] In 1991 the aggregate sales of the world's ten largest corporations totaled more than the aggregate GNP of the one hundred smallest countries of the world. The world's five hundred largest industrial enterprises, employing only 0.05 of 1 percent of the planet's population, nonetheless control 25 percent of the planet's economic output. The 1992 sales revenues of General Motors alone ($133 billion) was about the same as the combined GNP of eight countries whose combined population is one-tenth of the world's — Tanzania, Ethiopia, Bangladesh, Zaire, Nigeria, Kenya, Nepal, and Pakistan.[37] H. Maurer's description from the mid-1950s, with its solar-system metaphor, was exaggerated at the time. It no longer exaggerates at all:

> The large corporation might better be imagined as a gravitational body, around which revolve several planets in roughly established orbits. In the primary series of orbits are members of the corporate community itself: directors, managers, executives of various degrees, and wage-earners. In the secondary group of orbits are consumers, the public in general, competitors, dealers, distributors, sub-contractors, suppliers, and stockholders. In a third and somewhat erratic orbit lies the federal government, in com-

---

34. "When Money Talks, Governments Listen," *New York Times,* July 24, 1994, 4:3.
35. Hawken, *Ecology of Commerce,* 197.
36. Ibid., 108.
37. These statistics are taken from David C. Korten's *When Corporations Rule the World* (San Francisco: Berret-Koehler; West Hartford, Conn.: Kumarian Press, 1996), 220–21.

pany with the governments of foreign countries in which the corporation operates.[38]

The role of corporations is equally significant for the third revolution — the informational. But before that a concluding comment about industrialism is in order, and two stories.

An oft-heard thesis is that ours is a "postindustrial" society and age. The relative decline in heavy industry and factory production (smoke as progress) and the surge of new ventures and jobs in service and high-technology are trotted out as sure evidence. Some of this — the surge of new ventures — is part and parcel of the information revolution and is arguably "postindustrial." But the term and thesis are deceptive. Paralleling the agricultural revolution, the industrial revolution has seen more and more of the population supported in nonindustrial occupations. New services and livelihoods grow, and older ones are extended with the dividends of an expanding industrial economy. It is also true that industrial employment is going the way of agricultural employment. In that sense, too, we may be "postindustrial." In the middle of the nineteenth century in the United States, for example, a *majority* of the population was engaged in agriculture. By 1900 that had dropped to a third, by 1940 to a fifth, and today to just under 3 percent. Here the United States is typical of all developed countries. By 1990 no developed country had more than 5 percent of the workforce in farming. Moreover, with the exception of Japan, none of the developed countries was a heavy net importer of food. The workforce is way down; farm production is way up; and both of these have been achieved through the industrialization of agriculture. Agribusiness is in fact one of the most capital-intensive, technology-intensive, and information-intensive industries around.[39] The point is that the same pattern is developing in other industrial sectors. Between 1960 and 1990, production of manufactured goods of all types continued to rise, but the jobs required for that flow dropped by half. Industrial production grew faster in the United States than in any other developed country except Japan in the very same decades (1960–90) that the industrial workforce shrank faster than in any other developed country.[40] Services, an exploding sector during this season, generated the jobs to plug the largest part of the employment gap. (By 1990 perhaps as many as nine million jobs in the United States alone were in services. This is enough jobs to safely conclude that this sector saved us from utterly devastating unemployment.)[41] Services notwithstanding, on a *global* scale unemployment is at the highest levels since the Great Depression of the 1930s, somewhere in the neighborhood of eight hundred million unemployed and underemployed human

---

38. Herrymon Maurer, *Great Enterprise: Growth and Behavior of the Big Corporation* (New York: Macmillan, 1956), 167, as cited in Bas de Gaay Fortman, "Is Capitalism Possible?" State and Society Relations Paper 95–96 (Institute of Social Studies, The Hague, Netherlands), 10.

39. From Robert L. Heilbroner's foreword to Jeremy Rifkin, *The End of Work: The Decline of the Global Labor Force and the Dawn of the Post-market Era* (New York: Putnam, 1995), xii; and Drucker, "Age of Social Transformations," 54.

40. Drucker, "Age of Social Transformations," 62.

41. Heilbroner, foreword, xii.

beings.[42] Peter Drucker argues that one of the twentieth century's big stories is the rise and demise of the blue-collar worker. No class has risen faster, nor any fallen in the same way.[43]

In short, the worker exodus from agriculture and other industries has led to the claim that ours is a "postindustrial age." The deception is that the term "postindustrial" suggests less industrial production, when the truth is emphatically the opposite. In fact, industrialization is expanding greatly, as are the resources and energy consumed, significant efficiencies notwithstanding. The flow of natural capital into manufactured capital is *not* declining. It is increasing. Moreover, the call is for much more of the same, so that "developing countries" might become "developed" at high levels of goods and services. (Sustainable development stakes its global success on this, as we shall see.) Yes, certain communities, even societies and nations, experience not only new (informational) industries but *de*industrialization as well. Yet their response is not to assist it, but fight it. The struggle for industry, including heavy industry, *intensifies* in order to get economic growth "back on track," retain precious jobs, and make up for lost ones. Like agriculture, industrialization has changed significantly. But through the changes, it has always sought to expand and intensify, and has largely succeeded. Of the 5.4 million people employed abroad by U.S. corporations alone, 80 percent are in manufacturing.[44] Even in the United States, the steady rise in the use of industrial materials is dramatic. Four and one-half *b*illion metric tons of industrial materials were used in 1993, compared with one hundred *m*illion in 1900 and four hundred million in 1950.[45] This success, which is closer to hyper- than to postindustrialization, is the chief reason for both unprecedented wealth *and* the earth's distress. Ironically, and cruelly, this second human revolution has contributed the most *both* to human well-being and to unsustainability. Which is to say that this most generous of all revolutions cannot continue in the form we have come to know and depend upon. It speaks only a few syllables in the last sentence of the last volume of earth history to date. But they have been the most unsettling utterances of all.

The first of the two stories won't make immediate sense to people in industrial cultures. That is its point. It reveals the coming revolution from the perspective of a preindustrial culture. (The story was recorded more than four hundred years ago.) The words are of an elderly Tupinamba "Indian" as recorded by Jean de Lery, a Frenchman. The French at the time (the 1500s) were battling

---

42. Ibid., xv.

43. Drucker, "Age of Social Transformations," 55ff.

44. "U.S. Corporations Expanding Abroad at a Quicker Pace," *New York Times*, July 25, 1994, A1. Those who think, however, that this means U.S. blue-collar jobs have simply moved offshore, or that these industries will employ huge numbers of presently unemployed abroad, haven't been watching the technological trends and the establishment of jobless growth. The argument about exported jobs giving work to roughly parallel work populations elsewhere made some sense thirty years ago. It no longer captures the reality very well.

45. The President's Council on Sustainable Development, *Sustainable America: A New Consensus for Prosperity, Opportunity, and a Healthy Environment for the Future* (Washington, D.C.: U.S. Government Printing Office, 1996), 142–44.

the Portuguese and Dutch for control of the seas and the conquest of what was then called Pau-Brasil, now Brazil. (Pau-Brasil was the signature tree of Brazil. It is now largely extinct in areas Europeans developed.) Jean de Lery remembered this exchange:

> An old Indian once asked me: "Why do you come so far in search of wood to heat yourselves? Don't you have wood on your land?"
>
> I told him we had a lot, but not of the same quality, and that we did not burn it but extracted color tints. I added that in our country there were traders who possessed more cloth, knives, scissors, mirrors, and other merchandise than he could even imagine, and that one of these traders bought the whole of Pau-Brasil with the laden cargoes of many ships.
>
> "Ah," said the savage, "you're telling me tales of marvels...but this rich man you're talking about, won't he die?"
>
> "Yes, he'll die just like everyone else."
>
> "And when he dies, to whom will he leave all he owns?"
>
> "To his sons, if he has any."
>
> "Well," continued the Tupinamba, "I see now that you're all mad. You cross the seas, suffer all sorts of upsets, and work so hard to amass riches which you leave to your sons and those who live on. Don't tell me that the land you feed couldn't also feed you? We have our fathers, mothers, and sons that we love, but we believe that when we die the earth that has nurtured us will also nurture them. That's why we rest easy."[46]

The second story continues the first. It is also set in "Pau-Brasil," two and three centuries later. This is Brazil depicted in the huge murals of the Palacio Tiradentes, the lavish parliament building in Rio de Janeiro. The occasion for my presence there was the Earth Summit and the report of Stephen Schmidheiny, of the Business Council for Sustainable Development, to the Global Forum of Spiritual and Parliamentary Leaders on Human Survival. Before Schmidheiny's report, we all participated in a joint North American meditation led by a swami (from Boston), an imam (from Baltimore), a Native American chief (from upstate New York), and a Roman Catholic sister (from Manhattan). Together they shared an Omaha Indian prayer. Then Schmidheiny spoke of the global business community's responsibility for environmentally sustainable development. As he spoke, I pondered the murals at the base of the cupola and the portrait of the first assembly of Brazil next to them, both well above Schmidheiny. The first assembly was white, vested, well-mustachioed, and neo-European to a man. The murals pictured the procession of settlement and civilization come to the New World and were the more or less official history of Brazil. The Europeans were in white, arriving in long wagon trains. They were led by a priest processing with a crucifix. As they entered "Indian" villages

---

46. Marcos Terena, "Sing the Song of the Voice of the Forest," in *Story Earth: Native Voices on the Environment*, ed. Pablo Piacentini (San Francisco: Mercury House, 1993), 31–32.

evangelizing, they were eagerly welcomed. But what was striking to the wandering eye was high above Schmidheiny's podium and oblivious to his words. Just above the ornate arch and probably ten feet tall was a stark, agonized Christ, gray, skinny, dusty, suspended on a cross. Hanging dead center above the portrait of the first assembly and the murals depicting the neo-European procession, it was a stunning commentary on power. Most astounding of all, this tortured figure hanging amidst palatial splendor was considered the natural expression of the culture and faith of those who brought civilization and salvation to the indigenous peoples of the "New World." So at home was this incongruency that it was not, for its cultural supporters, incongruous at all.

Schmidheiny's report on business-led sustainable development was followed by a congenial panel representing corporations, science, the arts, news media, women, and youth, and then a nice earth song by John Denver.

INFORMATION AND RESOURCES

The basic assumptions of the information revolution don't wander far from those of the industrial revolution, at least not at first glance: knowledge is power, power is control, and growth is only limited by human imagination and the access to knowledge. It is particular knowledge, however — information about how things are "coded," work, and can evolve. What Frank Parsons was serious about in 1894 (see below) belongs to the hubris of the information age as much as the "sensible industrial system" he intended. The information revolution in fact extends industrial hubris by assuming that all things, including biological mechanisms, are raw material for social manipulation. There is nothing, biological life included, that human beings cannot "program" for human ends.

Yet there are new twists that may justify the word "revolution." Parsons in 1894:

> Life can be moulded into any conceivable form. Draw up your specifications for a dog, or a man, ... and if you will give me control of the environment, and time enough, I will clothe your dreams in flesh and blood. ... A sensible industrial system will seek to put men, as well as timber, stone, and iron, in the places for which their natures fit them, and to polish them for efficient service with at least as much care as is bestowed upon clocks, electric dynamos, or locomotives.[47]

"Timber, stone, and iron," together with "men," are not all that can be "polish[ed] ... for efficient service" now. The information revolution lets us combine the organic and the technological in potentially endless configurations.

---

47. Frank Parsons, *Human Engineer,* as cited in Donna J. Haraway, *Simians, Cyborgs, and Women: The Reinvention of Nature* (New York: Routledge, 1991), 43. No bibliographic data are given for Parsons other than the date — 1894.

With knowledge of life "codes," not least DNA and RNA, we might reinvent nature itself to solve the problems brought upon us through inadequate information, harmful practices, and nature's nasty irregularities.

The cutting-edge technologies here are communications technologies and biotechnologies. Through bioengineering, nano-technology, robotics, and computerization we might, for example, create molecules that feed on pollution and produce ozone, engineer foods with genes that delay decay or banish cholesterol, alter plants to fix their own fertilizer, banish bad genes that afflict us or other creatures with every successful reproduction, and patch ailing organisms and whole ecosystems with the right doses of creations tailored just for them. The organic is not opposed to the technological and artifactual here. It is the partner for the "one flesh" of a truly good marriage. Nature is the open-ended, unlimited resource for culture, and information culture is the resource for a new nature.

Decoding nature and recoding with human instruments to create new worlds means porous borders between forms of life and artifacts. When all is "information," on the one hand, and "resources," on the other, as it is for the information revolution, the distinction between the organic and technical is leaky. It is not clear who makes and who is made in computer programs that can mutate, or what is mind and what is body in thinking machines. Nor is it altogether clear what is organ and what is not when a defective body part is replaced by a crafted one that performs the same function and is permanently attached to other "body parts." It even seems that the ontology of life and nature is no longer what it has been. Anything and everything can be a matter of "systems" and "ecologies," and is. Everything can be "total" and "holistic" and simultaneously relative, mutable, and mobile. There are endless ways to "be" in a dynamic, relational mode. The decoded cannot only be recoded but transcoded as something that has not yet been. New life-forms can be created and patented. Technology, like nature, is the rest of *us*, as it has always been. But the lines between technology and nature blur as never before. If nature and technology are less discrete than ever, and borders more permeable, where is human power (and sustainability)? When everything is susceptible to uncoding and all heterogeneity "can be submitted to disassembly, reassembly, investment, and exchange,"[48] then surely the last barriers have fallen to human power as instrumental control — the power of some humans more than others, to be sure. In the Information Age, the world — all of it — is rendered a series of systemic problems and solutions. The people who count most, then, are those who understand systems and have access to information. The rest is important, but only as "resources." Power is "effective communication," whether the language is biology or engineering or both together.

Differently said, all the long-standing Western dualisms of mind and matter, mind and body, human culture and resistant nature, have finally been resolved. They have been resolved more successfully and unambiguously than the indus-

---

48. Haraway, *Simians, Cyborgs, and Women*, 164.

trial revolution ever managed. They have been unequivocally resolved in favor of human mind and culture as creators, controllers, and high-tech cowboys working "integrated circuits" of various kinds.

Relational, holistic, "ecological" thinking has also triumphed in ways the industrial revolution did not know. So has decentralized networking as an organizational style working across distant "nodes."

Peter Drucker thinks this is a massive social transformation in the making. Industrial society, he says, was still a rather traditional society in its basic relationships of production. The society based on knowledge and knowledge workers is not. "It is the first society in which ordinary people — and that means most people — do not earn their daily bread by the sweat of their brow. It is the first society in which 'honest work' does not mean a callused hand. It is also the first society in which not everybody does the same work, as was the case when the huge majority were farmers or, as seemed likely only forty or thirty years ago, were going to be machine operators." This is "far more than a social change. It is a change in the human condition."[49]

Drucker ponders this change in the human condition further. Information societies will be so mobile and dynamic that people will no longer have roots. They will not have neighborhoods and communities that do essential socialization. The welfare state and government constituted one go at a substitute for the essential socialization, but they have been "totally disproved"[50] as effective agents of citizenship and moral formation (the formation of human character and conscience). Drucker himself argued earlier, in *The Future of Industrial Man* (1942), that the big corporation should and could become the community successor to yesterday's village. The workplace would be the chief agent of socialization, he thought then. He now scoffs at the notion. Thus, some new social sector for "growing people up" (Richard Rohr) and creating human well-being is needed.[51] We don't know what it is, concludes Drucker; we only know what it is not — business and government. And we know what this new social creation must do — re-create community.[52] But we don't know how it will do so, "virtual community" via the Internet notwithstanding.

Basically the problem is that in an information society of borderless networks and megahertz pace, few are concerned with the common good and the cohesion of society. So all Drucker sees emerging is a kind of "pluralism of feudalism." Private entities are assuming public power in a proliferation of fiefdoms.[53]

Drucker's insights on the social transformations of the Information Age are many. But he fails where most analysts do. He does not ask how the information revolution appears from the experience of earth. He does not ask what configurations human/earth relationships take in the wake of this revolution. Like virtually all management theorists in this age of *Homo economicus*, he remains

---

49. Drucker, "Age of Social Transformations," 64.
50. Ibid., 74–75.
51. Richard Rohr, "Why Does Psychology Always Win?" *Sojourners* 20 (November 1991): 14.
52. Ibid., 76.
53. Ibid., 78.

trapped in "social-social" frames of mind,[54] apparently oblivious to the fact that all societies and their institutions are configurations of nature that never break loose from impacting earth and being impacted by it. So we must double back to the earlier discussion of mind and matter, biology and information, and take another look. The socio-ecocrisis is not, after all, the creation of ignorant folks. It is the product of educated minions who, like Drucker, have married knowledge to power on the assumption they could have a world of their own making, in accord with their own design.

Information societies *try* to break loose from earth and its distress. "Information" here is largely disembodied content, the "codes" of things abstracted from all that makes the codes living flesh. A certain docetism flourishes as "virtual reality." Information as coded, recoded, transcoded reality carries a certain contempt for being earthbound at all. It prefers avoiding the messy world of finite, limited, placed, dependent bodies.

Breaking loose from earth and its distress takes another form as well, a geographical one. Robert Reich's point in "Secession of the Successful" is that "community" in the United States now generally means one thing: similar incomes. That means similar educational backgrounds and roughly the same taxes and consumerist impulses. "Tell me someone's zip code," Reich quotes a direct mail company founder, "and I can predict what they eat, drink, drive — even think."[55] This means that lifestyle enclaves replace diverse and genuine community and tend to create "we" and "they" social divisions. With information revolution dynamics, it also means a new stratum linked by jet, modem, fax, satellite, and fiber-optic cable to the great commercial and recreational nodes of the world, a stratum that no longer need even pass through neighborhoods of urban decay on the way to and from work morning and evening. These fortunate secessionists can spend more and more time in the bucolic settings of their country homes, corporate headquarters, and research parks, worlds away even from the blue-collar service workers who travel to them from the city. In the name of "quality of life," they can choose their own personal Valhallas and leave the unfortunate to deal with crime, crowded schools, and toxic wastes. It's a soft and legal version of apartheid, and the information revolution only widens this gap between "the symbolic analysts" and most everyone else.[56] Earth's distress is at a great remove in the same moment that earth-abstracted data are handled in ways that mimic "ecological" patterns of complex interrelatedness.

This is another revival of a certain Platonism in which immaterial pure forms inhabit another world as true "reality," while the flawed one we live in is the illusory cave of shifting shadows. The earthbound is denigrated, the abstract and precisely mathematical is elevated. Important and formative reality is put out to travel the ether on a web of its own, unencumbered by the mundane.

The resolution of dysfunctional dualisms of mind and matter, body and spirit,

---

54. See the comment in the introduction above, p. 8.
55. Robert Reich, "Secession of the Successful," *Other Side* (July/August 1994): 20.
56. Ibid., 24–26.

mind and body, in favor of networks of disembodied "mind," information, and a certain earth-avoiding spirit and ethic, when joined to a social stratum of beautiful people in beautiful places, will certainly have as one outcome the systemic neglect of earth and those who inhabit its distress most directly. Most of the world will grow shabbier as a consequence; and in another instance of prophecy self-fulfilled, this will reinforce the contempt of the information strata for the world. "Our material world gets dirtier . . . and more brutal as we go along," says one commentator. "So we're attracted to this ethereal bill of goods we've been sold, this wonder world elsewhere."[57]

Perhaps little more need be said, in part because we are young into the information revolution and cannot assess a revolution that hasn't run its course. But for now the information revolution would seem both an intensification of industrial-era logics as well as a new turn. It removes humanity yet further from its sense of embeddedness in nature close up and as part and parcel of it. It seems to treat all nature, humans included, in even more abstracted and utilitarian ways. All are "resources" and "capital." It fosters the illusion that we control what we create and that what we can create knows no bounds. Nor does nature place claims upon us, moral claims included. Through intensification of complex global systems, it appears as well to push the consequences of decisions still further from sight and thus dim down personal responsibility so as hardly to be felt in the gut at all. Not least, it further advantages the already advantaged and disadvantages the already dependent.[58] And it intensifies the dynamics of a globalizing economy in ways that treat the planet as a commons for the taking, anywhere, anytime, night or day. The world, to repeat, is game and booty and landfill.

These are all extensions and intensifications of negative industrial-era dynamics. No doubt the information revolution offers much that is positive. But the point here is not, in the first instance, an assessment. It is to explain how we arrived at the state of *un*sustainability and where and how we begin to wrest sustainability from present conditions. From such a vantage point, the promise of the information revolution is far less apparent than its part in aiding and abetting unsustainability by extending industrialism in new, more "efficient," and ecologically comprehensive ways. Some of it is sheer extension, a wider reach. So at the very first meeting of the G-7 countries,[59] after the conclusion of the seven-year GATT (General Agreement on Trade and Tariffs) negotiations on trade and tariffs, the agenda turned chiefly to three areas not yet adequately covered: protecting "intellectual property" (patents, software, research, movies, videos); gaining more access for service industries such as insurance, banking, brokerage, and consulting; and lowering the barriers and securing more guarantees for the flow of billions of dollars (or yen, marks, or francs) of investment capital around

---

57. "Get on Line for Plato's Cave," *New York Times*, June 25, 1995, E5.

58. See "Haves and Have-Nots Revisited," *New York Times*, October 9, 1995, D4. The gist of the article is that while rich nations talk and increasingly live "high-tech," poor nations live "no-tech."

59. The United States, Japan, Great Britain, France, Germany, Italy, and Canada.

the world and especially to emerging economies in Asia and Latin America.[60] All of these are quintessential "information revolution" industries rather than "heavy" industry. Yet the economic dynamics and goals are the same.

So is the constellation and play of human power. When the cyberspace prophets and visionaries met in Aspen, Colorado, on "Cyberspace and the American Dream II," they agreed that the information revolution "will profoundly restructure the political, economic and social landscape."[61] Yet of the two hundred participants in this gathering of cyberspace leadership, the overwhelming majority were male, white, middle-aged, and already established and recognized as technological elites. Only two of the program panelists were women (both white), and only one of the men was black. The conference was organized by the Progress and Freedom Foundation, best known for its *Progress Report*, a cable television show whose cohost is Newt Gingrich. Gingrich's advisers and supporters were themselves prominent on the conference program.[62] From the viewpoint of a sociological profile, this Aspen tableau was the Palacio Tiradentes murals of Brazil's neo-European colonization, adjusted for a couple centuries of globalization.

If there is a difference from industrialism, it is use of "ecological," or holistic, thinking in the interests of colonization and conquest. But of course it is not truly ecological or genuinely holistic. "Earth" is largely absent. In its place are webs of abstracted information in world-spanning networks handled by major economic players.

Take present capitalism as an example of both an intensification of industrial-era dynamics and new twists and turns effected by the information revolution. Current capitalism has plural forms. The Europeans want to promote free markets vigorously, yet subordinate them to programs that place social welfare and regional interests above unfettered capitalism. The Americas are, with the strong push and pull of the United States, pressing the agenda of deregulation and maximum feasible laissez-faire. The Asians continue to experiment with state capitalism, where government invests heavily in business enterprises and works consciously from detailed industrial policies worked out jointly by government and business.[63] At the same time, these varied renditions of capitalism are all affected by the information revolution. The engine seems to be "blitz capital," which can be moved anywhere anytime in a twenty-four-hour economy. There is more to it than rapidly moving capital, however. Despite capitalist variations, the tendency is for fewer collective contracts, reduced power of unions and often of governments (except as allies for business), easy shifting of manufacturing to different regions, smaller size but coordinated establishments, emphasis upon

---

60. "At the Summit: New Leaders and Old, with New Agendas and Old," and "Group of 7 Summit Leaders Find There Is Little to Fix in Their Economies," *New York Times*, July 8, 1994, A8.

61. "Cyberspace Prophets Discuss Their 'Revolution' Face to Face," *New York Times*, August 23, 1995, A17.

62. Ibid.

63. See the discussion in "A Lingering Unease Despite Strong Growth," *New York Times*, January 3, 1995, C1ff.

short-term profitability over longer-term commitments and profitability, and less and less commitment and attachment to people and place.[64]

Befitting the topic itself, there is no summary or conclusion here. Instead a nagging question: What kind of knowing is it that we need? Is it "information," via the information superhighway or other routes? Are we dying for want of information? Is late twentieth-century life unsustainable because too few data come our way, and we just can't get enough of that wonderful stuff? Is earth's distress traceable to facts we need that are not yet known? While we certainly do need knowledge about the empirical world, and while no truly integrated earth science yet exists, the intuition runs deep that the issue is less knowledge (*scientia*) than it is understanding and wisdom (*sapientia*).[65] It is less "know-what" and "know-how" than "know-wherefore" and "why." It is not so much information of the kind the information revolution manipulates as it is the choices that ethics poses: What understanding do we lack in order to live with earth and one another, on terms enhancing for life in its many guises? What sort of existence and way of life do we need to learn so as to take responsibility for the home we have? What norms and values do we measure information itself by, as a moral guidance system for what kind of society? These are not questions for newborn gnostics and docetists, or for high-tech nomads and cowboys. They are questions for plain earth creatures who know they have nowhere else to go and whose (not altogether unhappy) destiny it is to be dumped on the beach by the cresting waves of agricultural, industrial, and information revolutions.

There is still another matter to put in front of us in this panning of history and sustainability: to be clear about the concrete social outcome of these human revolutions. If earth's distress is socio-ecological, how does that appear, not for "humanity" as some nebulous global whole, or for some half-imagined division of information elites and the great unwashed, but for varied human beings who presently occupy different strata in domestic and world social orders? How do the issues of power and sustainability play out "in the 'hood"?

We illustrate with one development, here tagged "environmental apartheid."

---

64. De Gaay Fortman, "Is Capitalism Possible?" 11.

65. See the discussion of Sandra Harding, *Whose Science? Whose Knowledge? Thinking from Women's Lives* (Ithaca, N.Y.: Cornell University Press, 1991).

Environmental Apartheid

In 1987 the United Church of Christ Commission for Racial Justice published the landmark study *Toxic Wastes and Race*.[1] Using U.S. census data and government data on the location of hazardous-waste sites, it documented what has come to be called "environmental racism," and it helped spark the environmental justice movement. The findings "suggest the existence of clear patterns which show that communities with greater minority percentages of the population are more likely to be the sites of commercial hazardous waste facilities."[2] Numerous other studies have since been conducted, including those of the Environmental Protection Agency (EPA). A distinct picture has been drawn.

All communities, it turns out, are not created equal. Some get dumped on more than others.[3] The *National Law Journal,* summarizing reports on the 1,177 EPA Superfund toxic-waste sites, found the following:

- Penalties under hazardous-waste laws at sites having the greatest white population were about 500 percent higher than penalties at sites with the greatest minority population. [That is, the government's environmental law enforcement is more stringent in white communities than in communities of color.]

- For all the federal environmental laws aimed at protecting citizens from air, water, and waste pollution, penalties in white communities were 46 percent higher than in minority communities.

- Under the giant Superfund cleanup program, abandoned hazardous-waste sites in minority areas take 20 percent longer to be placed on the national priority action list than those in white areas.

- In more than one-half of the ten autonomous regions that administer EPA programs around the country, action on cleanup at Superfund sites begins from 12 to 42 percent later at minority sites than at white sites.

- At the minority sites, the EPA chooses "containment," the capping- or walling-off of a hazardous dump site, 7 percent more frequently than the

---

1. The founding event of the movement was the gathering in October 1991, under the sponsorship of the United Church of Christ Commission for Racial Justice, of the First National People of Color Environmental Leadership Summit in Washington, D.C.
2. From Vernice D. Miller, "Building on Our Past, Planning for Our Future," in *Toxic Struggles: The Theory and Practice of Environmental Justice,* ed. Richard Hofrichter (Philadelphia: New Society, 1993), 128. Vernice Miller was one of the researchers for the UCC study and a cofounder, with Peggy Shepherd, of West Harlem Environmental Action (WHEACT).
3. See Robert D. Bullard, "Anatomy of Environmental Racism," in *Toxic Struggles,* 25ff.

cleanup method preferred under the law, permanent "treatment," which eliminates the waste or rids it of its toxins. At white sites, the EPA orders treatment 22 percent more often than containment.[4]

What do these statistics mean on the ground? The United Church of Christ study offered the details. Three in five African-Americans live in communities with abandoned toxic-waste sites. Three of the five largest commercial hazardous-waste landfills are in predominantly African-American or Latino-American communities. These account for approximately 40 percent of the nation's estimated landfill capacity. Garbage dumps themselves are not randomly scattered either. In Houston, from the 1920s to the 1970s, to cite one example, all the city-owned municipal landfills and six of eight of the garbage incinerators were in African-American neighborhoods. Three of four privately owned land-fills were also in African-American communities. (African-Americans comprise 28 percent of Houston's population.) South Central Los Angeles, the southeast side of Chicago, West Dallas, and West Harlem show similar profiles for waste siting. Thirty thousand to forty thousand cubic yards of lead-contaminated soil is now scheduled to be removed from West Dallas, twenty years after it was first reported. It is to be taken to the Magnolia Landfill in Monroe, Louisiana. Monroe is 60 percent African-American.[5]

Economic factors correlate with racial/ethnic ones. *Playing with Fire,* a 1990 study of hazardous-waste incinerators, found that the minority stratum of the population with incinerators already present is 89 percent higher than the national average. Communities where incinerators are proposed, rather than already in place, have minority populations 60 percent higher than the national average. Property values are 38 percent lower in communities where incinerators exist. They are 35 percent lower where incinerators are proposed. Average income in these communities runs 15 percent lower.[6]

Have the dots on the map changed the picture since 1987? In 1994, *Toxic Wastes and Race Revisited* was published. There *has* been some change. The summary of the study opens as follows:

Despite growing national attention to the issues of "environmental justice," people of color today are even more likely than whites to live in communities with commercial hazardous wastes than they were a decade ago. The disproportionate environmental impacts first identified and documented in the 1987 report *Toxic Wastes and Race in the United States* have grown more severe.[7]

---

4. Summary reported by Marianne Lavelle and Marcia A. Coyle, "Unequal Protection: The Racial Divide in Environmental Law," in *Toxic Struggles,* 137.

5. From Bullard, "Anatomy of Environmental Racism," 27–28.

6. Ibid.

7. Benjamin A. Goldman and Laura Fitton, *Toxic Wastes and Race Revisited: An Update of the 1987 Report on the Racial and Socioeconomic Characteristics of Communities with Hazardous Waste Sites* (Washington: Center for Polity Alternatives, 1994); no page number given for the executive summary.

Between 1980 and 1993, it turns out, the concentration of people of color living in ZIP codes with commercial hazardous-waste facilities increased from 25 percent to almost 31 percent of the average population around the facilities; in 1993, people of color were 47 percent more likely than whites to live near a commercial hazardous-waste facility; and like 1980, the percentage of people of color in 1993 is three times higher in areas with the highest concentration of commercial hazardous-waste facilities than in areas without commercial waste facilities.[8]

In short, the racial gap between communities with commercial toxic-waste sites and those without has grown even more in little over a decade. A separate study by Laura Fitton found that while poorer communities fared far worse than affluent ones, poorer people of color fared worse than poorer whites. Racial factors thus reinforced already effective socioeconomic ones.[9] Michel Gelobter has traced both race and class factors in urban core areas specifically, with similar results.[10] Since poor women and children, and especially poor women and children of color, fare worse than men in most communities, there are negative gender and generational correlations as well. In a word, race, class, and gender are all intersected by biased environmental practices. Different communities suffer different consequences. The wrong side of the tracks has always been more toxic.

As details like these connect the dots and draw the picture, it is clear that economic and racial/ethnic correlations combine for a "soft" — and usually legal — version of environmental apartheid. Apartheid of green and brown hue results from the way legal processes, the economy, and long-standing social and cultural patterns of discrimination work.[11]

Nor is this only a U.S. phenomenon. When the net is cast more broadly, parallels emerge. That is, poorer nations and poorer communities within nations, most often nonwhite societies and communities, bear disproportionate burdens in the trashing of the globe. This is true for earth both as "resource pool" (goods to be extracted) and as "sink" (waste and pollution sites). Social

---

8. Ibid.

9. Cited in ibid., 5 n. 2. Fitton's study was prepared for Cornell University in 1992. It is entitled "A Study of the Correlation between the Siting of Hazardous Waste Facilities and Racial and Socioeconomic Characteristics."

10. Michel Gelobter, "Urban Environmental Justice," *Fordham Urban Law Journal* 21, no. 3 (1994): 841–56. After a detailed treatment of this nexus as it appears in urban areas, Gelobter concludes his article by saying the following about groups organized to combat the negative outcomes: "All of these movements share a recognition that environmental justice must be linked to increased economic and political control over the life of the city. Sustainability cannot be achieved by existing urban power structures. Environmental justice activists are working not only for immediate justice, but also for new models of economic and environmental activity that will form the foundation of true justice and sustainability for cities around the world" (856).

11. John Hart's summary of what he calls "eco-racism" is apt. In their relationships with indigenous people, dominant societies have: "1. exil[ed] native peoples to what were perceived by a dominating group as being the least desirable lands. 2. diminish[ed] native lands when some desired resource was discovered on them. 3. [sought] to dump undesirable byproducts of industrial production on native lands. and 4. [sought] to locate industrial plants with harmful emissions and effluents on native lands" (see John Hart, *Ethics and Technology: Innovation and Transformation in Community Contexts* [Minneapolis: William C. Norris Institute, 1995], 129–30).

and environmental degradation are knit tightly here, not because these communities are disproportionately the *source* of the basic causes of earth's distress — they are not — but because they are most vulnerable. And they are most vulnerable because they are less powerful in the world of industrial and information revolutions and their ways. All socioeconomic and racial/ethnic groups resent the nearby siting of major facilities, but some strata can resist more effectively than others. And those with privilege, power, and advantage tend to be white rather than colored, from the North rather than the South, male rather than female, rich rather than poor, and from richer nations rather than poorer ones.

At the same time a flourishing brand of conforming "multiculturalism" is quickly being established as the newest phase of a globalizing economy, with "diversity" and market "pluralism" its leading exports. In fact multiculturalism, diversity, and market pluralism are being celebrated with gusto now. Ads show the multiracial and multinational faces of the people who work for the phone company or at computers busy with transnational contacts and contracts. Such diversity and multiculturalism, Michael Lind says, are about as threatening "as an avant-garde sculpture in the lobby of a bank."[12] They are in fact celebrated at the same time that real differences are being eradicated by mass-oriented consumption and communication. So while this is the face of diversity, it masks growing commercial conformism. The diverse, multicultural face masks the fact that class lines are hardening the contours of the soft apartheid already in existence. Only now it is western Africa and Latin America, and not just West Virginia, Alabama, and Louisiana, that import industrial waste and trash from the United States and Europe. The EPA itself has proposals to send sludge and other wastes from New York, New Jersey, and other East Coast states to Central and South America. "In my 20 years of working at customs I have never seen the quantities of industrial waste and trash coming into this country from the United States and Europe that we are seeing now," reports Carlos Milstein of the Office of Technology Imports of Argentina.[13]

Waste is not the half of it. It is only a small component and in the overall picture is hardly noticeable because it is only one item among the consequences flowing from current industrialism and global finance. The truly generative factors are historically unprecedented extraction of renewable and nonrenewable resources, the processing of these for global markets and their distribution and

---

12. Michael Lind, "To Have and Have Not," *Harper's* 290, no. 1741 (June 1995): 45.

13. "Latin Nations Getting Others' Waste," *New York Times*, December 16, 1991, A8. Probably the most notorious statement on related issues belongs to Lawrence Summers, the assistant secretary of the U.S. Treasury and the former chief economist of the World Bank. Summers wrote a now-infamous World Bank memo calling for concentration on Africa as a site for some of the downside of degradation caused by economic activity: "Underpopulated countries in Africa are vastly underpolluted, their air quality is probably vastly inefficiently low [in pollutants] compared to Los Angeles or Mexico City" (cited by Paul Hawken, *The Ecology of Commerce: A Declaration of Sustainability* [San Francisco: Harper Business, 1993] 174) Not that Summers is alone. The highly regarded journal *The Economist*, commenting on the Summers memo, argued that the rich have a moral duty to export their pollution to the poor. This will create economic opportunities the poor would otherwise forego (see "Pollution and the Poor," *The Economist*, February 15, 1992, 16–17).

consumption (waste is one environmental spin-off of this), and the competitive scramble to find cheaper natural resources and take advantage of low wages and less stringent environmental conditions. When these factors interact, they generate *cumulative* social and environmental consequences for peoples and the rest of nature in a given locale. This cumulative impact includes the fraying of community bonds, with its psychic and cultural consequences, as well as the depleting of community-directed material resources. It is the cumulative impact that has led participants in the environmental justice movement to speak of improved "quality of life" as their goal rather than "a cleaner natural environment"; and to pursue comprehensive urban and rural community improvement as the aim. Michel Gelobter's definition of environmental injustice captures the matter well. Environmental injustice is "a three-dimensional nexus of economic injustice, social injustice and an unjust incidence of environmental quality, all of which overwhelmingly assures the continued oppression of communities of color and low-income communities on environmental matters."[14]

The varied pressures on already poorer communities, and some not-so-poor ones, can be glimpsed in a *New York Times* article on commercial logging in Chile. There are cumulative effects for Chile as a whole, but different groups within Chile are affected in very different ways.

As of 1994, sections of thirty-five million acres of pristine Chilean forest were being cut for wood chips and precious hardwoods. Much of the cutting is illegal, but stronger laws have been beaten back by Chilean lobbyists with market interests abroad. (The lumber and paper industry is Chile's second largest exporter.) In any case, such laws as there are cannot be stringently enforced, since government cutbacks have left fifteen inspectors with twelve vehicles to stop illegal logging on an assigned territory of eight million acres. Seven hundred thousand rural poor in the region are affected. Many of them are desperate for work and have hired on with the loggers. In addition to present jobs, they have been promised future work. Monocrop, fast-growing eucalyptus, and pine for export are scheduled as the replacement forest and the source of these future jobs. Presently, Japanese firms are the chief buyers, turning mountains of wood chips into paper. Swiss firms purchase the precious hardwoods. Neither claim any special responsibility, however, since their business transactions conform to market practices, and they have simply offered a better price than competitors. "They say they are only buying the wood," Carlos Ritter, head of regional inspectors, explains. "But they have created so much demand, the peasant farmer cannot resist cutting his forest."[15]

Nor is this an anomaly. *Times* articles report identical dynamics in the Philippines and in Papua New Guinea for logging practices.[16] In a different case, but

14. Gelobter, "Urban Environmental Justice," 842.

15. "Vast Areas of Rain Forest Are Being Destroyed in Chile," *New York Times*, May 31, 1994, C4.

16. See "In Isolation, Papua New Guinea Falls Prey to Foreign Bulldozers," *New York Times*, June 5, 1994, A1, and "Filipinos Are Bereft: How Befouled Is Their Eden," *New York Times*, July 13, 1993, A4.

with parallel patterns, the Mescalero Apaches of New Mexico first voted against a U.S. government plan to store nuclear waste on the reservation; then, reconsidering the lucrative offer and its promise of high-tech jobs and investment opportunities, they voted for it.[17] It is another instance of "the peasant farmer [who] cannot resist cutting his forest."

Such dynamics repeated over and over again are what precipitated the environmental justice movement and what *Toxic Wastes and Race* unearthed in the case of dumping practices. More extended analysis shows all this furthered by so-called free-trade agreements that are really managed-trade agreements. The 160 countries under the GATT treaties, for example, cannot "discriminate between like products on the basis of the method of production." This means that most local environmental and social regulations and standards cannot be imposed on a product group in international trade when they impose restrictions upon the *process* of production. GATT ruled against the United States, for example, in what has been dubbed the *GATTzilla vs. Flipper* case. U.S. laws banning the import of tuna caught in nets that also snare dolphins were found to violate free-trade regulations.[18] *How* goods are produced cannot be subject to extensive transnational regulation.

To cite another case: when Austria imposed a 70 percent tariff on imported tropical lumber, as well as insisting that rain forest timber be labeled as such, Asian nations, through GATT, were able to prohibit this because the tariff and labeling were not required of temperate forest timber as well. "In other words," Paul Hawken says, "woods that are harvested in areas that destroy traditional cultures cannot be discriminated against."[19] Hawken also mentions that for the Uruguay Round of negotiations from 1986 to 1992 the advisers to the U.S. GATT negotiators were all from large multinational corporations—Nestlé, Pepsico, Phillip Morris, Monsanto, Dupont. No representatives were present "from small businesses, farms, churches, environmental organizations, or unions."[20] The same was the case for teams negotiating the North American Free Trade Agreement (NAFTA). Items important to sustainable local community—for example, small businesses, cultural diversity, and community concerns themselves—were thus not on the agenda. In fact, environmental matters themselves received only secondary, though very hard-fought, attention. Meanwhile, GATT has been institutionalized as the World Trade Organization (WTO), part of the

---

17. See "Nuclear Waste Dump Gets Tribe's Approval in Re-vote," *New York Times,* March 11, 1995, A6.

18. "When Money Talks, Governments Listen," *New York Times,* July 24, 1994, 4:3.

19. This discussion of GATT is taken from Hawken, *Ecology of Commerce,* 98. The precise language is that of Article XVI of the GATT agreement creating the World Trade Organization: "Each member shall ensure the conformity of its laws, regulations and administrative procedures with its obligations as provided in the annexed Agreements." The "annexed Agreements" are the multilateral agreements relating to trade in goods, services, and intellectual property rights. As presently structured, what this means in practice in that each member country must bring its laws into accord with the lowest international standard or face fines or sanctions. See David C. Korten's *When Corporations Rule the World* (San Francisco: Berret-Koehler; West Hartford, Conn.: Kumarian Press, 1996), 174ff.

20. Ibid., 97.

UN system. The earlier effort of the United Nations to regulate the behavior of transnational corporations through the UN Center on Transnational Corporations went by the board when the organization was disbanded not long before the 1992 Earth Summit. Membership in the new creation of the Earth Summit, the Business Council for Sustainable Development, is strictly voluntary.

What all this means is that companies with the wherewithal to roam the globe in pursuit of resources and profits can shop around for the best ways to externalize costs into human and natural communities. "Free trade" effectively means access to resources and commodities with minimal impositions from socio-ecological and community regulations.[21] Typically, then, poorer countries and communities and even relatively prosperous ones like Chile compete with one another to get in on "development" by lining up with negotiated land, labor, and environmental compromises. Some succeed and benefit. For the rest it is largely a slow race to the bottom. In time such plundering becomes so severe that "overshoot" appears, first in poor communities, then in poor regions that still offer something of value to the global economy, either as "resources" or as "sinks." The signs of overshoot, which profile environmental apartheid, are these:

- Falling stocks of groundwater, forest, fish, and soils.

- Rising accumulations of wastes and pollutants.

- Capital, energy, materials, and labor devoted to exploitation of more distant, deeper, or more diluted resources.

- Capital, energy, materials, and labor compensating for what were once free natural services (sewage treatment, flood control, air purification, pest control, restoring soil nutrients, preserving species).

- Capital, energy, materials, and labor diverted to defend or gain access to resources that are concentrated in a few remaining places (such as oil in the Middle East or water in several locales).

- Deterioration in physical capital, especially long-lived infrastructure.

- Reduced investment in human resources (education, health care, shelter) in order to meet consumption needs or pay debts.

- Increasing conflict over resources.

---

21. The *New York Times* editorial "Falsely Demonizing the Trade Pact" (July 23, 1994, A18), from which some of this information is taken, argues that the United States and all major industrial nations will, on balance, gain more than they lose from such open access. This may be correct, *unless* intensification of what we list as "overshoot" conditions is, in the end, bad business for virtually everyone. The *Times* editorial finishes by saying that "the broadest and best response to the critics is this: International accords achieve a common good by chipping away at everyone's insularity." Yet it is not "a common good" if the "chipping away" uses, not community-commons regulations, but minimally qualified free-market dynamics. These often result in "the tragedy of the commons" (G. Hardin) with its heightened degradation.

82   *Earth Scan*

• Less social solidarity, more hoarding, greater gaps between haves and have-nots, more crowding, more crime, less security, and larger-scale displacement of persons, both within and across borders.[22]

Some argue that while these are the remaining reverberations of a rapacious industrial order, the "second industrial revolution" (combining eco-efficiency and the information revolution) promises other results. The information revolution's widespread sharing of essential data, when tied to its democratizing and decentralizing tendencies, will yield greater community control. Instances of overshoot will diminish as a consequence.

That may yet prove to be the case. But not if initial studies are borne out. The Center for Media Education in Washington, D.C., together with the Consumer Federation of America and the National Association for the Advancement of Colored People, found that information superhighway wiring is happening first and foremost in affluent neighborhoods and in affluent sectors of both the domestic and world economy. "The pattern is clear — low-income and minority neighborhoods are being systematically underrepresented in these plans [of telephone companies to build the advanced communications networks]."[23] Robert Reich's "secession of the successful," with its digital apartheid, seems well underway.[24]

When company leaders were confronted with these data and accused of ignoring the poor and racial/ethnic minorities, the response was telling. All of them said their companies had "no intention" to bypass poor and minority areas, as though "intentions" chiefly determine market actions. But the most revealing comment was that of Jerry Brown of U S West Inc.: "To say that we're going to stay out of areas permanently is dishonest and ridiculous. But we had to start

---

22. These items are taken largely verbatim from the executive summary of Donella Meadows, Dennis Meadows, and Jørgen Randers, *Beyond the Limits* (Post Hills, Vt.: Chelsea Green, 1992), 3–4. A particularly chilling example of extreme overshoot has been reported in Robert D. Kaplan, "The Coming Anarchy: Nations Break Up under the Tidal Flow of Refugees from Environmental and Social Disaster," *Atlantic Monthly* (February 1994): 44ff. Kaplan was reporting firsthand on conditions in West Africa but his interpretive framework was heavily informed by important studies of thirty researchers in several locales under the direction of Thomas F. Homer-Dixon, Jeffrey H. Boutwell, and George W. Rathjens. Their work is entitled "Environmental Change and Violent Conflict" and was published in *Scientific American* (February 1993): 38ff. The first paragraph reads: "Within the next 50 years, the human population is likely to exceed nine billion, and global economic output may quintuple. Largely as a result of these two trends, scarcities of renewable resources may increase sharply. The total area of highly productive agricultural land will drop, as will the extent of forests and the number of species they sustain. Future generations will also experience the ongoing depletion and degradation of aquifers, rivers and other bodies of water, the decline of fisheries, further stratospheric ozone loss and, perhaps, significant climate change" (38). The remainder of the report discusses the social consequences of environmental overshoot as pressed by needed economic growth and growing numbers of people, focusing especially on the contributions of these to violent conflict. The final two sentences are: "To prevent such turmoil, nations should put greater emphasis on reducing such scarcities [of renewable resources]. This means that rich and poor countries alike must cooperate to restrain population growth, to implement a more equitable distribution of wealth within and among their societies, and to provide for sustainable development" (45).

23. "Data Highway Ignoring the Poor, Study Charges," *New York Times*, June 24, 1994, A1.

24. See the discussion of Robert Reich in the previous chapter, p. 71.

building our network someplace. And it is being built in areas where there are customers we believe will use and buy the service. This is a business."[25] Intentions, it turns out, aren't much required. Nor are they the point. Brown himself may not be personally racist at all, or disregarding of the poor. He may in fact be African-American, and U S West Inc. may have a strong affirmative-action hiring policy. All that is required — this is the point — is simply to be in business as business has evolved over the past few hundred years. It's the system's ways that matter, and the system regularly overrides intentions out of sync with market pressures and practices. ("We had to start building our network someplace. And it is being built in areas where there are customers we believe will use and buy the service.")

HISTORY

Jerry Brown is simply riding his own history and culture, like the information revolution rides the industrial. Were he able to step outside both the world in his head and the one around him, he might understand "this-is-a-business" less naively. Perhaps in the manner of Willa Cather.

Willa Cather, writing in 1927, couldn't resist an aside clearly unessential to the plotline of *Death Comes for the Archbishop*. In foraging the history necessary for the book, Cather found her own culture in sharp contrast to native peoples'. She describes "the Indian manner":

> It was the Indian manner to vanish into the landscape, not to stand out against it....In the working of silver or drilling of turquoise the Indians had exhaustless patience; upon their blankets and belts and ceremonial robes they lavished their skill and pains. But their conception of decoration did not extend to the landscape. They seemed to have none of the European's desire to "master" nature, to arrange and re-create. They spent their ingenuity in the other direction; in accommodating themselves to the scene in which they found themselves. This was not so much from indolence, the Bishop thought, as from an inherited caution and respect. It was as if the great country were asleep, and they wished to carry on their lives without awakening it; or as if the spirits of earth and air and water were things not to antagonize and arouse. When they hunted, it was with the same discretion; an Indian hunt was never a slaughter. They ravaged neither the rivers nor the forest, and if they irrigated, they took as little as would serve their needs. The land and all that it bore they treated with consideration; not attempting to improve it, they never desecrated it.[26]

---

25. "Data Highway Ignoring the Poor," A1.
26. Willa Cather, *Death Comes for the Archbishop* (New York: Vintage Books, 1990 [1927]), 233–34.

But desecrated it is now, and a short version of the deeper historical experience that created environmental apartheid and gave rise to the environmental justice movement is a necessary insertion. It goes like this.

The wealth generated by European and neo-European globalization is inconceivable apart from the Americas. The enslavement of African peoples brought to North America, South America, and the Caribbean was an essential element of this. While the profitability of American slavery over two and one-half centuries can be debated, there is no doubt slave labor helped build and maintain the "New World." Slaves built buildings, cleared forests, plowed lands, planted seeds, kept livestock, mined ore, tilled and harvested crops, laid track and roadways, loaded and off-loaded cargo, nursed and raised children, cooked, cleaned, and tended households, both their masters' and their own; all this and more as chattel who received no pay.[27] The indentured servitude of tens of thousands of others, chiefly Europeans, added more labor in an age when most everything economic was labor-intensive.

Peoples already native to American soils North and South were just as essential to the transformations that issued in the modern world, transformations in both agriculture and industry. Indians mined the gold and silver that made European capitalism possible, which is to say, made the industrial revolution possible. With African slaves, they supplied cotton, rubber, dyes, and essential chemicals for the new systems of production. Through them corn, potatoes, cassava, peanuts, and other "New World" foods entered European diets and made for major strides in agricultural production and human nutrition. Drugs found in the Americas and learned about from Indians became essential to the emergence of modern medicine and pharmacology. The colonizers themselves followed trails already used by native peoples, adding to the infrastructure and, in some cases, impressive cities. Not least, the wealth generated for the emerging commercial and industrial economy of Europe, together with agricultural transformations and improved nutrition and medicine, were the crucial factors in the population explosion that led to the "Sweet Betsy" "swarmings" of Europeans.[28]

Jack Weatherford, from whom the above summary is taken,[29] opens his account of these matters with a description of a day in the life of Rodrigo Cespedes. Cespedes is a modern Quechua and Aymara Indian who rises at five-thirty, eats two rolls, and drinks a cup of tea before leaving the town plaza in a bus headed up Cerro Rico, the mountain that towers over Potosi, Peru. Cerro Rico was first mined in 1545. After disembarking the bus near the abandoned church, Heart of Jesus, Cespedes climbs two hours and, in a succession of twenty generations of Indian miners, works the rocks for eight hours for traces of tin his mining cooperative will send to the Chilean port of Arica. Weatherford chose Cerro Rico because it became the symbol of the fabulous wealth

---

27. For one account, see Vincent Harding, *There Is a River: The Black Struggle for Freedom in America* (New York: Freedom House, 1983).

28. See the discussion in the chapter above entitled "Sweet Betsy and Her Avalanche."

29. See Jack Weatherford, *Indian Givers: How the Indians of the Americas Transformed the World* (New York: Fawcett Columbine, 1988).

of the Americas. It is the richest mountain ever discovered on earth. With only mildly intended exaggeration, Indian miners said it produced enough silver for the Spaniards to build a sterling-silver bridge from Potosi to Madrid!

But Weatherford describes Rodrigo Cespedes on Cerro Rico to show what happened to the natural wealth of the Americas and the peoples who worked it once Western globalization began.[30] As the book closes, Weatherford muses about factors that left Cespedes and native peoples a largely decimated lot in the late twentieth century. Given their own civilizational sophistication, Indian intellectual and cultural inferiority could not be the reason. Indian civilizations were the match of any anywhere. So Weatherford looks elsewhere. Indians built their elaborate civilization on human energy. By the time Europeans came, they (the Europeans) had already exploited animal energy and even begun tapping inanimate energy sources (the key to the industrial revolution). The latter, in the form of ships and sail, windmills and waterwheels, guns, gunpowder, and cannons, made Europeans better soldiers with better instruments of invasion and war, as well as able settlers breaking and domesticating new lands. Too, the Europeans brought what we earlier called "the advantage of their diseases." Horrendous epidemics devastated native peoples. "By the time the Europeans arrived in Tenochtitlán or Cuzco or on the plains of North America," says Weatherford, "their microbes had preceded them and thoroughly decimated and weakened the native population."[31] The face of disease, together with advantageous technologies and raw, mounted strength, favored the Europeans.

Weatherford might also have cited the cultural differences noted by Cather and collective European confidence about a messianic mission.[32] Not least important is the fact that the very centuries in which Europeans discovered the power of knowing as mastery in a relationship to nature (a relationship in which nature was seen as other) are the same centuries when a new set of masters (Europeans) found a new set of slaves (Africans, Indians). Nor is it coincidental that the European conception of these populations was as peoples "closer to nature," more "bestial," and "less human."[33] George Tinker puts it strongly: "Genocide seems all too often to accompany ecocide."[34]

Weatherford and others need not be cited further. The history is familiar enough that, once recollected, it can be seen as continuous with the present. The point is that some peoples have always known it, from the inside out. To them Chris Kiefer and Medea Benjamin's opening of the final chapter of *Toxic*

---

30. Ibid., 1ff.

31. Ibid., 252.

32. See the important volume by George E. Tinker, *Missionary Conquest: The Gospel and Native American Cultural Genocide* (Minneapolis: Fortress, 1993).

33. A longer discussion of this last-named cultural and ideological factor can be found in G. Whit Hutchison, "The Bible and Slavery: A Test of Ethical Method" (Ph.D. diss., Union Theological Seminary, New York, 1995). The present reference is to pp. 49–50 of Hutchison's work.

34. George Tinker, "An American Indian Theological Response to Ecojustice," in *Defending Mother Earth: Native American Perspectives on Environmental Justice*, ed. Jace Weaver (Maryknoll, N.Y.: Orbis Books, 1996), 153–76.

Struggles: The Theory and Practice of Environmental Justice comes as no surprise
at all. Jerry Brown's naive "this-is-a-business" doesn't surprise them either.

> The last five hundred years of global history tell a sorry tale of the theft
> and destruction of land and resources. Colonialism, slavery, the intro-
> duction of export agriculture, and the unequal control of financial and
> informational flows have resulted in the passage of livelihood-sustaining
> resources into the hands of elite cultures and classes.[35]

TOWARD SUSTAINABLE DEVELOPMENT

The "passage of livelihood-sustaining resources into the hands of elite cultures
and classes" summarizes a lot of history and brings us into the present.[36] We
can thus close down and summarize the narrative analysis of the previous few
chapters. One of those chapters began with Bruce Rich's remark that economic
development is now the self-evident organizing principle for almost every society and nation on the planet, though this was not always so; we noted also that
the centuries of the planet's busiest life thus far (the eighteenth to the twentieth)
parallel the centuries of the greatest alteration of the global environment, including its most extensive degradation. We deepened these observations by tracking
the historical forces that have cumulatively made for unsustainability over the
long haul. Unsustainability is the systemic consequence of the agricultural, industrial, and informational revolutions. These fundamental human revolutions
reconfigured society by way of transformations of nature, with devastating effect
for some populations of humans and otherkind. They all reconfigured nature as
well, by way of the social transformations. All told, the double-sided transformations bore very different outcomes for different peoples and their communities,
as they did for other communities of life. Succinctly stated, social and environmental degradation follows from inequalities of power and can only be corrected
by equalizing power.

While this highly variegated but joint transformation of nature and culture together is the chief matter here, well beyond the selective narrative of
these chapters, the day will likely come when even those of us who have most
benefited will stand stunned. We will stand stunned at five hundred years of
expanding trade and empire; at the conviction, despite William Blake's "dark
satanic mills," that smoke is progress; at the Enlightenment construal of science, technology, and culture itself as the dynamics of mastery and control, with
nature, society, and psyche itself as the objects; at the spread of urban-based
civilization worldwide; and at the utter explosion of human numbers. And if

---

35. Chris Kiefer and Medea Benjamin, "Solidarity with the Third World: Building an International Environmental-Justice Movement," in *Toxic Struggles*, 227.

36. This history is also a history of resistance. "There has always been resistance to this plunder," Kiefer and Benjamin go on to say, "not only by victims but by outside sympathizers as well"
(ibid., 227).

not we, then certainly future generations will stand stunned at the collective arrogance and pathology of ordinary Western ways. These generations will shake their heads at the hubris of our intellectual heroes. Thomas Paine writing in the famous tract of the revolution, *Common Sense* (1775): "We have it in our power to begin the world over again. A situation similar to the present hath not appeared since the days of Noah until now. The birthday of a new world is at hand."[37] Francis Bacon (1571–1626) claiming that through scientific technology nature can be "disected" and "forced out of her natural state and squeezed and molded" so that "she" "takes orders from man and works under his authority." René Descartes arguing in his famous *Discourse on Method* (1637) that through knowledge of nature and human craft we can "render ourselves the masters and possessors of nature."[38] Future generations will also shake their heads at the decimation of whole peoples and cultures because their primal vision of the universe was as a sacred community of subjects, rather than a collection of objects for transformation and use, and because their ways were attuned to cycles of nature many Westerners literally had no time for. Future generations will stand stunned at the demented "success" of massive extraction from the earth, at a pounding rhythm robustly celebrated as a virile way of life, at violence against nature as progress itself, "crashing through [our] world like a bulldozer in an herb garden."[39] "A vast hole seems to open in nature," writes Loren Eiseley, "a vast black whirlpool spinning faster and faster, consuming flesh, stones, soil, minerals, sucking down the lightning, wrenching power from the atom, until the ancient sounds of nature are drowned in the cacophony of something which is no longer nature." Rather, this is "something...loose and knocking at the world's heart, something demonic and no longer planned — escaped, it may be — spewed out of nature, contending in a final giant's game against its master."[40]

Future generations will also stand stunned that what began with the slave trade and continues as the clearing of the rain forest is the same history and story, as is so much in the parallel treatment of nature and women. They will stand stunned as they look back on the fetishes of production and consumerism and a notion that land has reached a "developed" state when it is home to only one species, *Homo sapiens*, cars, and a few houseplants. They will stare in disbelief at the illusions that science and technology, in the interests of development,

---

37. Cited by Thomas Berry in *The Dream of the Earth* (San Francisco: Sierra Club Books, 1988), 41.

38. Citations from both Bacon and Descartes are taken from Carolyn Merchant, *Radical Ecology: The Search for a Livable World* (New York: Routledge, 1992), 46–47, drawing from the *Collected Works of Francis Bacon* and the *Philosophical Works of Descartes*. See Merchant for further citation and bibliographical sources.

39. Erazim Kohák, *The Embers and the Stars: A Philosophical Inquiry into the Moral Sense of Nature* (Chicago: University of Chicago Press, 1984), 74.

40. Loren Eiseley, *The Firmament of Time* (New York: Atheneum, 1960), 123–24. Gandhi seems to have gauged the matter about right at midcentury: "As human beings our greatness lies not so much in being able to remake the world — that is the myth of the 'Atomic Age' — as in being able to remake ourselves" (as cited from James J. Farrell, "The Politics of Insignificance" [keynote address, annual Nobel Peace Prize forum, St. Olaf College, published in *St. Olaf* [March/April 1994]: 13]).

yield the power to work our will on nature and bend the environment ever more to our needs, without requiring reciprocity.[41] And they will stare in disbelief that we insisted upon continuing material economic growth and furthering a consumption-driven economy even when what we added to the economy was outweighed by what we were removing from present and future generations.

Lastly, they will wonder over and again why it took so very long to realize that any adequate discussion of sustainability mandates understanding society in nature and nature in society, together, inseparable, each changing the other while being changed by it; why it took so long to grow ears for Karl Wittfogel's question of 1929: "How does a society's interaction with nature lead to its own restructuring, to its evolution from one form to another?"[42] And why it took so long to ask the equally important obverse: How does society's interaction with nature lead to nature's own restructuring, its evolution from one form to another? And how do the two together lead to or away from sustainability as *a socio-ecological state?*

This said, we turn to the discussion of sustainability in its most prominent and presently contested form, "sustainable development." Sustainable development is the most widely recommended antidote to the global social, economic, and ecological crises — and the most varied in meaning.[43] At the same time and by another measure, sustainable development is the latest in the long line of efforts to address the most tenacious problem since human settlements began in earnest: namely, how to find the means of extracting from the environment food, shelter, clothing, and other goods in a way that does not subvert earth's capacity to support its most transforming species. (Here is the first grand issue of sustainability itself.)[44] Damage and degradation are, in some degree, inevitable in the processes of sustaining human life. Enhanced environments are also possible and many have been realized. Yet both degradation and enhancement turn not only on the fine-tuned fit with natural surroundings. They turn on the nature of the societies transforming their own environments — and other peoples' in the process. (Here is the play of intrahuman power, the second grand issue of sustainability.) Thus while "sustainable development" is a phrase new to the human vocabulary, and now takes its place as the newest global recommendation, it is but the most recent version of the oldest issues of them all, issues that began with Gowdy's findings and that continue in escalated form as today's global patterns.

---

41. See the discussion in Langdon Gilkey, *Nature, Reality, and the Sacred: The Nexus of Science and Religion* (Minneapolis: Fortress, 1993), 161ff.

42. Donald Worster, *The Wealth of Nature: Environmental History and the Ecological Imagination* (New York: Oxford University Press, 1993), 33. The particular wording of Wittfogel's question is Worster's.

43. Task Force on Sustainable Development, the Reformed Faith, and U.S. International Economic Policy, "Hope for a Global Future: Toward Just and Sustainable Human Development" (Presbyterian Church [U.S.A.]), 67 of the draft dated August 8, 1995.

44. This is the conclusion in Clive Ponting, *A Green History of the World: The Environment and the Collapse of Great Civilizations* (New York: Penguin, 1991), 407. Ponting's last sentence, written in view of the impact of industrializing and human population levels, is this: "Past human actions have left contemporary societies with an almost insuperably difficult set of problems to solve" (407).

The history that makes for unsustainability behind us, then, we turn to the constructive proposal. The task is to get from here (unsustainability) to there (sustainability).

This first go at the constructive task (more will come in part 3) is an analysis of sustainable development together with an alternative, sustainable community. Because there is no quick fix for either, and a different way of thinking, even imagining, is needed, any effort at either sustainable development or sustainable community requires a basic reorientation to the complex of society and culture inevitably together. The next chapters, "Ecumenical Earth" and "Creation's Integrity," offer that.

Ecumenical Earth

Ships. Ships and symbols. Recall the triptych and the logo of the WCC, the Good Ship Oikoumenē. The center of that ancient symbol — *oikos* — has the feel of new truth. It is a way to picture the vision and interrelated elements of sustainability. It is a way into rounded ways of thinking and acting, and into cultures very different from industrialized ones to date.

EARTH AS *OIKOS*

The numbers are stunning, both the big ones and one very small one. There may be fifty billion galaxies. Each has billions and billions of stars. Planets circle some, perhaps many. But so far as we know to date, earth is the only life-form in the universe. Its thin, living envelope is its distinctive signature. Apparently good planets are hard to find.

Oikoumenē means "the whole inhabited world" or "the inhabited globe." All things belong to an all-inclusive form upon which the life of each depends. Humankind and otherkind are fit together in an undeniable, if precarious and sometimes mean, unity of life and death. We are not so much *at* home *on* earth as we *are* home *as* earth.[1] This is the first truth of earth ecumenics and earth ethics — the integrity of creation. It renders the ecumenical task basic and clear: to help life not only survive but thrive together indefinitely; that is, it means sustainability.

Since we are using the WCC discussion of *oikos* to introduce constructive thought on sustainability, its theological conviction about the earth as life's rare habitat is crucial.

Many religious traditions, Judaism and Christianity included, understand the inner secret of creation to be the *indwelling of God* within it. Creation is the place of God's presence. The very purpose of the Shekinah (a Jewish way of speaking of God's indwelling) or the work of the Holy Spirit (a Christian way of speaking) is to render the whole creation the very "house" of God. The phrase "economy of God" (*oikonomia tou Theou*) is an ancient way of speaking about the redemptive transformation of earth for its fulfillment. For the moment, however, we simply register the underlying notion of *oikos* that earth (and all creation) is the house of God's presence and indwelling. God is "home" here, as are we.[2]

---

1. See the nice discussion of Shannon Jung, *We Are Home: A Spirituality of the Environment* (Mahwah, N.J.: Paulist, 1993); see esp. chap. 4, "The Earth: God's Home, Our Home."
2. See the volume of Jürgen Moltmann, *God in Creation: A New Theology of Creation and the Spirit of God* (San Francisco: Harper and Row, 1985).

The word *oikos* itself is at home in both classical Greek and the "street Greek" of the New Testament (Koine). It was recently revived by Daniel Bell. Bell ended his study of the world coming apart, *The Cultural Contradictions of Capitalism*, in search of a unifying vision.[3] He sought direction for society and culture. *Oikos* offered this social vision. We will return to it after noting something basic that eluded Bell and most others who have discussed the theme.

What is most decisive for the whole inhabited earth (*oikoumenē*) is that it be habitable! The first and basic meaning of *oikos* is simply "Habitat Earth." Habitat is the core meaning of all *eco* words: economy, ecology, ecumenicity itself. *Oikos* — earth as a vast but single household of life — means the capacity for survival, that is, sustainable habitat. It means space and the means for the living of all living things. Without adequate hospitable habitat, nothing lives. Not only humans, but all life-forms need carefully fitted habitats.

This foundational fact — that we all need a fitted space for the basics — is the one great element of "democracy" in life, a kind of first equality. Whatever we choose to call it, the point is that without that which good space provides — productive land, a hospitable atmosphere, safe water, the numberless forms of life that provide sustenance for one another in astounding ecosystems — *none* of the other "household" goods we treasure, including educational, artistic, and spiritual ones, is possible. This is finite, bounded space, to be sure — the earth is curved, well-wrapped, a closed sphere, an *oikos*. But the boundaries are those of life itself and what life requires to stay in place.[4]

Habitation in this closed space (sunlight is the *only* life element not permanently resident within the biosphere) rests on further understandings. *Oikos* carries these. As noted, it is the root and common unity of economics, ecology, and ecumenics. Economics is *eco* (habitat as the household) + *nomos* (the rules or law). Economics means knowing how things work and arranging these "home systems" (ecosystems) so that the material requirements of the household of life are met and sustained. The household is established as hospitable habitat. The basic task of any economy, then, is *the continuation of life*, though no economist has put it that way for ages. In fact, the kind of economics generating earth's present distress made three decisive moves *away* from *oikos* economics.

One was to consider nature as interchangeable parts and machinelike rather than organic and communitarian.

Another was what Guy Beney calls the "genius" of the West always "to flee towards the exterior": that is, the propensity to generate affluence by expanding to new worlds until the globalized West became a full "planetary wave." (Beney also notes that this wave now doubles back on itself, having come up against the closed nature of the *oikos* and "fusing" with it so that what we have left are

---

3. Daniel Bell, *The Cultural Contradictions of Capitalism* (New York: Basic Books, 1976); see Bell's last chapter.

4. See the discussion of Konrad Raiser, *Ecumenism in Transition* (Geneva: WCC Publications, 1991), 84–91.

vulnerable systems with little room for error.)[5] This is frontier economics in a postfrontier world. In a contracting space it is destructive.

The third departure was to shift economic attention from the household and its community to the firm or corporation. The founders of modern economics chose the corporation or firm rather than the household (*oikos*) as the basic unit of economic agency. The long-term interests and perspectives of corporations or firms are very different from those of households and the communities that households constitute and that in turn sustain households. The household seeks to maximize the quality of life and benefits of its members. Corporations' and firms' measures of success are maximized profits and market share. The aggregation of household interests approaches the collective interest of the community. The aggregation of corporations' and firms' interests has very little to do with the interests of even their own geographical area, however, and represent instead the interests of those who own, control, and decide about capital.[6] Because firms' and corporations' perspectives are potentially global, attachment and loyalty to households and communities run a very poor second, third, or fourth.

These three changes in economics in the modern period — nature as parts, globalization, and the firm or corporation as the basic unit of economic agency and analysis — all revise radically the inherent eco-economics of *oikos* in a contracting world, even when "household and community" take on wildly different dimensions, from a single household in a given locale to earth community itself as an all-inclusive life-form.[7]

In New Testament usage, the one who knows the house rules and cares for the life of household members is the *oikonomos,* the "householder." English speakers customarily translate *oikonomos* as "steward." "Steward" is derived from "sty warden," the keeper of the pigs — a fine image, though at first blush somewhat less attractive. Whether as steward or householder, *oikonomos* signals trusteeship. It means broad responsibility for the world we affect, including the deep and far-reaching impact on nature later discussed as "appropriated carrying capacity." It means wise management of life's household, including knowledge of human limits, together with the rest of nature's.

---

5. Guy Beney, "'Gaia': The Globalitarian Temptation," in *Global Ecology,* ed. Wolfgang Sachs (London: Zed Books, 1993), 181–82.

6. Herman E. Daly and John B. Cobb Jr. (*For the Common Good* [Boston: Beacon Press, 1990]) note this shift as one from *oikonomia* to *chrematistics.* The distinction between the two is Aristotle's, but both refer to the discipline of economics. *Chrematistics* casts the work of economic practices as the "manipulation of property and wealth so as to maximize short-term monetary exchange value to the owner." *Oikonomia,* by contrast, "is the management of the household so as to increase its use value to all members of the household over the long run." With the shift in industrial economies from the household to the firm came also the shift from *oikonomia* to *chrematistics* as the substance of economic ends. Daly and Cobb go on to say that "if we expand the scope of household to include the larger community of the land, of shared values, resources, biomes, institutions, language, and history, then we have a good definition of 'economics for community'" (138). This is the direction the discussion above takes — *oikos* as the symbol for an understanding of sustainable development as "economics for community."

7. See the discussion of David Korten, "Sustainable Development," *World Policy Journal* (winter 1991–92): 183–84.

"Ecology," another *oikos* word, is *eco + logos* — "the logic of the house" or knowledge of the structures and dynamics of the household, how it has been configured and runs. "Ecology" thus means knowing, from inside, the interrelated dynamics that make up the total life of the household and the requirements for living together. This is to respect creation's integrity and live in accord with it. English usage is recent, apparently initiated by Ernst Haeckel in 1870. "By ecology," he wrote, "we mean the body of knowledge concerning the economy of nature — the investigation of the total relations of the animal both to its inorganic and to its organic environment; including, above all, its friendly and inimical relations with those animals and plants with which it comes directly or indirectly into contact."[8] In short, ecology as a dimension of *oikos* is the knowledge of life-systems necessary for good home economics. *Oikos* members are themselves *oikeioi*, household dwellers. Their tasks together comprise what the Apostle Paul called simply "the mutual upbuilding of community." In the full vision of *oikos*, however, building up community includes all that belongs to *oikodomé*, yet another *oikos* word. *Oikodomé* is the continual upbuilding of the *oikos* as a whole,[9] what we might term global citizenship and earth patriotism, with all the attendant duties of "choosing life" (Deut. 30:19) and living in accord with the choice.[10] This includes restoration for damage suffered and done, a clear dimension of *oikodomé*. Yet for all the attention to global stewardship, *oikodomé* never loses its focus on the particular community at hand and its well-being. The global does not substitute for the local. Just as the understanding of "ecumenical" in the early church designated the *whole* church in *each* place, so *oikodomé* refers to the global in the local community. The local is the basic unit of the global. Microcosm and macrocosm live within one another. They share the same dwelling.

Oikodomé, moreover, is understood by the first Christians as forging and sustaining a specific *moral* culture precisely at a turning point in history, a time when a "new age" was taking shape, one that required moral and religious communities attuned to it.[11]

This Christian notion mirrors a Stoic one. Stoics, too, considered themselves citizens of a moral world that transcended their immediate locale. Yet transcen-

---

8. This is cited from Charles Birch and John B. Cobb Jr., *The Liberation of Life* (Denton, Tex.: Environmental Ethics Books, 1990), 27. Birch and Cobb cite this date and definition but do not give the source in Haeckel's writings.

9. This discussion of various meanings of *oikos* and its derivatives is in part indebted, as noted above, to Raiser, *Ecumenism in Transition*, 102–11.

10. I was reminded of this, not by slow social-environmental degradation, but by the historical reality of genocide. At the U.S. memorial to the Holocaust there is a hall of remembrance. It is a stark, open, octagonal, marbled space that gathers only silence and has only an eternal flame, an inscription about remembrance, and two biblical verses, one high to the left, the other high to the right. To the left, Gen. 4:10: "And the Lord said: 'What have you done? Listen; your brother's blood is crying out to me from the ground!'" To the right, Deut. 30:19: "I call heaven and earth to witness against you today that I have set before you life and death, blessings and curses. Choose life so that you and your descendants may live." Life is increasingly a matter of human choosing, as is death.

11. This discussion of *oikodomé* is indebted to Geiko Müller-Fahrenholz, *God's Spirit: Transforming a World in Crisis*, trans. J. Cumming (Geneva: World Council of Churches; New York: Crossroad, 1995).

dence of the local was not in contempt of it, or a way of leaving it behind. Rather, it was a way, through local responsibility, of gathering the whole world together as community over time. For the Stoics, humans, gods, animals, and vegetation were all included and understood by way of the *theologia naturalis* — knowledge of the essence of things.

It is important to distinguish this relationship of local and global from others. *Oikos* as a vision is not the Enlightenment project of an earth ethic grounded in the universally human as some core that can be stripped of particularity and exist independently of differences generated by race, gender, class, and culture. There is no core human nature so solid that differences are normatively insignificant. Nor is the *oikos* an earth ethic held together by universal norms and procedures secured through the power of shared universal reason as a capacity that some-how leaves the peculiar treasures and passions of time and place behind. In the Enlightenment tradition, the provincialities of communities and cultures, of lo-cal time and space, are held in veiled contempt as tribal residues impeding a developing cosmopolitanism. By contrast, earth as *oikos* sees these provincialities as the diverse and finite units *of* a greater community to which they intrinsi-cally belong. *Oikos* is also at odds with the kind of "systems" ecologies of the information revolution, removed as they are from the grime of earth's distress. And, as we shall see in more detail soon, *oikos* is at odds with the particular kind of globalism of the present globalized economy, even an ecologically sensitized global economy. Such global thinking has disregard for local community loyal-ties and needs and, by virtue of globalism itself, oversimplifies, often with cruel results. *Oikos* earth ethics is instead closer to Wendell Berry's contention that "properly speaking, global thinking is not possible. Those who have 'thought globally' (and among them the most successful have been imperial governments and multinational corporations) have done so by means of simplifications too extreme and oppressive to merit the name of thought."[12] Global thinking, Berry goes on, can only be statistical. So unless one is willing to be destructive on a very large scale, one cannot do something except locally, in a small place. Global thinking can only do to the globe what a space satellite does to it: reduce it, make a bauble of it. "If you want to *see* where you are, you will have to get out of your spaceship, out of your car, off your horse, and walk over the ground. On foot you will find the earth is still satisfyingly large and full of beguiling nooks and crannies."[13] Berry's is an *oikos* citizenship *through* local in-habitation of Habitat Earth.

The ancient *oikos* vision includes instruction for all this. *Peri oikonomias* (on household management) was a steady theme of philosophers and rhetoricians from Plato's time until the waning of the Roman Empire. Essentially practical moral instruction for the conduct of different family members, the circle was often enlarged to include societal duties and roles as well. *Peri oikonomias* was

---

12. Wendell Berry, *Sex, Economy, Freedom, and Community* (New York: Pantheon, 1993), 19–20.
13. Ibid.

the articulation of such duties. (Here is Bell's quest for a coherent culture and the basic moral education necessary for life together.)

Early Christians appropriated this instruction in household management. Their new communities were originally conceived as *oikoi* themselves ("households of faith" in this case), partly because the early church was literally a house-church movement. Community instruction came in numerous forms, some of them demanding and lengthy. Catechumens, for example, normally studied for two years and were subjected to rigorous examination leading to the rite of baptism and their first eucharist. The purpose of instruction was to learn the way of life and belief of these communities, including the respective roles and responsibilities of different community members. It was concrete instruction for *oikos* living. It was socialization and education together.

"Ecumenics" was another early Christian *oikos* notion. It meant recognizing the unity of the household — all belonged to the same family — and nurturing this unity, both within each community and across the collective "household of faith" scattered on three continents around the Mediterranean Basin. The *oikoumenē* included a conscious effort to stand for the whole church in each place.

The Stoics, too, had their take on this venerable tradition of instruction. They taught *oikeiosis* — "appropriation" (in the sense of making something one's own). *Oikeiosis*, like Christian catechesis, included practical moral instruction and its supporting rationale. To become what one might as a human being, the Stoics taught, meant care of the self and nurturing the divine spark that by nature lived within each person. The primary means was cultivated reason, employed to tutor and order human appetites and actions. This was not care of the abstracted individual, however, but each person as a member of the *oikos*. Such membership might refer to membership in the familial household, society, the human race collectively, or the *cosmos* (world) as a whole. Underlying all was the Stoic conviction that we owe one another reciprocal duties that express the law of nature. (Western natural law traditions have deep roots in the Stoic *theologia naturalis*.) Taking these duties on and living them out "realized" one's true nature. Appropriation (*oikeiosis*) both followed from and issued in right living.[14]

But *oikos* also includes "homelessness," if only as a way to mark that which violates the vision and its implicit earth ethic. The forms of homelessness are at least two — deprivation and alienation. Each in turn has multiple expressions, but for the moment our attention is given to deprivation and alienation through the destruction of home as habitat and the economic, cultural, and spiritual uprooting of people from their homes.[15]

In a manner anticipating the Chiapas episode soon to be discussed, testimony to homelessness in our time was part of a public hearing of the World Commission on Development. The Krenak people, a living continuation of early

---

14. See Alasdair MacIntyre, *Whose Justice? Which Rationality?* (Notre Dame, Ind.: University of Notre Dame Press, 1988), 147–50. Also Wayne Meeks, *The Origins of Christian Morality* (New Haven, Conn.: Yale University Press, 1993), 77–79.

15. Maria Mies and Vandana Shiva, *Ecofeminism* (London: Zed Books, 1993), 103–4.

agricultural peoples, were protesting a resettlement from their traditional home in the valley of the Rio Doce, which the government had designated for development. "The only possible place... to weave our lives is where God created us," their elder explained to blue-ribbon commissioners. "It is useless for the government to put us in a very beautiful place, in a very good place with a lot of hunting and a lot of fish. The Krenak people... die insisting that there is only one place for us to live." The elder finished by saying: "I have no pleasure at all to come here and make these statements. *We can no longer see the planet that we live upon as if it were a chess board where people just move things around. We cannot consider the planet as something isolated from the cosmic.*"[16] First colonization, then development and commerce, spelled alienation and deprivation for the Krenak. The prospect of relocation meant homelessness of both spirit and place, a homelessness different but no less real than any encountered on the streets of New York or Calcutta. Many peoples long at home in their world, like the Krenak, have been rendered strangers in their own land as globalizing intruders from elsewhere decide how their earth space, their habitat, their *oikos,* would be developed. The Krenak did not understand that space and traditional borders mean nothing to the porous ways of an economy that renders land and its native treasures as commodities. Chief Seattle's claim in 1854 that you cannot "buy the sky" or "the warmth of the land" or "own the freshness of the air and the sparkle of the water" turns out not to be true at all. But the Krenak did not comprehend the culture of the commercial and industrial globalization of the past few centuries, even when they sensed clearly enough what it asked of their bodies, spirit, culture, and religion. Thus even another "very beautiful place" "with a lot of hunting and a lot of fish" was no substitute. Homelessness is alienation from the land of one's forebears, where one's own spirit still lives with those who have gone before in near-biblical fashion — generation by generation by generation.

Masses of people are homeless now, and not only indigenous peoples and the poor. *Oikos* as the experience of belonging somewhere intimate to one's bones eludes most moderns. The homeless mind, Peter Berger argues, is a condition of modernity itself, accompanying development and modernization as their sure offspring.[17] Even apart from homelessness as a matter of mind, it is quite literally displacement of one kind or another. The highly mobile rich, living from hotel to suburban lot to condo and hotel again, hardly have an enduring community they consider their own and, even less, a binding commitment to a neighborhood. All the advertisements and job descriptions, the trade and tourism, declare the world their "oyster." Every location is at their disposal. But no particular locale is home in a deep, settled sense.

The uprooted but formerly settled poor are another case. They also move about, in search of new places to sustain them or their families "back home" via money and the mail. The lucky ones live in places alien to them, the rest

---

16. As reported in ibid., 104.
17. See the argument in Peter Berger, et al., *The Homeless Mind* (London: Pelican, 1981).

in refugee camps. The biggest issues of the coming decades may in fact include refugees and internally displaced persons near the top of the list. "It's a privilege to be exploited now" is the humor of the 1990s. The alternative to bad employment is none; the alternative to exploitation is exclusion; and exclusion is homelessness. Homelessness, then, is both a physical and psychic condition, a reality spun off the global marketplace and free-wheeling "development." It affects both rich and poor, though hardly in the same way.[18]

Still, alienation and unsustainability are not the primary subject; sustainability is. Nor is *oikos* the subject, except as an orientation to earth as a single household of life whose sustainability needs to be understood in light of the planet's unforgiving unity and on terms of community. Yet this ancient picture of earth serves its purpose as a basic approach to sustainability itself, one that has a positive vision for earth, even earth as a closed and crowded sphere.

That sphere and its task were nicely pictured by Martin Luther King Jr. in an essay, "The World House," published the year he was assassinated in the book whose title remains the right question: *Where Do We Go from Here? Chaos or Community.*

> Some years ago a famous novelist died. Among his papers was found a list of suggested plots for future stories, the most prominently underscored being this one: "A widely separated family inherits a house in which they have to live together." This is the great new problem of mankind. We have inherited a large house, a great "world house" in which we have to live together — black and white, Easterner and Westerner, Gentile and Jew, Catholic and Protestant, Moslem and Hindu — a family unduly separated in ideas, culture and interest, who, because we can never again live apart, must learn somehow to live with each other in peace.[19]

Such is the kind of ecological thinking needed to take up this task as our own. It is described in fuller measure as "the integrity of creation."

---

18. Mies and Shiva, *Ecofeminism*, 98.
19. Martin Luther King Jr., *Where Do We Go from Here? Chaos or Community* (Boston: Beacon, 1968), 167.

Creation's Integrity

A replica of planet earth was part of closing worship at the World Convocation on Justice, Peace, and the Integrity of Creation held in Seoul in 1990. Borne to the center of the assembly by children of different colors, languages, and cultures, the earth was the object of the children's message to the delegates: "Treat earth gingerly; pray and plead for its future; pray and plead for ours." Then the entire assembly, representing most of historic Christianity[1] and every race and region, was invited to come forward and lay hands quietly on the earth. They came and touched, in gratitude and blessing.

Given two millennia of Christian ambivalence about earth loyalties and strong taboos about pagan nature religions, did those delegates fully understand the significance of placing earth at the center? Did they comprehend the ritual they performed with such sincerity and the commitments they prayed with such earnestness?

"The integrity of creation" was not a phrase new to Seoul. It first surfaced at the Vancouver assembly of the WCC seven years earlier (1983). There it was added to long-standing themes of justice and peace. The early 1980s were dense with Cold War fear, and dual threats hung heavy in the air. Nuclear apocalypse was a possibility. And even if the worst did not happen, preparation for it was exacting too high a price around the world. Another comprehensive threat loomed as well. Slowly, incrementally, planetary life-systems were under assault.

"The integrity of creation" gave a name to the comprehensive good that delegates at Vancouver sensed stood under threat; namely, nothing less than life itself, abundant or otherwise.

The new phrase soon resonated with church members everywhere. The fate of the world seemed an ultimate concern to them and the turn to earth an imperative of Christian faith itself. "To turn toward God (*theo*centrism)" was simultaneously "to turn toward God's beloved world (*geo*centrism)."[2]

Still, these high stakes were only intuited as the rainbow of children bore the globe into the center of the Seoul assembly. Delegates may have identified with the poster of earth in space labeled "Our Town." They may even have experienced what the Saudi astronaut did: "The first day or so we all pointed to our countries. The third or fourth day we were pointing to our continents. By the fifth day we were aware of only one earth."[3] They nonetheless continued

---

1. Protestantism, the Orthodox communions, Roman Catholicism.
2. Douglas John Hall, *Professing the Faith: Christian Theology in a North American Context* (Minneapolis: Fortress, 1993), 309.
3. Sultan Bin Salman al-Saud, in *The Home Planet*, ed. Kevin W. Kelley (Reading, Mass.: Addison-Wesley; Moscow: Mir, 1988), 82. (The volume has no page numbers as such. The

to puzzle, as Vancouver delegates did, about what precisely "the integrity of creation" meant.

We often act, ritualize, and worship our way into new perspectives, knowledge, and commitments before we think our way there, so this is not a surprise. We gradually learn the world in the course of dealing with changes we have already effected, and much of that learning begins with a picture, a few hunches, a couple analogies, and a song or two. The question nevertheless pushes for an answer. What *does* "the integrity of creation" mean? What does it mean for an earth ethic and the habits that either sustain us or do us in?

Undaunted by great mysteries, theologians were the first to reply. Jay Mc-Daniel's sentences summarize nicely: "The integrity of creation" refers to "*the value of all creatures in themselves, for one another, and for God, and their interconnectedness in a diverse whole that has unique value for God. To forget the integrity of creation is to forget that the earth itself is a splendid whole.*"[4]

The theological work has often been insightful, and we will return to it. But there are other approaches, from different angles with different results. Most of these are descriptions of nature and how it works, the slow reward of detailed science.

Empirical accounts are as important to an ethic as theological and philosophical claims. No earth ethic will be viable if its morality does not issue from the heart of reality itself. Yes, specifying the nature of the "heart of reality" is itself a metaphysical and philosophical or theological judgment in the end. But such an ethic will not command allegiance if there is no resonance with the universe and the world as we experience, know, and describe them. An understanding of things in their integrity, knowing them the way they are with all else, accounting for them in ways true to where they are tending, is elementary for any ethic.

So what *does* "the integrity of creation" mean? What does it mean as a description of nature, and what does it mean for sustainability and an ethic for earth?

MEANINGS

The integrity of creation has six dimensions.

1. The integrity of creation describes *the integral functioning* of endless

---

photographs are numbered, however, and astronauts' commentaries placed beside them. I cite the photograph numbers for purposes of locating these comments.)

4. Jay McDaniel, "'Where Is the Holy Spirit Anyway?' Response to a Sceptical Environmentalist," *Ecumenical Review* 42, no. 2 (April 1990): 165. Charles Birch, an Australian biologist and an experienced hand in WCC circles, puts it this way: "The Sixth Assembly of the World Council of Churches in Vancouver in 1983 called for a program on justice, peace, and the integrity of creation. The phrase *integrity of creation* was not then spelled out. The phrase needs to be given precise meaning. Integrity of creation should refer to the recognition of the integrity of the intrinsic value of every living creature and the maintenance of the integrity of the relations of each creature to its environment" (Charles Birch, "Christian Obligation for the Liberation of Nature," in *Liberating Life: Contemporary Approaches to Ecological Theology*, ed. Charles Birch, William Eakin, and Jay B. McDaniel [Maryknoll, N.Y.: Orbis Books, 1990], 61).

natural transactions throughout the biosphere and even the geosphere. These comprise the specific exchanges and cycles from which all nature lives, ourselves included atop a complex food chain. They have an integrity that must not be violated if life is not to fall from its perch. Mess too much with the composition of finely tuned greenhouse gases and watch interlocking dynamics of warmer temperatures, changed growing seasons, different patterns of rain- and snow-fall, rising seas, altered ocean currents, dying coral reefs, and interrupted food chains — for starters.[5] The behaviors of one domain are integrated with those of others. These exchanges and cycles in their totality are integrated. They have an integrity *all together.* Earth is a dynamic but closed system.

One example suffices. The $150,000,000 Biosphere 2 in the Arizona desert is a self-sustaining world that is to last one hundred years. (Owners and managers graciously consider earth Biosphere 1.) Bio 2's initial trial period of two years suffered a major flaw, however, and large amounts of extra oxygen had to be supplied from outside, thus negating the goal of the project — self-sustainability. Scientists discovered the reason. The creators, in order to assure rich organic soils for food production, put five to ten times more organic matter in the soil than earth normally provides. The glut of the organic matter like peat and com-post set off an explosion of oxygen-eating bacteria. These in turn produced high levels of carbon dioxide as a result of bacterial respiration. There are thirty thou-sand tons of soil in Biosphere 2. Evidently most of it will have to be replaced to keep the atmosphere balanced for human life. But how to do that and yet have sufficient amounts of rich soils in a confined space in order to grow all the food and feed necessary is the problem. (In an interesting editorial the scientist who discovered the flaw called the action of enriching the soil "arrogant.")[6]

2. The integrity of creation refers to nature's *restless self-organizing dynamism.* While creation is finally one, its internally connected condition is dynamic. The integrity of creation assumes this dynamism and its endemic creativity. There is a persistent restlessness in nature, but this novelty is self-organizing. Nature is comprised of charged and changing stabilities. The sources of novelty seem to be the sources of order and vice versa. Differently said, while all life is a restless adventure, it exists within vital boundaries and depends on elements working together as an astounding community. Sometimes the changing relationships are harmonious, sometimes symbiotic, sometimes predatory, sometimes a mix of these. Destruction is certainly part of nature's creative process. Biodiversity itself seems as dependent upon traumatic disturbances as upon smooth continu-ities. Yet there are patterns and limits, even when they shift and change. Even chaos has a theory now! The integrity of creation means recognition and re-

---

5. The point is not that all these things definitely will happen. The point is rather that we don't know exactly what will transpire and *cannot know* in such a huge, complex, dynamic system of systems until after the fact. For practical purposes, it is then often too late to return to more desirable conditions.

6. "Too Rich a Soil: Scientists Find the Flaw That Undid the Biosphere," *New York Times,* October 5, 1993, C1.

spect for these boundaries and integrated patterns. They are integral to nature's dynamism itself.

Integrity thus refers to the dynamic, internally connected, and organized condition of creation. "Matter" is not "parts," chunks hammered together by God or chance. The elements of life are not related to one another like monads putting out contracts. They are all *inter*related from the outset, at the core of being itself. With the exception of sunlight, nothing gets out or away. Everything here stays here, in changing forms. Recycling is the fundamental earth dynamic. "Waste" in one phase normally becomes food or habitat in another, via cyclical and spiral processes.

Yet the natural recycling can be altered, and in ways destructive of the processes themselves. This occurs when the integrity of their organic character is not respected or their calibrated balance is ignored. Destroy the ozone shield high above earth's skin and see what happens to marine phytoplankton at the beginning of the food chain, to immune systems, to human skin and eyes.[7] The atmosphere, it turns out, is to life even more than the fur is to the cat or a nest to a bird. While not living itself, it is made by living creatures and protects all of them against the deadly cold of space as well as fierce radiation.[8] But more than that, it literally lets us live and breathe within this thin envelope. So care must be taken not to brew the chemistry in destabilizing ways. Or, to recall Bio 2, enrich the soil inordinately and watch the chain set in motion by those wildly promiscuous bacteria. The whole has a dynamic and inclusive integrity. Earth is one and on the move. Stasis does not describe it; bounded, integrated dynamism does.

The implication is utterly clear: "the fundamental 'law of nature' is that human activities must stay within the bounds of nature."[9] Given nature's own exuberance *in the form of humans,* this law is always being transgressed. The distinctiveness of modern times, however, is the scale of the transgression and the consequences, altering the rest of nature in dangerous and unforeseen ways. This, we shall see, pertains with special force to certain current economies simply because all economies are "inextricably integrated, completely contained, wholly dependent sub-systems of the ecosphere."[10] Denying that changes it not a whit.

In religio-moral terms, the error is the failure to submit human power to earth's finitude and inhabit earth on terms it can accept. Course correction means disciplining human nature and curbing human power. Norman Maclean's ac-

---

7. Again the point here is not sure prediction. We don't know exactly what ripping the ozone shield means. We do know it will not be insignificant. The far-reaching and dynamic relationships of the atmosphere with the rest of life rule out insignificance.

8. Al Gore Jr., *Earth in the Balance: Ecology and the Human Spirit* (Boston: Houghton Mifflin, 1992), 264, in a discussion of the Gaia hypothesis of James Lovelock.

9. James Nash, "'Toward a Just, Sustainable, and Frugal Future,' Part One: Confronting the Economics-Ecology Dilemma," *Earth Letter* (March 1994): 6.

10. William E. Rees and Mathis Wackernagel, "Ecological Footprints and Appropriated Carrying Capacity: Measuring the Natural Capital Requirements of the Human Economy," in *Investing in Natural Capital: The Ecological Economics Approach to Sustainability,* ed. A-M Jannson et al. (Washington, D.C.: Island Press, 1994), 365.

count of his father in the opening pages of *A River Runs through It* says it
nicely:

> To him, all good things — trout as well as eternal salvation — come by
> grace and grace comes by art and art does not come easy. . . . As a Scot and
> a Presbyterian, my father believed that man by nature was a mess and had
> fallen from an original state of grace. . . . I never knew whether my father
> believed God was a mathematician but he certainly believed God could
> count and that only by picking up God's rhythms were we able to regain
> power and beauty. . . . If you have never picked up a fly rod before, you will
> soon find it both factually and theologically true that man by nature is a
> damn mess. . . . Until man is redeemed he will always take a fly rod too far
> back [and lose] all his power somewhere in the air . . . since it is natural for
> man to try to attain power without recovering grace.[11]

3. The integrity of creation not only points to nature's dynamism and internally bounded order. It also refers to earth's treasures as *a one-time endowment*.
As we have noted, nature is by nature a creative borrower. The planet is self-
renewing in ways seen and unseen. This totality is immensely rich, varied, and
dynamic. It is also finite, limited, vulnerable, and subject to subversion and exhaustion. This came as a surprise to a German astronaut: " . . . a thin seam of
dark blue light — our atmosphere. Obviously this was not the ocean of air I had
been told it was so many times in my life. I was terrified by its fragile appearance."[12] But the chief point is that all the central goods of the biosphere were
in place long before humans arrived, and none of them — oceans and sky, sun
and forests, photosynthesis and hydrologic cycles, rivers and mountains, DNA
and food chains — can be created by humankind.[13] No human power, or any
other, can bring back an extinct species. Yet we live as if the endowment were
boundless and eternal.[14]

We cannot imagine society as we know it functioning without fossil fuels, for
example. But not only is the drawdown of oil well underway; the atmosphere
holds a limit to expanded burning of coal, oil, and natural gas. Jeremy Rifkin's
generalization is correct: "The most important truth about ourselves, our artifacts and our civilization is that it is all borrowed."[15] Even the molecules in

---

11. This excerpt from Maclean's *A River Runs through It and Other Stories* (Chicago: University
of Chicago Press, 1976), 2–4, is taken from David C. Toole, "Farming, Fly-Fishing, and Grace:
How to Inhabit a Postnatural World without Going Mad," *Soundings* 76, no. 1 (spring 1993): 87.

12. Ulf Merbold, in *The Home Planet*, 58.

13. See the discussion of Holmes Rolston III, "Wildlife and Wildlands: A Christian Perspective,"
in *After Nature's Revolt: Eco-Justice and Theology*, ed. Dieter T. Hessel (Minneapolis: Fortress, 1992),
126–31.

14. The Russian astronaut Yuri Glazkov responded to the lifelessness of space in this way: "Nature has been limitlessly kind to us, having helped humankind appear, stand up, and grow stronger.
She has generously given us everything she has amassed over the billions of years of inanimate development. We have grown strong and powerful, yet how have we answered this goodness?" (from
The Home Planet, 83).

15. Jeremy Rifkin, *Declaration of a Heretic* (Boston: Routledge and Kegan Paul, 1988), 97.

our face and body are there only temporarily, on their journey to and from the environment. We are forever borrowing from the environment. We transform resources from nature into utilities in order to create and maintain our way of life. Yet everything we transform must eventually end up "back" in nature "after we have expropriated whatever temporary value we can." *All* our activity is an economics of borrowing. We are indebted "to the core of our being."[16]

In short, the integrity of creation refers to a one-time endowment that can be and is being jeopardized. What is to be made of this for earth ethics will be considered later. But there is wisdom for the journey in the underlying attitudes of another era. Consider William Penn (1644–1718) in *Fruits of Solitude:* "[W]e have nothing that we can call our own; no, not our selves: For we are all but Tenants, and at Will, too, of the great Lord of our selves, and the rest of this great Farm, the World that we live upon."[17] The conclusion of many religious communities is that earth is a commonwealth — the world a great farm in which we are all but tenants, an *oikos* we are to till and tend but not own.[18]

4. The integrity of creation as a one-time, dynamic, natural endowment that can and is being jeopardized is related to another dimension, *the integral relation of social and environmental justice*. Perhaps better said, there is an internal connection in the redress of "the three great instabilities" of "injustice, unpeace, and creation's disintegration" (Charles Birch). Earth as *oikos* — a vast world house — is a *shared* home. "The whole inhabited world" displays this: all, humans and otherkind alike, are relatives of one another as a consequence of basic relatedness on a curved and closed planet. We live into one another's lives and die into one another's deaths.[19] Earth knows this. The implication is that justice for people and for the rest of the planet are knotted together.

Cause and effect of the great instabilities can run in all directions. The earth in a survivalist mode cannot tolerate masses of people straining at survival's edges any more than it can the assault of affluence. Both poverty and affluence threaten and degrade basic life-support systems, albeit in wildly different degree.

---

16. Ibid.

17. J. Philip Wogaman, *Christian Ethics: A Historical Introduction* (Louisville: Westminster/John Knox, 1993), cites this from William Penn, *Some Fruits of Solitude*, ed. Charles W. Eliot, Harvard Classics 1 (New York: Collier, 1937 [1909]), no page given.

18. In this view, absolutist notions of private property and justification of sharp inequity on the basis of property rights are usually rejected. The Diggers, a movement contemporary with Penn's Quakers, included in their 1649 manifesto: "And the Earth, which was made to be a Common Storehouse for all, is bought and sold and kept within the hands of a few, whereby the Great Creator is mightily dishonoured, as if He were a respecter of persons, delighting in the comfortable livelihood of some, and rejoicing in the miserable poverty and straits of others" (cited in Wogaman, *Christian Ethics*, who is citing "The True Levellers Standard Advanced," from Lewis H. Beren's *The Digger Movement in the Days of the Commonwealth* [London: Holland Press and Merlin Press, 1961 (1906), 96]). Not that such religion belongs only to the past. The papal encyclical *Laborem exercens* (1981) of Pope John Paul II says, on moral grounds, that "the right to private property is subordinated to the right to common use, to the fact that goods are meant for everyone." From Wogaman, *Christian Ethics*, 244, citing John Paul II, *On Human Work: Encyclical Laborem Exercens* (Washington, D.C.: U.S. Catholic Conference, 1982), 6.

19. See the discussion of George E. Tinker, "Creation as Kin: An American Indian View," in *After Nature's Revolt*, 144–53.

(Affluence is responsible for roughly 70 percent of environmental degradation.) These life-support systems beg for attention to their intrinsic requirements for health and well-being. At the same time human rights and well-being cannot be realized in an utterly devastated environment. Even something as basic as human health itself is the complex outcome of the interaction of our genes and the environments we are exposed to and part of throughout life. Social justice and health in any degree thus require renewable, "reasonably" unpoisoned material resources and systems.[20] Because we can appropriate other peoples' sustainability for our own, and transport their earth resources into our habitats, we lose sight of the dense integral connections here. But earth does not. Earth is the ultimate bookkeeper, and eventually all pay, though not the same price or in the same way. In any event, "the integrity of creation" names this ecumenicity and the moral imperative of a whole-earth justice. Earth and its distress are indivisible. So are earth and its health. The whole has this kind of integrity.

This internal linkage and its complexity were hard-won insights for the Seoul delegates, even with earth at the center of the closing worship. Like so much of the world, the WCC has its passionate advocates of social justice and environmental awareness. But they weren't the same crowd in Seoul, and the advocates of a just integration of society found themselves yelling at the advocates of a healthy integration of nature, both parties assuming the very dualism they were creating. That the blood and arguments flowed hot is understandable. Indeed, 1.3 billion people live in absolute poverty, and another 1.5 lack access to even rudimentary health care. This poverty enervates, grinds, destroys. Epidemics, uncontrolled migration, markets too poor to attract investors and absorb exports, elemental missing supplies of adequate food, shelter, and employment — all these contribute to a vicious undertow for the poor. Small wonder that the environmentalist agenda, whenever it smacked of conservation and preservation of nature for the pleasures of the affluent, was suspected of moral and religious violation. Yet when the gavel finally went silent, the Seoul delegates said something that in some ways the wider world has not yet quite understood. In the words of the official document:

> The integrity of creation has a social aspect which we recognize as peace with justice, and an ecological aspect which we recognize in the self-renewing, sustainable character of natural ecosystems.[21]

Note: *both* the "social aspect" and the "ecological aspect" are dimensions of the integrity of creation. While the internal linkages are not made explicit in this passage, the whole document belongs to a significant refusal issued by the

---

20. See the discussion of James Nash, *Loving Nature: Ecological Integrity and Christian Responsibility* (Nashville: Abingdon, 1991), 171–76.

21. World Council of Churches, *Now Is the Time: The Final Document and Other Texts from the World Convocation on Justice, Peace, and the Integrity of Creation* (Geneva: WCC Publications, 1990), 18.

WCC: a refusal to separate human from nonhuman worlds, *together with* a simultaneous refusal to separate any one human world from another in the tangled skein of earth relations today. The integrity of creation here is a way of thinking that stands opposed to what we earlier tagged "apartheid thinking." Apartheid thinking assumes that some worlds can be separated from other worlds (other humans' and otherkind's), and these can pursue separate development because they are not internally related. Against this, the integrity of creation assumes that life today carries a fateful "integrity" from the inside out. Effects may and do vary widely. Some are life-giving while others are deadly. But none is without effect, socio-environmentally.

5. For millions and probably billions, the integrity of creation also names a *divine source and a certain intrinsic dignity*. It gives theological voice to faith's conviction that creation in its totality and its differentiation is good and the work of a life-giving God. "Integrity" is a word for the preciousness integral to creation's being. West Virginia hills in October, where trees color the morning mist like a mad artist, or the bright eyes of a child running with the latest daycare creation to the parent waiting in the doorway — these bespeak God and a primal goodness that reaches toward eternity.

In fact, the very passage from Seoul cited above is preceded by these sentences: "We affirm that the world, as God's handiwork, has its own inherent integrity; that land, waters, air, forests, mountains and all creatures, including humanity, are 'good' in God's sight. The integrity of creation has a social aspect."[22] "Creation" is thus a theological word rather than a scientific one. It means all things together, in, with, and before God, all things in their totality and in their differentiation as an expression of the divine life. It may be used as a synonym for nature, but nature understood in a certain way and bearing a sure dignity. It is nature understood in the manner of aboriginal religion — nature as a luminous expression of sacred power, life, and order. Langdon Gilkey says that nature in this understanding embodies powers and values that "we associate with God and therefore respect, revere, and cherish: infinity, power, life, order, uniqueness or individuality, and self-affirmation."[23] Nature, then, has its own integrity as "*independently good* expressions of the divine love... expressed in God's creative activity."[24] Nature bears traces of the divine.

Such is the theological meaning with which we began. It was given specifically Christian articulation in a WCC consultation in Norway (1988): creation "possesses an inner cohesion and goodness" by virtue of its origin in "the will and love of the Triune God." Moreover, it is not we who "integrate" creation. Its integrity "is prior to our concern, prior to our participation."[25] One implication is

---

22. Ibid.

23. Langdon Gilkey, *Nature, Reality, and the Sacred: The Nexus of Science and Religion* (Minneapolis: Fortress, 1993), 152.

24. Tom Regan, "Christianity and Animal Rights: The Challenge and the Promise," in *Liberating Life*, 81. Regan is speaking of (nonhuman) animals, but both his argument and ours properly extend this to the rest of creation.

25. *Integrity of Creation — An Ecumenical Discussion*, Granvollen, Norway, February 25–March 3, 1988 (Geneva: World Council of Churches/Church and Society, 1989), 3.

clearly moral, since that which has integrity as an embodiment of a goodness in and before God cannot be mere means. It cannot be "the material furniture of a (human) region"[26] and little else. It cannot be simply subject to other creatures' whims and even needs (namely, ours),[27] at least not without taking account of its own life value and requirements. Traces of the divine demand more.

6. Yet a true understanding of the integrity of creation carries more than the common conviction of Judaism, Christianity, and Islam about the interior goodness of creation as God's. As intimated, it also carries the *specific ethical freight of its religious content*. We must elaborate this as a basic dimension of its meaning.

Creation is not an abstract mass, an undifferentiated sea in which we might immerse ourselves in romantic, ecomystical fashion. As noted, creation is the gathering of "*independently good* expressions of the divine love." This means respect for otherkind *as* differentiated other. Initially it means the rightness of the other's presence and respect for the fact that each creature ought to be able to strive to be and do what it is created to be and do — to be a lady slipper or daffodil, a mushroom or red oak, a chipmunk, squirrel, or blackbird. Any loss here is genuine loss — to the creature, of course, but also to earth and, it can be argued, to God. The sparrow's fall is a significant event.

More deeply, the integrity of creation here means *being-with* as creation's own way of being. Existence is coexistence; creation is "social" by nature; life requires mutuality. What never works is going it alone, fleeing relatedness, acquiring utter autonomy, being-above rather than being-with, bending all in one's own direction without reciprocity, without mutuality, without consequence or responsibility.

Take the case of the attine ants, an impressive tribe that has worked out a pact with certain fungi. The ants and the fungi are "symbionts," dependent upon one another, with each evolving characteristics that contribute to mutual survival. The ants gather huge amounts of plant materials from ant real estate and deposit it conveniently on fungi real estate. It's food the sedentary fungi could not possibly hustle on their own. The fungi mulch this biomass and grow so fat that their tips swell with precisely the nutrients, sugars, and proteins attine ants need. It's an impressive arrangement. Equally impressive is the life of this contract to date. This has been going on, says E. O. Wilson, for something on the order of fifty million years.[28]

In short, what Douglas John Hall calls "the ontology of communion"[29] as creation's own *logos* (logic) merits respect for difference and independent goodness within creation's matrix of interrelatedness and common ground. And it means a consideration of otherkind's well-being as more than an extension of ours. The

---

26. Erazim Kohák, *The Embers and the Stars: A Philosophical Inquiry into the Moral Sense of Nature* (Chicago: University of Chicago Press, 1984), 69.

27. Gilkey, *Nature, Science, and Religion*, 175.

28. See "Ant and Its Fungus Are Ancient Cohabitants," *New York Times*, December 13, 1994, C1,4.

29. See Hall, *Professing the Faith*, 317ff.

"integrity of creation" carries moral freight of this kind. It resists our natural anthropocentrism from the inside out. We are all symbionts. Evolution is the history of an extended family.

MORAL IMPORT

From this initial discussion of the integrity of creation, and from an orientation to earth as *oikos*, what follows for an earth ethic?

James Nash's discussion begins an answer. It is only a beginning. The rest of this book is the rest of the answer. "All creatures are entitled to moral consideration."[30] This is Nash's way of making morally explicit the respect for nature I've discussed. It means that all life, not just human life, shares a common moral universe. Aboriginal traditions have assumed this. Most modern ones have not.

Common moral citizenship does not mean that all creatures and their claims carry "the same 'moral significance.'"[31] All may bear intrinsic value as independently good expressions of divine activity. But intrinsic value does not ipso facto mean *equal* intrinsic value. All creatures may even have "rights" in a moral sense (a much disputed issue in ethics). But when rights conflict, not all carry the same weight. Some rights are more basic than others and override them when rights conflict. Nash's own "graded model" argues, for example, that other things being equal, sentient creatures' claims are stronger than nonsentient ones in conflict situations. Since all species necessarily use other species for survival and flourishing, choices are unavoidable, including the taking of life itself. Yet the larger point is Nash's initial one. For earth ethics, the integrity of creation means that all creatures are entitled to moral consideration even when that does not issue in moral equality. The crass moral dualism in which only one species occupies the moral universe — us — while the rest of creation lacks any moral consideration at all is flat out repudiated.

From this basic moral respect and consideration, Nash creates an ethical infrastructure. At one pole, an adequate ethic "*must posit respect for individual lives, not simply aggregates like species.*"[32] In our language, the integrity of creation pertains to all lives. The other is more than a nameless instance of the whole. Otherkind is differentiated. Of course, as a matter of sheer practical logic (and moral sentiments aside), it is also the case that a species cannot be a species apart from the individuals who comprise it. So a derivative case for attention to individual lives is that species welfare itself depends upon it. The integrity of creation, however, posits respect and value directly, and not only derivatively. And it does so for all. "Listen, God love everything you love," Shug says to Celie in *The Color Purple*, "and a mess of stuff you don't."[33] Before God all creation has standing.

---

30. Nash, *Loving Nature*, 181.
31. Ibid.
32. Ibid., 179.
33. Alice Walker, *The Color Purple* (New York: Pocket Books, 1982), 178.

At the other pole, an adequate ethic *"must be holistic, concerned about collective connections."*[34] This is not identical to respect for species, though that is included. Rather, Nash means moral consideration of the systems that together comprise the biosphere. He means ecosystems, including not only their living creatures but also their essential abiotic elements (for example, minerals and gases). The proper language here is "Earth Community," and the proper morality is the morality of community life and well-being. But it is community as the whole biotic community. Aldo Leopold's ethical guideline is invoked: "A thing is right when it tends to preserve the integrity, stability, and beauty of the biotic community. It is wrong when it tends otherwise."[35] This is one way to formulate the moral import of the relatedness at the heart of being itself. We have named it "the integrity of creation."

Within this bipolar framework of moral consideration Nash discusses "biotic rights." Biotic rights are only one dimension of earth ethics, and the following is little more than a taste and a tally. It is nonetheless a fitting way to close a first run at the ethical implications of "the integrity of creation" as itself a first contribution to an earth ethic.

These rights are held against humans only. Flora and fauna do not make moral claims in relation to one another, since matters of justice or charity or equal treatment or other expressions of moral consciousness do not belong to otherkind's sense of what appropriate behavior is. The character of human beings as moral and religious creatures thus means that biotic rights belong to human consideration of nonhuman life, not flora's and fauna's treatment of one another. A salmon cannot claim the grizzly violates salmon rights as the grizzly sweeps it from the fast-running waters. Humans can, however, make moral decisions and exercise moral responsibility about human treatment of salmon and grizzly habitat and their right to exist insofar as human actions affect that.

With this as moral framework, Nash proposes the following biotic rights:

1. *The right to participate in the natural dynamics of existence.* This is a right to flourish as nature provides this, without undue human alteration of the genetic or behavioral "otherness" of non-human creatures.

2. *The right to healthy and whole habitats.* The right to flourish on nature's terms and contribute to the common ecological good assumes and requires that otherkind enjoy the essential conditions which appropriate habitat provides.

3. *The right to reproduce their own kind without humanly-induced chemical, radioactive, hybridized, or bioengineered aberrations.* This right asks human respect for genetic integrity, evolutionary legacies, and ecological relationships. By implication, it defends biodiversity.

---

34. Nash, *Loving Nature,* 183.
35. Cited in ibid., 185, from Aldo Leopold's *A Sand County Almanac* (New York: Ballantine, 1970), 262.

4. *The right to fulfill their evolutionary potential with freedom from human-induced extinctions.* Extinctions are a natural part of evolutionary process, but human-induced extinctions are unjust. Humanity's exercise of its power ought not to undermine the existence of viable populations of non-human species in healthy habitats until the end of their evolutionary time.

5. *The right to freedom from human cruelty, flagrant abuse, or profligate use.* Minimal harm to otherkind within necessary usage ought to characterize human treatment of non-human life.

6. *The right to reparations or restitution through managerial interventions to restore a semblance of natural conditions disrupted by human abuse.* Because of human abuse of natural environments in the past, interventions are often necessary to enable a return to an approximation of previous ecosystemic relationships.

7. *The right to a "fair share" of the goods necessary for individuals and species.* "Fair share" is, of course, a vague criterion. Yet it is possible to determine ways in which human populations can coexist with viable populations of humanly unthreatened species and thereby preserve for them a fair share of the shared ecological good.[36]

It is important to recall Nash's framework, including his graded model of rights corresponding to graded valuations of species. It is also important to note that these rights are not moral absolutes. They are prima facie claims. Like most principles and norms in ethics, including most human rights, biotic rights are not inviolable. As prima facie rights, they are to be honored unless there is stronger moral justification for violation than for compliance: that is, unless a persuasive case is made for morally justified exceptions. Making the case for exceptions will need to appeal to the same basic ground that underlies such rights in the first place; namely, respect for life and the ecosocial common good.

CONCLUSION

The integrity of creation issues in a full-court press for a life-centered earth ethic. It could be formulated in Hans Jonas's rendition of Kant's categorical imperative: "Behave in such a way that the effects of your actions are always compatible with the permanence of nature and of human life on the earth."[37] A life-centered earth ethic could also adapt Mother Ann Lee's version of Shaker

---

36. I have excerpted Nash's formulations from "Biotic Rights and Human Ecological Responsibilities," *Annual of the Society of Christian Ethics* (1993): 154–57. The italicized sentences are Nash's verbatim. The remainder is my summary of his commentary.

37. Hans Jonas, as cited by Leonardo Boff, "Social Ecology: Poverty and Misery," in Hallman, *Ecotheology,* 244.

sustainability: "Do all your work as though you had a thousand years to live on earth, and as you would if you knew you must die tomorrow."[38]

For some, earth ethic inspiration will be religious. If so, it will recall the convocation in Seoul: a turning to God is simultaneously a turning to earth. But recalling that convocation, we might ask again whether the delegates fully understood the earth patriotism and earth ethic they sang and prayed and ritually enacted as they bore earth, blessed it, and committed themselves to "justice, peace, and the integrity of creation." They understood in varying degree, no doubt. Perhaps they understood little more than the symbolic power of earth as the living jewel of life in space. Perhaps as a start, even a moral and religious jump-start, that is enough.

Yet more is required if we are to understand what the integrity of creation means for sustainability. For that we must listen well — and anew — to earth itself at the point of the biggest rub. The biggest rub is the grind of the human economy against the economy of nature.

---

38. As cited by Wendell Berry in *Sex, Economy, Freedom, and Community* (New York: Pantheon, 1992), 111.

The Big Economy and
the Great Economy

Like Sherlock Holmes's case of the dog that didn't bark, the clue is in the silence that shouldn't have been. The blank spaces, the missing sounds, tell the story.

Take classic texts in the study of economics. Take macroeconomics, especially, the branch of economics that analyzes mass structures, dynamics, and scale. Take a look at the index and note what isn't there. Typically not there is an entry for "environment." Or for "natural resources," "pollution," or "depletion." There is no entry for "sustainability" either, or even the hot topic, "sustainable development." These spaces are blank. These dogs don't bark. These clues give away the case.

But what's the case? The case is a violation of creation's integrity, located in a huge mismatch between the Big Economy (the present globalizing human economy) and the Great Economy (the economy of nature).[1]

And where's the mismatch? Thomas Berry explains, for starters: "The difficulty of contemporary economics is its effort to impose an industrial economy on the organic functioning of the Earth....[H]uman economics is not integrated with the ever-renewing economics of the natural world. This ever-renewing productivity of the natural world is the only sustainable economics. The human is a subsystem of the Earth system."[2]

Berry slides easily into species talk: "[*H*]*uman* economics is not integrated....."; "the *human* is a sub-system...." Yet his opening line doesn't refer to all human economies, but a particular one, "an industrial economy."

So what's the bind, the big rub between our present huge, globalizing economy and "the ever-renewing economics of the natural world"? We won't know until we stroll among the details awhile and inspect the plants. We need to see up close how the Big Economy and the Great Economy work.

The place to begin is with one of the basics of the integrity of creation: every human economy of whatever sort in any time and place is necessarily

---

1. "The Great Economy" is a phrase used first by Wendell Berry for that complex economy which sustains the whole web of life. See "Two Economies" in Berry's *Home Economics* (San Francisco: North Point Press, 1987), 54–75. Berry writes that his term of personal preference is "the Kingdom of God" but that it is limited to biblical traditions. A more culturally neutral term is therefore desirable. He goes on to say, however, that while we can "name it whatever we wish, ... we cannot define it except by way of a religious tradition....The Great Economy, like the Tao or the Kingdom of God, is ... the ultimate condition of our experience and of the practical questions rising from our experience, and it imposes on our consideration of those questions an extremity of seriousness and an extremity of humility" (56–57).

2. Thomas Berry, "The Meadow across the Creek" (unpublished manuscript), 128. Used with permission.

a subsystem of the Great Economy, the economy of earth. Just as there can
be no postagricultural society,[3] there can be no economic order that is not totally dependent upon the planet's ecosystems and the biosphere and geosphere
as a whole. Human economies have considerable latitude, to be sure, and new
resources can often substitute for depleted and exhausted ones. But human economies dare not exceed tolerable environmental margins or violate requirements
for renewing and replenishing nature. Economic production and consumption,
as well as human reproduction, are unsustainable when they no longer fall within
the borders of nature's regeneration. So the Bottom Line below the Bottom
Line[4] is that if we don't recognize that the laws of economics and the laws of
ecology are finally the same laws, we're in deep doo-doo. Eco/nomics is the only
way possible.[5]

Differently said, an expanding human economy that issues in a diminishing
earth economy commits suicide by increments.

Where are we presently? A glimpse is offered by William Rees and Mathis
Wackernagel. They want to measure "the natural capital requirements of the
human economy" (the present Big Economy) in order to judge what earth's
"carrying capacity" might be. Their project leads them to consider conventional
economic assumptions and rationality. It is here that we stumble on the reasons
for the missing entries in the economics index. Far more important, it is here
that we stumble on the reasons for the sustained unsustainability at the heart
of earth's distress.

Rees and Wackernagel contrast "the expansionist vision" of the Big Economy
with a still largely nameless ecological perspective. The former remains the prevailing vision of global economic reality, as led by the world's richest industrial
powers. The latter is the minority report of a new, mutated subspecies called
"ecological economists." With little help from anyone, they are still trying to find
full voice and be heard in the auditoria of business, government, and academe.[6]
Lawrence Summers, former chief economist of the World Bank before becoming undersecretary of the treasury in the Clinton administration, is a devotee of
the expansionist perspective: "There are no ... limits to carrying capacity of the
Earth that are likely to bind at any time in the foreseeable future. . . . The idea
that we should put limits on growth because of some natural limit is a profound

---

3. The point is nicely made by Timothy Weiskel, as cited in Lester R. Brown, *State of the World,
1991* (New York: Norton, 1991), 15.

4. The image is Thomas Berry's, "Meadow across the Creek," 131.

5. William Ashworth uses this neologism to signal the necessary synthesis of ecological and
economic laws. He asks for a discipline called "eco/nomics" (see *The Economy of Nature: Rethinking
the Connections between Ecology and Economics* [New York: Houghton Mifflin, 1995], 306).

6. An ecological economy may in fact have found one name — "the restorative economy" —
and its voice in Paul Hawken and the policies he offers. We have cited his *Ecology of Commerce:
A Declaration of Sustainability* (San Francisco: Harper Business, 1993) often. What is
distinctive about the book is that it is so distinctive! Hawken is one of the few persons in business
who could have, or would have, written it. Ecological economics remains a minority report and
activity.

error."[7] Curiously, the contrasting ecological view also belongs to a former World Bank economist, albeit a renegade one, Herman Daly.

But understanding *either* expansionist economics *or* ecological economics requires a knowledge of the Great Economy's principles.

PRINCIPLES OF EARTH'S ECONOMY

For all nature's wonder and awesome complexity, its economic principles are few and simple. For our purposes, three are most determinative.

First, "waste equals food."[8] As noted earlier, "recycling" is fundamental to nature's generation and regeneration. Tossed-off material from some life-forms becomes nourishment for others or builds up essential habitat. Everything goes somewhere and contributes something, with a minimum of energy and inputs. What is taken in is changed in a way that can be used by another living body. Like Aldo Leopold's picture of Wisconsin as a "round river," there is "the stream of energy which flows out of the soil into plants, thence into animals, thence back into the soil in a never ending circuit of life." "Dust unto dust," he adds, "is a desiccated [dry] version of the Round River concept."[9] If such a principle were applied to the human economy, all waste would have a place in, and value to, further production. Everything would be reclaimed, reused, or recycled and would, from the outset of production, be made to do so.

(Paul Hawken's insight is worth including here. We call ourselves consumers, he says, "but the problem is we do *not* consume."[10] On average, for example, each person in the United States generates twice his or her weight per day in household, hazardous, or industrial waste and an additional half-ton of gaseous wastes such as carbon dioxide.[11] This is *not* used up or consumed but released, with the hope the rest of nature will deal with it. The rest of nature knows how to consume, we don't.)

Second, "nature runs off current solar income."[12] Sunlight is the one and only input into earth's economy from outside, the only resource not already contained within earth's closed system. Everything else belongs to the one-time endowment. What this means for human economies is far-reaching but cannot be

---

7. Cited from William E. Rees and Mathis Wackernagel, "Ecological Footprints and Appropriated Carrying Capacity: Measuring the Natural Capital Requirements of the Human Economy," in *Investing in Natural Capital: The Ecological Economics Approach to Sustainability*, ed. A-M Jannson et al. (Washington, D.C.: Island Press, 1994), 363. Rees and Wackernagel cite as their source S. George, "Comment on the World Bank's 'World Development Report 1992' and the June 1992 Earth Summit," from *The Globe and Mail*, Toronto, May 29, 1992.

8. Hawken, *Ecology of Commerce*, 12.

9. Aldo Leopold, *A Sand County Almanac* (New York: Ballantine, 1966), 188.

10. Hawken, *Ecology of Commerce*, 12; emphasis added.

11. Ibid.

12. Ibid. Hawken has overlooked internal geological income. Radiated heat from the breakdown of deep radioactive minerals drives much of earth's geology — plate tectonics, volcanism, earthquakes. This in turn impacts geography, weather patterns, and so on. My thanks to Dan Spencer for bringing this to my attention.

taken up here except to note that this second principle reinforces the first. If sunlight is *the* only true addition to what is already "in-house" (*oikos* again!), re-claiming, reusing, and recycling need to qualify and characterize all production and consumption. If a given entity cannot be reclaimed, reused, and recycled, don't make it. If it cannot be reproduced without deleterious, degrading effects, don't grow it. If the probable consequences of its use cannot be reasonably known, tracked, and paid for, don't venture it.

Third, "nature depends on diversity, thrives on differences, and perishes in the imbalance of uniformity. Healthy systems are highly varied and specific to time and place. Nature is not mass-produced."[13] This is the economic version of Swimme and Berry's universal principle of differentiation. Diversity, variation, complexity befitting local and changing conditions, and disparity to take advantage of them — earth's economy lives by such. This is why biodiversity is utterly crucial. It is the means for adaptability, evolution, and survival in complex and only partially stable environments. Diversity breeds stability and sustainability. When things do go badly awry in the economy of nature, *as they can and do,* the problem is precisely that "mass production" and uniformity overwhelm diversity and the fine-tuned differences that make for life. (Old Pharaoh's experience with the hypernature of those Genesis plagues was sufficient, if begrudging, example for him.)

Our present economic scheme of globalized mass production runs in other directions entirely. It doesn't care where resources come from and what "niche" roles they play in and for the Community of Life in home locales. Thus it not only contributes to destruction of social-environmental ecologies but generates one of the great periods of species extinction in history (and the first at human hands). The present globalizing economy also amplifies the negative impact of ignoring nature's third economic principle of production and consumption sufficiently varied and specific so as to continue life in each place. That is, it does not incorporate adaptability and sustainability into the very fabric of locality. Differently said, the Big Economy prefers globalized "development," sustainable or otherwise, to complex local and regional sustainable society and community. It thereby runs against the grain of nature itself and creation's integrity. Nor does the Big Economy live off solar income. It extracts from the one-time endowment that is earth in ways that, on a crowded planet with high consumption rates in rich quarters and an exploding population, threaten sustainability for present and future generations. Yet present economic theory and practice don't have routine ways to measure carrying capacity, largely because the biological life central to the economy of nature is absent in the equations of the Big Economy. Much modern economic theory and most of its determinative practices are still ecologically empty, while nature's economy is necessarily "full."

This sketch in hand, we turn to some details of dominant economic theory and culture.

---

13. Ibid.

OUR PLIMSOLL LINE IS MISSING

An expanding world needs to know the total activity the biosphere can toler-ate and yet renew itself indefinitely. It needs to know the biosphere's "carrying capacity" for the long haul. Scale matters immensely where scale refers to the physical size of human economic presence and its impact on the biosphere. Sus-tainability is at issue. Yet we noticed a curious fact when we looked to standard macroeconomics to help determine carrying capacity: the missing entries for "environment," "natural resources," "pollution," "depletion," and "sustainability" itself. Standard economic practices have few indicators for these.[14]

Why? Because the economy of nature isn't a subject of economics. The thin envelope atop the earth's crust that is its one and only life-system, thinner to earth than an apple's skin to the apple, is omitted from modern economic prac-tices even though every item and action in the human economy utterly depends upon it. So one would never know from economists and entrepreneurs, say Rees and Wackernagel, that "our ecological relationships to the rest of the ecosphere are indistinguishable from those of the millions of other creatures with which we share the planet."[15]

What happens is that standard economic theory and business practices abstract human beings and commerce from the ecosphere. There is an eerie "oth-erworldliness" here. Not in the sense of *ascetic* otherworldliness, of course. After all, material consumption and its increase drive the system. Ever-growing ap-petites are encouraged. But otherworldliness in a more basic, profound sense — a disregard of earth's rhythms and requirements, sometimes even masked dis-dain for them. The economy is thus strangely docetic.[16] The body it lives in is not fully real as an earthly body. Biological life is absent in economic equa-tions.[17] The equations are airy, ghostly, removed from real referents. Even Nobel Laureates say things that must leave other creatures scratching their crania and twitching their antennae. Robert Solow, for example: "If it is very easy to sub-stitute other factors for natural resources, then ... the world can, in effect, get along without natural resources, so exhaustion is just an event, not a catastro-phe."[18] Even when Solow later acknowledges that "there is no reason to believe in a doctrinaire way" that "the goal ... of sustainability can be left entirely to the

---

14. This is not to say that standard economic theory *couldn't* move a long way toward including the economy of nature in its analysis, much of it by adaptation of categories it already uses. It can, and one of the chief tasks of "ecological economists" is to make that adaptation. There is now rather vigorous debate about this.

15. Rees and Wackernagel, "Ecological Footprints," 364.

16. Docetism is the view, declared heretical by early Christians, that Jesus Christ merely *seemed* to inhabit an earthly body but was not fully human.

17. Hawken, *Ecology of Commerce*, 57.

18. Rees and Wackernagel, "Ecological Footprints" (365), are citing Solow's "The Economics of Resources or the Resources of Economics," *American Economics Review* 64: 1–14. The specific page is not given.

market,"[19] we are left wondering how we are to know when threats to carrying capacity and sustainability are being broached.

We need, says Herman Daly, something like the Plimsoll line drawn on commercial ships. When the water hits the Plimsoll line on the hull, the edges of safe carrying capacity have been reached. Innumerable kinds of cargo can be carried, and the cargo can be allocated in various ways for balance and efficiency, but at some point added cargo will push the Plimsoll line below the water line. Beyond that, Daly quips, "optimally loaded boats will sink under too much weight — even though they may sink optimally!"[20]

But we don't have a Plimsoll line for "our biospheric ark."[21] The Good Ship Oikoumenē is without one. In fact, not only does the index of a standard text in macroeconomics not have an entry for "carrying capacity." Such notions as evolution, biological diversity, and the health of the commons are alien and absent. Economics surpasses theology as a docetic science.

How did this huge gap in economic theory and practice develop? Why this ozone hole over the continent of modern economics? Why no real attention to the lifelines tying the human economy to the biospheric one, or regard for the principles of the Great Economy? The answer is complex. For the present the following must be said.

BLINDNESS FOR GOOD REASON

We can avoid thinking about the size and impact of the human economic subsystem so long as it seems small *relative to* the total biosystem and the Great Economy. We are thus lulled into believing that nature offers both infinite resources and ample natural "sinks" for our wastes. Accordingly we — some more than others — have ridden across the globe like cowboys on endless plains, moving herds from resource to resource, never worrying about reseeding or recycling. Overall scale is negligible if there are other pastures. Intricate requirements for the ongoing life of pastures need not command much notice when pastures are plentiful. "Pastures" are fewer and more complicated now, and a growing cabal of economists and businessmen and women do voice concern about the environment and carrying capacity. People sense, and speak of, limits. Yet expansionist economics remains firmly in place, led by such as NAFTA (North American Free Trade Agreement), the 160 nations of the GATT treaty (General Agreement on Tariffs and Trade), and the establishment of the WTO (World Trade Organization). Even the blue-ribbon Brundtland Commission, appointed by the

19. Rees and Wackernagel, "Ecological Footprints," citing Solow's *Sustainability: An Economist's Perspective* (Woods Hole, Mass.: Woods Hole Oceanographic Institution), no page number given.

20. Herman Daly, "The Economist's Response to Ecological Issues," in *Sustainable Growth — A Contradiction in Terms? Economy Ecology and Ethics after the Earth Summit*, 42 (a report of the Visser't Hooft Memorial Consultation, organized under the auspices of the World Council of Churches, and held at the Ecumenical Institute, Chateau de Bossey, June 14–19, 1993; available from World Council of Churches, Geneva).

21. Ibid., 48.

United Nations for the Earth Summit on Environment and Development, says that sustainable development must include "more rapid economic growth in both industrial and developing countries." "A five- to ten-fold increase in world industrial output can be anticipated by the time world population stabilizes sometime in the next century."[22]

Evidently, then, Lawrence Summer's dismissal of limits has more reasons going for it than the cowboy vision of earth as flat and endless. Even when cowboy assumptions are dumped, and environmental alarm is sincere, economic growth tenaciously hugs the saddle.

One reason could be the runoff of piecemeal knowledge and partial vision. We know the local manifestations of environmental degradation — depleted soil, falling water tables, a diminished forest, subverted wetlands, toxins affecting the immune systems of newborns. We may even know some global developments — the rip in the ozone, for example, or global warming. But, in Paul Hawken's helpful analogy, viewing these is still like peering up close at gray dots on a sheet of paper. We examine each dot in detail and see those clustered close around. Yet only with some distance and a broader view do we recognize they are the face of Abraham Lincoln. The fact is, we do not have a full picture of what is happening to the planet. We don't have all the dots, and we don't have the means to see them all in relation to one another. Lacking a full picture, the sense of urgency about fitting the Big Economy inside the principles of the Great Economy is lacking.

Another reason, and the most compelling one, is the sensed lack of alternatives. Millions of livelihoods depend upon the present globalized economy. It *has* to be defended. And given its institutionalized clout, it most certainly will be. In the United States alone, the four hundred companies profiled in *Everybody's Business Almanac* employ or support a quarter of the U.S. population. The largest one thousand companies account for over 60 percent of the gross national product.[23] So while we lack the sense of progress that the industrial-technological process once carried like the damsel cradling the ship in the New Amsterdam Theater, we simply cannot imagine how otherwise to retain affluence, address poverty, and secure livelihoods. And if we cannot imagine an alternative, it doesn't exist. The dream drives the action.[24]

Dreams or not, an impressive track record reinforces loyalty to the going arrangements and feeds lost dreams of the resurrection of the goddess Progress. Despite convulsive decades, wars and rumors of wars, and times of inflation, depression, and unemployment, there has been a two-hundred-year boom. Unparalleled material well-being has come about for unprecedented numbers of people whose ancestors often lived with little and died with less. A whole panoply of institutions has grown up around the taming and control of na-

---

22. Cited by Rees and Wackernagel, "Ecological Footprints," 363–64.

23. From Hawken, *Ecology of Commerce,* 8.

24. This is the argument of Thomas Berry. See especially the chapter "The Dream of the Earth: Our Way into the Future," 194–215, in his *The Dream of the Earth* (San Francisco: Sierra Club Books, 1988).

ture and managing society for expanding abundance. Social problems themselves have been solved by taking not from the already affluent but from the dividends of "growing the economy." No wonder many are still in the grip of this exuberant memory.

In short, we are at a loss for alternatives to what a massive economy delivers, even when we fear its consequences for the very life-support systems it depends upon.[25]

As if this weren't enough, positive incentives to mimic the principles of earth's economy are minimal. The "free ride" or "free lunch" at nature's expense is so deeply institutionalized in capitalism that most businesses that might voluntarily do the right thing by earth would suffer in the marketplace if they did. Given present systemic dynamics, doing the right thing might even put them out of business. There is a reason, after all, why no American corporation has supported a single major piece of progressive environmental legislation while most all of them have rushed to congressional offices to help write deregulative legislation. Satisfying shareholders and protecting markets and profits would be made more difficult by internalizing more costs and meeting more regulations in a globally competitive environment.[26] Privatizing profits while socializing costs to vulnerable human and other communities is a deeply ingrained capitalist practice, and a very bad one.

There are still other reasons why standard economic life ignores the economy of nature. They belong to the conceptual environment deep inside modern economic culture. A world within matches the world without and reinforces it.

REGNANT NOTIONS

One reason we don't live by earth's economic principles is the notion of nature in the Big Economy. As creatures of language and culture, humans don't simply live *in* nature. We live within our understandings and images of nature, and we live them out. Objective nature may set basic and unforgiving terms. But our actions accord with our understandings of what nature is and how it works. What we define as real is real in its consequences.

Modern economic notions and practices are based on concepts and methods borrowed from Newtonian mechanics. Without quite saying so, the basic assumption of the economics that forged the modern world to the beat of Sandburg's "Chicago" seems to be that all physical things consist of the same indestructible matter and that with sufficient capital this eternal matter can be recycled in an endless variety of sometimes useful, sometimes frivolous, forms. Given sufficient capital, neither shortages nor physical limits need appear.

Such a notion of nature "lacks any representation of the materials, energy sources, physical structures, and time-dependent processes basic to an ecological

---

25. See the more extensive discussion in Bruce C. Birch and Larry L. Rasmussen, *The Predicament of the Prosperous* (Philadelphia: Westminster, 1978), 19–41.

26. Hawken, *Ecology of Commerce,* 31.

approach."[27] Little attention is paid the varied details, needs, and dynamics of the earth economy that sustains the whole Community of Life over time. Why? Because, like the advertisement selling chicken parts in various combinations, the Big Economy sees nature as basically "parts." Parts that can be assembled, disassembled, reassembled, exchanged, and substituted. Nature resembles mechanism more than organism in standard practices and mind-sets. It's the enduring hangover of Descartes's contention that because there is no obvious vital principle integrating and guiding the living world, the living world is best understood by the human mind as mechanism. Mind understands matter, and matter is malleable.

Where this view prevails, as it still does in most capitalist practices, business is an open, linear system. With technology, resource extraction, and mobility, growth will always be possible, assuming capital. No inherent limits to expansion exist.[28] Bizarre as it may seem, economic theory and practice do not conceive nature as genuinely alive and finite here. Nor do they entertain the notion that earth is round and thus closed. The Big Economy still lives by the flat-earth policy that serves it so effectively.

Its location of value is yet another reason the expansionist version of economics lives on. Value resides not in nature but in human creativity and use. Both Adam Smith and Karl Marx regarded nature as essentially "free goods" whose value emerges as humans do something creative to them. And John Locke (in *Two Treatises on Government*) and Pope Leo XIII (in *Rerum novarum*) both argued that the value of property derived from nature is established not by earth's eternal labor but by humans' labor, understood somehow apart from earth. In the pope's nearly infallible words: "Now, when man thus spends the industry of his mind and the strength of his body in procuring the fruits of nature, by that act he makes his own that portion of nature's field which he cultivates — that portion on which he leaves, as it were, the impress of his own personality; and it cannot but be just that he should possess that portion as his own, and should have a right to keep it without molestation."[29]

Nor need one consult only the dead. Peter Drucker argues that entrepreneurs create values and resources. Before entrepreneurs get hold of them, "every plant is a weed and every mineral is just another rock."[30] Commodities and production, distribution and consumption processes thus count as the economy. Nature doesn't — except as "raw material." So land is "empty" when "nothing is on it." In that pitiful state, it is "undeveloped" and passively awaits human investment and labor to "make something of it." Until humans acquire and transform it, land is thus devoid of value. Five thousand species may live in this unoccupied

---

27. Rees and Wackernagel, "Ecological Footprints," 365, citing P. Christensen, *Ecological Economics: The Science and Management of Sustainability*, ed. R. Constanze (New York: Columbia University Press, 1991), no page given.

28. Hawken, *Ecology of Commerce*, 32–33.

29. Pope Leo XIII, *Rerum novarum*, 7, as published in *Seven Great Encyclicals* (Glen Rock, N.J.: Paulist, 1963) and as cited by J. Philip Wogaman, *Christian Ethics: A Historical Introduction* (Louisville: Westminster/John Knox Press, 1993), 212.

30. Peter Drucker, "The Age of Social Transformations," *Atlantic Monthly* (November 1994): 30.

earthpatch, and its life-support capacity may be very full indeed. But modern economics does not see this and does not understand in either theory or practice that land is embodied energy with its own complex life, its own complex actions and needs, and its own economy. So an economist, much less a developer, never thinks to ask, for example, how these limited places under the sun might be shared with other species on reciprocal terms.[31] The very notion of nature that conventional economists and people of commerce hold in their heads — passive resources awaiting human transformation — doesn't let the question slide over the brain cells. It is disallowed because the notion of nature we think *with* as we think *about* things of the earth categorically disallows it. The problem, it would seem, is not with the limits of human ingenuity and creativity per se. The problem is with the limits of human perception that spark and channel these. Regnant notions themselves are askew.

Speaking of unallowed questions, we need to circle back to Rees and Wackernagel's concern for "carrying capacity." That omission in standard economics is yet another reason so much present theory and practice are ecologically empty.

APPROPRIATED CARRYING CAPACITY

For ecologists, carrying capacity refers to "the population of a given species that can be supported indefinitely in a defined habitat without permanently damaging the ecosystem upon which it is dependent."[32] It is thus a notion that is territorially bounded, intrinsically tied to local ecosystems and their populations. The ecological notion cannot presently pertain to humans, however, because of the way national, regional, and global economies work. Trade, technology, consumption patterns, and urban settlements extract from and impact *widely scattered* ecosystems. We are utterly dependent on near *and* distant ecosystems.

This dependence notwithstanding, we fail to recognize or account for it as we put the groceries in the bag and the car, turn on the furnace or air conditioner, or place an order for seltzer or underwear. We are not alone. Economic theory and practice fail to as well. Prices reflect relative market scarcity. Prices do not reflect what is happening to disparate ecosystems.

Somehow, Rees and Wackernagel conclude, we must reinterpret carrying capacity. The measure cannot be the health of our local ecosystems, even for the ecologically attuned economies Rees and Wackernagel advocate. Rather, carrying capacity must be understood "as the maximum rate of resource consumption and waste discharge that can be sustained indefinitely without progressively impairing the functional integrity and productivity of relevant ecosystems *wherever the latter may be.*"[33]

---

31. See the discussion in Herman E. Daly and John B. Cobb Jr., *For the Common Good* (Boston: Beacon, 1990), 190–208.

32. Rees and Wackernagel, "Ecological Footprints," 369.

33. Ibid., 369–70.

Take ecosystems in Germany, for example. A concerned population, instructed and agitated by "Green" activists, has increased its attention to nature's health on and in German soil. Does good health there then mean that Germans and Germany are living within nature's carrying capacity? Not when tropical forests in Brazil and other South and Central American countries have been razed to grow soybeans that are fed to cows in Germany that produce surplus butter and cheese for U.S., Brazilian, Costa Rican, and other markets.

Germany is only one example. Industrial societies all reach beyond their borders to appropriate carrying capacity elsewhere for themselves.[34] Ecosystem health within their own borders is thus only one indicator of their impact. And in rich countries with long arms it is a very deceptive one. Germany's own national carrying capacity is "increased" *because* those of Brazil and other nations have been lowered.

Or take cities. Half the human population now lives in urban centers. Cities graphically illustrate the error of focusing on local ecosystem health as a measure of the human economy's relationship to earth's, because cities appropriate carrying capacity from all over the globe. They appropriate it from the future as well. But this appropriated carrying capacity from elsewhere or from future generations is not seen. Both urbanization and trade distance these populations physically *and* psychologically from the scattered ecosystems that sustain them.[35] "However brilliant its economic star," Rees and Wackernagel conclude, "*every city is an entropic black hole*."[36]

Lewis Thomas's notes on cities — and nations — is probably *too* graphic, joining Rees and Wackernagel's exaggerated "entropic black hole": "[Cities] defecate on doorsteps, in rivers and lakes, their own or anyone else's. They leave rubbish. They detest all neighboring cities, give nothing away. They even build institutions for deserting elders out of sight."[37] And nations are no better: "Nations...bawl insults from their doorsteps, defecate into whole oceans, snatch all the food, survive by detestation, take joy in the bad luck of others, celebrate the death of others, live for the death of others."[38]

This is hyperbole. Yet Thomas, Rees, and Wackernagel make a valid point. Cities and nations *do* "snatch their food from elsewhere," "defecate into whole oceans" as well as "rivers and lakes," and "leave rubbish" or export it. They draw resources from everywhere into their vortex and scatter consequences far from their own streets and alleys. Trade systems allow them to be "entropic black holes." Their own sustainability requires that they import the wherewithal of others' sustainability.

More than trade systems depend upon distant ecosystems. Nature's own *direct*

---

34. Hawken, *Ecology of Commerce*, 25.
35. Rees and Wackernagel, "Ecological Footprints," 377.
36. Ibid., 370.
37. Lewis Thomas, *The Lives of a Cell: Notes of a Biology Watcher* (New York: Bantam, 1974), 129.
38. Ibid., 129. I must add that Thomas is not discussing carrying capacity, so this quotation is itself "appropriated" out of context. But he is discussing the behavior of cities and nations, and his words describe behaviors we are discussing as appropriated carrying capacity.

"goods and services" work in a similar manner. Urban dwellers North and South depend on the tropical forests that form a broken band around the middle of the globe. The forests not only supply extracted products but also supply the "sink" for urban-generated carbons, function to effect global heat transfers and regulate climate, and generate life-giving and life-preserving oxygen. Human populations are directly dependent on distant ecosystems and sustained by them, even when humans are apparently "doing nothing" to them.

Still, for all this, appropriated carrying capacity is simply not an issue for dominant economics. Its docetic theory and practice have no routine place for it, so it is not noticed or asked about. Genuine benefits are delinked from true costs. By contrast, ecological economics tries to ascertain the total "ecological footprint" human settlements leave upon the earth. How to do so is the stickler, since the impact cannot be ascertained by measuring local or even regional impact alone. The right measure of carrying capacity is that which finds the point at which further human economic activity degrades and destroys the host, whether the destruction and degradation are local *or* distant. But it is this measure we do not have, at least not internal to economic activity itself.

Our immediate concern, however, is not to determine how ecological footprints might be traced to their maker but to expose the reasons for the structural blindness of dominant economics to the requirements of creation for its own integrity and well-being (and thus to our own well-being). For further explanation, we return to the workings of trade, technology, and consumption.

TRADING PLACES

The argument here is not against trade per se. Trade has been a vital human activity since the first exchange of tools and good advice around the fire tens of thousands of years ago. The issue is the way trade works in the Big Economy. Much trade today is not, in the first instance, undertaken to play its classic role; namely, to facilitate the exchange of goods and services a particular culture could not produce itself or find a substitute for. Rather, trade is to further economic growth by increasing demand and consumption. Nike, L A Gear, and Adidas are not selling shoes worldwide because baby needs new shoes. They are selling fashion. In fact they do not *want* folks to wear out one pair before buying another.

But let's forgo commentary and take trade on the terms current economic theory itself uses to justify it.

The role of trade in reigning economics, the texts say, is to relieve local constraints. Each locale plays its "comparative advantage" and offers what it can from its relative abundance of resources, including human resources. Greater economic efficiency all around is the aim and sometimes the result. Because products of nature can now be so easily transported, this scheme makes sense. What happens? Essentially this process intensifies production of specialized commodities for export. Benefits thus largely redound to international capital, transnational corporations, local elites tied to these, and that stratum of the population benefiting from an export-oriented economy. What also happens is the consolidation of land and its resources for consumption abroad, usually by way of high-energy, monocultural agriculture or concentrated extraction of goods by way of mining, logging, and so on. This removes land and resources from their vital place in providing domestic staples for local markets in local communities. Not infrequently it also means erosion, desertification, deforestation, depletion of resources, or other severe negative outcomes for both culture and agriculture.[39] In short, the effective pressures do not make for local sustainability over the long haul. They promote export-oriented profit in the shorter term within a system of fierce international competition. Local production from local resources for local use by those with a life investment in these is vitiated.

Other elements of expansionist, flat-earth economics worsen this. William Ashworth explains what he calls a "frontier paradigm" and what we have named the expansionist vision. The basic tenet of the frontier model is that it "always assume[s] ... there is plenty of the [desired] good available over the next hill."[40] For practical purposes, resources are assumed to be infinite, and scarcity is assumed to be a local phenomenon. When scarcity is experienced, the response is not to look for a substitute for the resource. What is "substituted," rather, is *"the same good, taken from more distant supplies."* So when you cut down most of the local timber, you don't quit building wood-frame houses and turn to stone. You substitute the timber over the next hill and the next after that, ad infinitum. Distant goods are substituted for different goods. "One continues stripping hills one after another until the whole world is bald as a nut," Ashworth writes. The stripping simply proceeds outward as local sources are jeopardized. Fierce competition in this paradigm — our present one — assures that the search will go far and wide to stay competitive in the market. In the process, local communities will be offered as little as necessary to exploit their resources. They will, in fact, compete with one another to offer the "best" terms, so as to secure jobs and income. So Louisiana Pacific signs a deal for a million acres of Venezuelan timber while other U.S. companies sign on for logging operations in Chile, New Zealand, Indonesia, and Malaysia. Competition also means that local resources

---

39. The Rees and Wackernagel work includes the marshaling of data I have omitted for reasons of length. But the reader may wish a sample. On the point here made, the authors include these trends pertaining "to the state of global natural capital: encroaching deserts (6 million hectares/year); deforestation (11 million ha/yr of tropical forest alone); acid precipitation and forest dieback (31 million ha damaged in Europe alone); soil oxidation and erosion (26 billion tonnes/yr in excess of formation); soil salination from failed irrigation projects (1.5 million ha/yr); draw-down and pollution of ground water; fisheries exhaustion; declining *per capita* grain production since 1984; ozone depletion (5 percent over North America [and probably globally] in the decade to 1990); atmospheric and potential climate change (25 percent increase in atmospheric CO_2 alone in the past 100 years)." Rees and Wackernagel add: "This partial list shows that far from growing with the expansion of the urban world, the resource base sustaining the human population is in steady decline. It should also remind us of an important corollary to Liebig's law of the minimum — carrying capacity is ultimately determined not by general conditions but by that single vital factor in least supply" (Rees and Wackernagel, "Ecological Footprints," 380).

40. Ashworth, *Economy of Nature*, 133.

in areas already affected will need to keep their own resource prices as low as possible to remain in the game. The combination thus depletes local community twice over, while steadily enlarging the area of annexed communities. Ashworth's closing lines are these: "The frontier paradigm obviously is not going to die easily. As long as forests stand in Siberia and substitution-by-distance remains a viable fiction, there is not a stick of timber in America that is completely safe."[41]

For all these reasons, consumers in such an economy do not readily see the undermining of local sustainability, despite it happening right under their feet and nose. And the dynamics of highly mobile business do *nothing* to make them aware of it. In fact, competition and bottom-line results work at virtually every point to press local limits in the interest of a net transfer of wealth *and sustainability* from one locale to another and then portray this, via mass advertising, as the good life for all! Such a scheme thus reveals nothing about ecosystem health, the necessary requirements for it, or the remaining stock of natural capital available for future generations. The integrity of creation is simply ignored, even when violated. Still, somewhere in its brain cavity dominant economics *must* be aware that manufactured capital can be sustained only by flows of natural capital. It nonetheless seems oblivious to the long-lasting consequences that count most — local ones. Preserving intact communities and their healthy natural and social systems is simply not on the agenda of the "creative destruction" a globalized economy lives by.[42] So Carol Johnston's basic question is never asked: "Why is the goal of economics growth in production of goods and services, instead of growth in the health of persons in communities?"[43]

The striking analogy of this trading system is the economy of colonies. Except, the colony is nature — *everywhere*. In a colonial economy the true costs to the colony are kept off the books of the colony's holder. The consequences of extractive economics for both the ecosystems in the colony and its communities and social structures are out of sight of the consumers "back home." Colonial rule directly appropriates extraterritorial carrying capacity. As Rees and Wackernagel put it, colonial rule may have ended now, but many of its same resource flows and local consequences for the planet continue in the form of commercial trade. "What used to require territorial occupation is now achieved through commerce!"[44]

41. Ibid., 133, 134.

42. The dynamism of capitalism as creative destruction is the argument of Joseph Schumpeter in his famous work, *Capitalism, Socialism, and Democracy,* 2d ed. (New York: Harper and Brothers, 1947). See especially part 2: "Can Capitalism Survive?" (59–164).

43. From the abstract of Carol Johnston's Ph.D. dissertation, "A Theological History of Economics: How Basic Assumptions and Value Choices Combined to Work against Communities and the Land" (Claremont Graduate School, 1994).

44. Rees and Wackernagel, "Ecological Footprints," 380. A parallel instance of the colony analogy is seen in the comments by wealthy communities within nations and by rich nations themselves about their local environment. Commonly they convey that their way of life is environmentally friendlier than that in locales where poverty destroys society and environment together. They imply thereby that they are more responsible about their use of the earth than are the poor. This parallels colonizers' frequent contempt for the peoples and places they colonized, all the while using the colonial economic system to generate the wealth to care for their own environment. Until the well-

In any event, trade in the present system works hard and methodically to postpone the day when peoples, especially affluent peoples, face up to living within nature's bounded capacities. By separating the costs from the benefits of exploitation of nature, trade makes them more difficult to compare and include. It thereby increases the tendencies for economies — and societies — to land in the condition ecologists call "overshoot."[45]

In sum, the logic of present trade and market patterns fails to ask about the requirements of environment, culture, and social relations for ongoing existence, except as "externalities" that need to be addressed for the sake of public relations and sales. It is especially difficult for the comfortable to see the impact of their lives elsewhere, not least because advertisements for the same goods they consume never show the stress and deterioration their acquisition has effected. On the contrary, the ads all portray color, song, sunshine, and clear running water.

CONCLUSION

What shall we say, then, about the mismatch of the present human economy and the economy of nature? First, the expansionist vision is "ecologically empty economic theory."[46] Second, because it is so, it deflects attention from the fact that the present human economy is becoming "increasingly coincident with the ecosphere" as economic expansion continues apace. A closed and round, rather than flat and endless, biophysical world intrudes on daily life more and more. Yet the globalized economy, or rather the way we think of it, continues to buffer the materially well-off from the fact that all of us everywhere share the same ecosphere and therefore ultimately "the same macro-ecological fate."[47] Third, "globalization of the economy means globalization of the ecological crisis."[48] Economic specialization; concentration of capital; the distancing of natural capital; modern industrial-technological processes of production, distribution, and consumption; minimally restricted access to resources; freer markets; and expanded trade — all these (the hailed markers of the tested route to greater prosperity!) seem paradoxically to yield joint ecological and socioeconomic impoverishment, at least at the periphery, where an increasing majority live. The misery of masses of people and of so much of the rest of nature can continue awhile. But on a finite and contracting planet it cannot continue indefinitely any more than apartheid can. And the signs that true carrying capacity is being exceeded do seem more and more with us. In many places the increased activity of economic growth is degrading and destroying the host.

---

off fully account for the positive effects to them of appropriated carrying capacity from elsewhere, such perspectives are disingenuous (and usually self-deceiving as well).

45. Herman Daly, "Free Trade and the Environment," *Scientific American* 269 (November 1993): 53. The conditions that make for "overshoot" are listed above, p. 81.

46. Rees and Wackernagel, "Ecological Footprints," 387.

47. Ibid.

48. Ibid., 386.

What are the consequences of this principled neglect of nature's finite endowment and of its absorptive and regenerative capacities, its organic dynamics and requirements, its integrity? They are essentially practical and pose essential questions. How do we readjust or otherwise change our ways of production, reproduction, distribution, and consumption radically enough that unforgiving natural boundaries are not violated and the integrity of creation is respected? How do we attune our changing cycles to fit within the rest of nature's changing cycles? How do we assure safe margins for local, regional, and global carrying capacity? How do we live as though we and all our kind for generations to come, and all other kind for a similar stretch, belong here? When we say "economic growth" or even simply "economics," how do we get ourselves to *think* "the economics of borrowing"? How do we fashion a culture attuned to the principles of the Great Economy and meet material human needs in ways that live within creation's uncompromising integrity? How do we get day-to-day transactions to account for "the profound ethic of biology"[49] embedded in the Great Economy and expressed as the integrity of creation?

Differently said, how do we attain sustainability?

The proposed answer is "sustainable development." To discussion of it, including the controversies, we can now turn.

---

49. Hawken, *Ecology of Commerce*, 8. Freeway-and-mall monoculture resists this ethic of biology. As William Ashworth (*Economy of Nature*, 299–307) explains, the sameness of infrastructure, housing developments, malls, and commerce nurtures the illusion that a geographic identity must exist here as well. But of course it doesn't. Take land, for example. It "differs in stability, in rock chemistry, in angle of repose. It differs in ambient temperature, in annual rainfall, in wind exposure. Soil depth and fertility, vegetal cover, and proximity to water vary greatly." Yet we build freeways much the same for both Phoenix and Boston, and, for that matter, we "industrialize" our agriculture so that we grow cotton in Arizona and rice in California. If need be, then, we'll change the land to fit the mold of geographic identity, whether for agribusiness, buildings, or transportation. It should not come as a surprise, then, that some Floridians *were* surprised when their "Cape Cod–style" homes went down like paper cartons before Hurricane Andrew, to a greater extent than some other styles. The owners blamed shoddy construction practices, and it never occurred to them to ask whether houses designed for the cool woods and stern morainal soils of New England were symbiotic with the palmetto scrub and shifting sands of the Florida sandbar. Such is only one small instance of Hawken's far-reaching comment that a "profound ethic of biology" underlies things, and our economics barely acknowledges it at all. Instead, single-minded environmental manipulation carries the day — until the Great Economy takes its revenge.

Message from Chiapas

Sustainability is the capacity of natural and social systems to survive and thrive together indefinitely. It is also a vision with an implicit earth ethic. Both sustainability and its earth ethic follow from creation's integrity and a picture of earth as *oikos*. This chapter and the next two unravel what "sustainability" can now mean. If previous chapters helped explain how we "got here from there" (the history and notions that leave us with unsustainability), these work at "getting there from here" (wresting a sustainable future from a precarious present).

Sustainability can be approached via discussions of sustainable development. Many of these are heated. Parties to the debate only lightly mask deep differences and no little confusion. International organizations such as the United Nations, the World Bank, the governmental delegations that negotiated the General Agreement on Tariffs and Trade (GATT), and transnational corporations, arrayed on the one bank of a wide river, and nongovernmental organizations (NGOs) and peoples' movements, including resistance movements, on the other, compete in the crowded scene this side of Sweet Betsy's avalanche. Among the latter, the World Council of Churches (WCC) occupies a special place, for reasons that will become apparent.

These chapters try, then, to tease out differing approaches to sustainability. One large and varied camp understands sustainability as a crucial qualifier of development economics on a global scale. Development "qualifiers" or the "relaunchers" of development take this path. Another large and disparate camp begins not with economics but with ecology, and not with a global reach but with a local and regional one. This is the work of development "dissenters." Unraveling the knotty discussions about sustainable development and sustainable society or sustainable community is the agenda.

Sustainability can also be approached as a practical reality to be achieved with crucial guidance from science about nature's behavior. Scientists' testimony, rather than development debates, is the source for this. That discussion, plus conclusions about sustainability, follows these chapters.

NORTH, SOUTH, AND CHIAPAS

Consensus about sustainability and how to achieve it did not happen at the UN summits over the period 1992–96. Instead, a ragged North/South abyss cut across others, with different views from different angles. Divisions have deepened and sharpened in complicated ways. Perhaps the best entry to the matter is a vignette from the largest gathering of development agencies under one

127

umbrella, the Society for International Development (SID), at its meeting in Mexico City in April 1994. In the closing plenary entitled "Building Partnerships and Collaboration towards Global Transformation," a representative of the indigenous peoples of the Mexican province of Chiapas was recognized. David Korten's account goes like this:

> For the first and only time during the conference, we were hearing an authentic voice of the world's poor and marginalized, specifically a voice from a group that only a few months earlier had declared war on the Mexican government as an expression of its discontent. Without accusation or rancor, he spoke as a plain and simple man of the desire of his people to have the opportunity to free themselves from poverty. He spoke of foreign aid that had never reached the poor. He spoke of the love of his people for the land, the trees, and the ocean. He spoke of their desire to share their ideas as fellow human beings, to have their existence recognized, to be accepted as partners in Mexico's development. He spoke of the people's call for a new order in which they might find democracy for all.[1]

Korten speaks of this as a defining moment in debates about sustainable development because it bespeaks a sharp alternative to the conventional view that runs, as we shall see, from Harry Truman and W. W. Rostow through the Brundtland Commission Report and beyond, albeit with significant modifications. The Chiapas rebellion was distinctive among guerrilla struggles in that it did not seek to seize state power. Instead, it aimed to win the right of people to govern themselves within their own communities. It did not call upon other Mexicans to take up arms for a new national social agenda, but for the space and means to elect popular, democratic movements tied to particular locales.[2] One commentator, Gustavo Esteva, called it "a new kind of movement" and the "first revolution of the twenty-first century." By that he meant the feisty manifestation of a growing struggle of people around the world for economic and political survival and sovereignty within their own communities. Despite the history of peasant rebellions and revolutions in Mexico and the Zapatistas own name, this wasn't simply an updated version of previous identities. It wasn't a Marxist guerrilla group, for example. It had no clear-cut socialist ideology or political platform and no one leader. Nor was it a fundamentalist or messianic group. Its members came from different Indian groups, professed different religions, spoke different languages, and were explicitly ecumenical. They were Tzeltales, Tzotziles, Choles, Tojolobales, Zoques, and Mames, with their own tongues, cultures, traditional lands, and faiths.[3] As mentioned, its goal was not

---

1. David C. Korten, "Sustainable Development Strategies: The People-Centered Consensus" (paper of the People-Centered Development Forum, May 17, 1994), 1.

2. Ibid.

3. The linguistic, cultural, religious, and other differences among indigenous peoples have in fact been major reasons it has been difficult to organize diverse peoples, even when they share common causes in response to common violations.

to seize power to govern the country but rather *to reclaim the community*. It did not eschew, but used, modern means of communication and a strategy of networking varied coalitions of dissent. Perhaps most strikingly, it did not call upon the government for cheaper food, more jobs, more health care, and more education. Rather than trying to find its niche in Mexico's efforts to solve its problems by strengthening its role in a global economy organized around the needs and wants of consumer society, it sought to order its own world around the organic needs of community. In Esteva's words, it was not a revolt in response to a *lack* of development but a response that Chiapas was being "developed to death." People "opted for a more dignified way of dying."[4] This more dignified way consists of a "commons" the community carves out for itself in response "to the crisis of development"; "ways of living together that limit the economic damage and give room for new forms of social life"; "life-support systems based on self-reliance and mutual help, informal networks for the direct exchange of goods, services and information"; and "an administration of justice which calls for compensation more than punishment." Esteva's conclusion is that the revolt was against conventional development as played out and planned in Mexico: "To challenge the rhetoric of development, however, is not easy. Mexico's economic growth, the promise of prosperity tendered by the IMF and the World Bank, the massive investment in modernity as an integral element of the war against poverty — these have been cast as truths beyond question." Nonetheless, the Indian rebels "announced to the world that development as a social experiment has failed miserably in Chiapas."[5]

The background to this "announcement" is as follows. Chiapas is among the richest of Mexico's provinces — a hundred thousand barrels of oil and five hundred billion cubic meters of gas per day; dams that supply better than half of Mexico's hydroelectric power, even though 33 percent of Chiapan households are without electricity and 40 percent without running water; one-third of the national production of coffee; and a considerable percentage of the country's cattle, timber, honey, corn, and other products of the land. To connect these riches with the capital, a freeway is being built through the El Ocote forest to Mexico City.

Yet Chiapas is among the poorest of Mexico's provinces. Consequently, it is a chief target of PROSONAL, the national war on poverty. One-third of the 3.5 million population is Indian. Large numbers of these have been displaced by dams, oil rigs, and cattle ranches and have been pushed into the Selva Lacandona forest (the largest tropical forest in North America), where they have contended with loggers and ranchers. Thirty thousand Indians died in 1993 from hunger and disease; malnutrition remains the number one killer; and on average there is one doctor for every fifteen hundred citizens. Yet the entire social budget of Chiapas is a fraction of the cost of the freeway to Mexico City, despite PROSONAL.

---

4. Gustavo Esteva, "Basta!" *The Ecologist* 24, no. 3 (May/June 1994): 84–85.
5. Ibid., 84.

The (now former) governor of Chiapas, González Garrido, carried out a three-pronged program. He protected armed landowners and cattle raisers against Indians' resistance to the taking of their lands, supported the creation of an Indian stratum to control the villages, and gave free rein to the police to combat drug trafficking. All this was called "modernization" and "development" and was highly lucrative for those who governed Chiapas and plied its economy. A success, Garrido was appointed Mexican minister of the interior in 1993, part of the successful economic growth in Mexico in the early 1990s.[6]

Garrido's program continued a certain history of governance, just as the rebellion continued a history of resistance. From the time of Spanish colonization onward, governments and landlords pushed Indians and other peasants from the land and consistently ignored human rights and legal claims for redress of losses, despite periodic revolt. In 1991 and 1992 indigenous peoples' groups organized a series of nonviolent demonstrations, including a seven-hundred-mile march from Palenque to Mexico City to demand agrarian reform and human rights anew. This march and other actions pressured officials into negotiation, some of them sympathetic to indigenous causes. An accord was reached. It was when the government failed to meet the negotiated pledges that the rebellion broke out.

The timing, however, points to other dimensions of the story. Chiapan Zapatistas began their rebellion in January 1994, two hours after the North American Free Trade Agreement (NAFTA) went into effect. Their appeals were "for an end to 500 years of oppression and 40 years of 'development.'"[7]

Behind this call to end "development" as an extension of oppression was a worsened situation traced in considerable part to President Salinas de Gortari's promotion, in 1991, of a constitutional amendment. The amendment was to Article 27. Article 27 had allowed the establishment of *ejidos,* the communal landholdings on which the indigenous peoples live and an outcome of land reform pursued in the 1930s. Article 27 had also permitted, through a system of legal petition, the reclamation by Indian villages of lands encroached upon by private owners in the nineteenth and twentieth centuries.

The amendment to Article 27 effectively repealed these arrangements by permitting outside corporations and individuals to buy up the *ejidos* and exploit their resources. Indigenous peoples were not driven from the land per se. But with little chance themselves to buy it and gain title, they were pitched against the wealthy in government-backed competitive privatizing of lands and resources.[8]

The amendment to Article 27 was one element of the Mexican economic reforms of President Salinas de Gortari and the ruling party. The larger plan pursued a national economic strategy congruent with the logic of what be-

6. Ibid., 83–84.
7. Ibid., 83.
8. The account of Chiapas here draws on two sources in addition to Esteva: José Luis Morin, "Indigenous Struggle for Justice," *Covert Action Quarterly* 48 (spring 1994): 38–43; and an unpublished research paper by Gary Matthews, Union Theological Seminary, 1995, "What Happened in Chiapas?" 20–35.

came NAFTA. In turn, NAFTA followed the general logic of liberalized global economic development.

Pursuing this plan, Mexico borrowed from the world financial system and attracted investment to develop its domestic resources, build industry and infrastructure, and provide jobs. Mexican products were then sold on the global market, with revenues plowed back into the Mexican economy for continued economic growth and — cross your fingers — prosperity. NAFTA pursues this free flow of capital and goods by lowering or eliminating barriers to it across the borders of Canada, the United States, and Mexico.[9] In Mexico itself the process consistently favored the economic power and access of corporations, domestic and foreign bankers and investors, and large landowners.

The choice of NAFTA's inaugural date as the rebellion's own could hardly have drawn a sharper contrast between the Mexican Indian peoples' understanding of sustainable development and their government's program, the latter in partnership with transnational institutions and framed by NAFTA. The contrasts can be summarized by saying that while conventional development revolves around economies and their growth in the form of free markets and economic globalization, the Indian notion of sustainability focuses on local land and the health of communities and societies tied more closely to self-reliance, indigenous social movements, culture, low-impact agriculture, sustainable energy use, environmental balance within locality and region, and community economic, social, and political accountability. Differently said and more broadly applied, conventional sustainable development qualifies economic growth with a view to *ecological sustainability* while the alternative vision tries to increase local economic self-reliance within a framework of *community responsibility and ecological balance.* The latter's attention is to webs of social relationships that define human community, together with ecosystem webs and the regenerative capacities of both human and ecosystem communities. More starkly put, conventional development is globalization from above, as led by "developed" sectors around the world. The Chiapas alternative is globalization from below, led by "underdeveloped" sectors.[10] Or, in Nicholas Hildyard's more helpful distinction: local peoples' notion of sustainable development views the environment as "what is around their homes." Government, business, and international organizations view the environment as "what is around their economies."[11]

This vignette is necessary to understand the conflicts of North and South. As the global integration of economies progresses, power is shifting from both communities and nations to transnational capital and the institutions that wield and regulate it. Accompanying this is an increase in the gap between rich and poor within and between nations, and a parallel diminishing of local power and of-

9. This succinct summary of the logic of NAFTA and global economics is taken from Matthews, "What Happened?" 25–26.

10. See the discussion in Korten, "Sustainable Development Strategies," 6. I draw heavily from the accounts of Korten and Esteva ("Basta!").

11. Nicholas Hildyard, "Foxes in Charge of Chickens," in *Global Ecology,* ed. Wolfgang Sachs (London: Zed Books, 1993), 23.

tentimes even national power to do much about it. Which is to say that "North" and "South" are increasingly defined by *class and culture rather than geography*.

To be sure, "North" still refers to nations and regions in meaningful ways. After all, the preponderance of advanced industrial nations are in the Northern Hemisphere and wield great power as players in the globalized economy, with its goal of sustainable economic growth on the capitalist model of liberalized international exchange.

"South" also refers to nations and regions in part, but on the basis of whether they possess an "expanding frontier" or a "shrinking" one (the terms are Tariq Banuri's).[12] An expanding frontier means a nation judges that it has the where-withal to successfully engage the global economy on competitive terms. It has expanding possibilities of growing income and wealth because it can command not just national but regional resources. Its economic frontier does not stop at its borders. It can appropriate carrying capacity from beyond its borders. Thailand, Malaysia, Singapore, and Indonesia all see themselves in this light, to use examples from Asia. A shrinking frontier belongs to those nations for whom a set of limitations on growth exists because of resource constraints and little leverage in a globalized economy. Pakistan has no more water and no more land and limited access to world markets and prosperity. Bangladesh and, in a more complex way, India face shrinking rather than expanding frontiers, to continue Asian examples. Southern nations themselves split along these lines now. Thus they often assume different postures in treaty and trade debates aimed at hammering out the terms of sustainable development.

All this noted, the more salient fact is that "North" now stands for classes and cultures everywhere in the world who are aligned with transnational capital and who see their own welfare or demise tied to the success or failure of the world economy. This is a diverse, multinational stratum. "South" designates those, whether they be in New York, Mexico City, Lagos, or Ankara, who are pressed more and more toward the margins in the globalization process. They are also diverse and multinational. Oftentimes this means, as in the case of indigenous peoples and subsistence farmers, seeing their land appropriated for export-driven agriculture with little attention given to either their own regenerative needs or the land's. And it means, for blue- and white-collar workers and the unemployed of urban megalopolises, highly mobile capital and not-as-mobile labor, accompanied by depressed wages, corporations shedding workers, and people displaced from their homes and means of livelihoods. In all cases, it means the stark absence of attention to the essential importance of a healthy spiritual connection to nature, place, community, and culture, as well as an "enchanted" or sacramental view of earth and the Community of Life. (This absence is somewhere near the heart of the Chiapan quest for an alternative.)

In sum, North and South correlate less with geography than with those who, on the one hand, pursue sustainable development in tandem with growth-driven economies as part and parcel of a globally integrated economic culture and those

12. Tariq Banuri, "The Landscape of Diplomatic Conflicts," in *Global Ecology,* 56–57.

who, on the other, grope for a local and regional alternative.[13] Or, in terms used at the outset, the divide is between the development qualifiers (sustained global economic growth within sustained environments) and development dissenters (global schemes rejected in favor of local and regional ones that do not transfer and deplete the carrying capacity of local communities and their ecosystems).[14]

In the end, the prestigious Brundtland Commission Report (*Our Common Future*), prepared for the 1992 Earth Summit, probably best reflects where dominant — and morally sensitive — thinking and practice are on sustainability and sustainable development. It documents clearly and forcefully the devastation of the environment at human hands. It singles out the limited capacity of nature to continue absorbing the waste products of world energy consumption and shows how economic growth has been deleterious toward the same resources upon which it is dependent. It records the brutal statistics of wasteful overconsumption in some quarters and crushing poverty in others. It casts its eye to the welfare of future generations and warns the present about the deprivations it is creating for them. It faces the realities of exploding populations head on. On the prescriptive side, it calls for major reductions in arms expenditures as well as major reductions in its chief target, poverty. It commends multilateral cooperation on all these border-crossing problems and, together with the Business Council for Sustainable Development, says that sustainability demands attention to the entire life-cycle of goods and requires the internalizing of all costs in production, distribution, and consumption, including environmental ones all along the way. In short, it shows a possible path for continuing development and rendering it sustainable.[15] Without doubt it is a pathfinding international analysis and proposal.

But after all this, which is purported to challenge radically standard development thought and practice, *Our Common Future* goes on to say, first, that "if large parts of the developing world are to avert economic, social, and environmental catastrophes, it is essential that global economic growth be revitalized."[16]

---

13. This discussion of North/South is indebted to David C. Korten's in "Sustainable Development," *World Policy Journal* (winter 1991–92): 173–74.

14. We might add parenthetically that if the UN Earth Summit had winners, the North won. All the treaties protected access to resources for conventional sustainable development, and none of them required less consumption and diminished standards of living in the North in order to address environmental constraints and poverty reduction. Ambassador Robert Ryan, head of the U.S. delegation, said at presummit meetings, as did President Bush, that the standard of living of U.S. citizens simply was "not up for negotiation" at the Earth Summit. And it wasn't. The South, and many Northerners with them, did succeed, however, in putting poverty reduction and environmental protection squarely on the global agenda *for everybody*. (These are written into the World Bank's mission statement, for example.) Yet prevailing economic relationships and dynamics were left in place, with the basic policy decisions of business and finance falling outside Rio's reach altogether. (This is the case with corporations and the World Bank as well, to continue the example.) Which is to say, Southern "victories" were written into the Northern pattern of ecoqualified development. See Maximo T. Kalaw Jr., "The Response of the South to the Justice and Ecology Debate," in *Sustainable Growth: A Contradiction in Terms?* report of the Visser't Hooft Memorial Consultation (Geneva: WCC Publications, 1993), 54.

15. Korten, "Sustainable Development," 160–61.

16. Ibid., 161, citing one of the conclusions of *Our Common Future* (Oxford: Oxford University Press, 1987).

Then, in its key recommendations, it calls for global economic growth at a level five to ten times the current output. Furthermore, the stimulus for this is increased consumption in the North since greater Northern consumption will create greater demand for Southern products.[17] This is oddly askew of the earlier analysis, where growth and consumption in richer regions are the problem. Now they are suddenly key to the solution. Or, to nuance this in the way Brundtland does: more eco-efficient production, distribution, and consumption, along with social changes, will render the necessary five- to tenfold economic increase ecologically benign.

One can only sympathize with Brundtland. A bulging world cannot develop in a way that meets basic, escalating needs without relying substantially on economic growth. Chances for meeting exploding needs while suffering declining national incomes are slim, by any standards. Yet how growth-driven economies are made compatible with biospheric sustainability in a world where the scale of human activity relative to the biosphere is already much too large is anything but clear, and may be impossible.

In a word, sustainable development in this scheme simply bundles contradictions without resolving them or showing the way forward, a judgment now plaguing those working with *Agenda 21*, the UN's master plan for sustainable development signed by all heads of state at the Earth Summit. There is even a sense of desperation here, though it is usually cast, cheerleader fashion, as the next great challenge. Stephen Schmidheiny, chairman of the Business Council for Sustainable Development, was interviewed by *Neue Züricher Zeitung* in December 1990 and was confronted with the statistical likelihood (read: unlikelihood) of successfully combining economic growth and environmental protection on the necessary scale. "Isn't this combination just a dream?" the reporter asked. "It brings together things that don't match." Schmidheiny replied, "For the time being, that's true,"[18] and went on to call for the development of ways to refute this truth as a lasting one.

We will return to the critique of sustainable development in subsequent treatments. What is presently imperative is understanding whence it came. And key here is the development of "development" itself, since it remains the noun. "Sustainable" is the modifier, not the thing itself.

DEVELOPMENT'S HISTORY

In his inaugural address as president of the United States in 1949, Harry Truman spoke of "a major turning point in the long history of the human race."[19] With fascism defeated, the world's chance for democratic freedom and prosperity had arrived. In the interests of this freedom, Truman offered a "bold

17. Korten, "Sustainable Development," 161.
18. Reported by Paul Ekins in "Making Development Sustainable," in *Global Ecology*, 92.
19. Cited by J. Ronald Engel, "Sustainable Development: A New Global Ethic?" in *The Egg: An Eco-Justice Quarterly* 12, no. 1 (winter 1991–92): 4.

new program for making the benefits of our scientific advances and industrial progress available for the improvement and growth of underdeveloped areas."[20]

This designation — underdeveloped areas — soon became standard vocabulary. Distinctions were drawn between developed and underdeveloped nations, with "developed" a clear-cut notion and the telos for the underdeveloped world. Truman himself supplied the criterion in his inaugural for moving from an underdeveloped to a developed state: "Greater production is the key to prosperity and peace." He went on to explain that "the key to greater production is a wider and more vigorous application of modern scientific and technical knowledge." "The United States," he noted, "is preeminent among nations in the development of industrial and scientific techniques."[21]

Later the terms would modulate somewhat to "developed" and "developing," but the model remained the same: development meant the way of life of capitalist democracies as defined by modern economic progress and advanced science and technology. Wolfgang Sachs, writing in 1992, sums up the legacy from 1949 onward: "The degree of civilization in a country could from now on be measured by its level of production. This new concept allowed the thousands of cultures to be separated into the two simple categories of 'developed' and 'underdeveloped.' Diverse societies were placed on a single progressive track, more or less advancing according to the criteria of production."[22] In short, developing countries were aspiring junior versions of developed ones, and developed ones were affluent industrialized democracies.

Of course, from another slant there is little new in this "new concept." It remains essentially the nineteenth-century European and American one of successful industrialism: higher levels of material consumption and a heightened ability to alter the natural world for human benefit hold the key to progress. Progress is by definition beneficial, something all societies should shoot for, and associated most of all with economic growth and democracy.[23] If there is something new here, it is the application of these postwar distinctions of "developed" and "developing," or "more developed" and "less developed," to all societies everywhere. Things have gone global.

Our attention, however, is to the notion of sustainability accompanying this understanding of development. Initially a strictly economic notion, presented among other places in W. W. Rostow's widely used *Stages of Economic Growth*, sustainability meant the critical threshold over which an economy passes so as to enter *long-term, continuous growth*.[24] To this day, and for all but a few analysts (though their number is increasing), development is quite unimaginable apart from sustained growth, even when sustainability is no longer a narrowly

---

20. Ibid., 5.
21. Ibid.
22. Wolfgang Sachs as cited by Wesley Granberg-Michaelson, *Redeeming the Creation, the Rio Earth Summit: Challenge to the Churches* (Geneva: WCC Publications, 1992), 1.
23. See the discussion of Clive Ponting, *A Green History of the World: The Environment and the Collapse of Great Civilizations* (New York: Penguin, 1991), esp. chap. 8, "Ways of Thought," 139–60.
24. The Rostow material is passed along by Engel, "Sustainable Development," 5.

economic notion. "Sustainable" has now come to mean rising national incomes and consumption together with environmental preservation. (Both the Brundt-land Commission Report, *Our Common Future*, and the UN master plan, *Agenda 21,* define it this way.) Sustainability is essentially, then, sustainable economic growth in a double sense: sustained production and consumption to meet expanded human wants and needs (the Truman-Rostow notion of sustainability) *and* production and consumption that nature can sustain (the Earth Summit's crucial qualification).

In some ways this means great change. Talk of "paradigm shifts" and a "second industrial revolution of ecoefficiency" is the language used by Stephen Schmidheiny, the Swiss millionaire who organized the Business Council for Sustainable Development. "The single biggest problem" within "the larger challenge of sustainable development," Schmidheiny says, is the requirement "for clean, equitable economic growth everywhere."[25]

"Clean, equitable economic growth everywhere" would indeed be a paradigm shift, especially "clean" and "equitable" "everywhere"! In any case, the focus of sustainability in this view, with paradigms shifting or not, is upon economies and their growth. This rests in three tenets David Korten dubs the "modern theology" of development: (1) sustained economic growth is both possible and the key to human progress; (2) integration of the global economy is the key to growth and beneficial to all but a few narrow special interests; and (3) international assistance and foreign investment are important contributors to alleviating poverty and protecting the environment.[26]

This may sound attractive enough, conventional enough, or perhaps simply banal enough to garner broad consensus. But it did not. In fact the Chiapas Zapatistas came to be true disbelievers. They rejected this faith and its promises. Our purpose at this point, however, is not to record the reception of sustainable development so much as trace approaches to sustainability and how it might be realized. And what has been sketched thus far is the history of sustainability on the part of those who begin with the massively institutionalized means of development regnant since the end of World War II and who move from there to qualify this now-global apparatus in a way that renders both growth and the biosphere "sustainable."

With this as the dominant approach to sustainability and the reigning notion of sustainable development, we turn in the next chapters to another approach. The Chiapas rebellion was a glimpse. But it says too little. It says too little because the sources of this alternative approach include even more than the

---

25. Stephen Schmidheiny, "The Business Logic of Sustainable Development," *Columbia Journal of World Business* 28, nos. 3 and 4 (fall/winter 1992): 21. Schmidheiny's appeal to business is straight-forward: "The bottom-line demand of the average business person is, 'Ethics and all that aside, give me one good business reason why I should care about the environment.' We [the Business Council] offer two. First, if you don't, your business will lose out in the coming environmental shake-out. Second, as the division between environmental excellence and economic excellence blurs, there will be increasing profitability and competitiveness in ecoefficiency" (22).

26. Korten, "Sustainable Development," 158–59.

long history of resistance to Western-instituted globalization and the invasionary power of the neo-Europes. The sources include established and growing middle-income ranks distant from marginalized peoples. Many of the established affluent, too, are convinced the current economic course is unsustainable, even when environmentally sensitized. What these varied sources hold in common is *a reverse ordering of economy and ecology.* The issue for many in both North and South (used now in a geographical sense) is not how to alter environments so as to serve the economy and yet be sustained but how to alter economies so as to serve comprehensive environments ordered around healthy communities. The difference is not subtle. Against the background of two centuries of industrial economies, it is the difference between an economic approach that begins with a notion of an "open," even "empty" and basically unlimited world, and an ecological approach that begins with a notion of a "full" and limited world that can only operate on a principle of borrowing. It is the difference between a mobile world (including mobile homes!) and a world of place and roots. It is the difference between viewing the whole world as sets of industrial and information systems that need to be managed globally as human and natural capital, and local and regional communities attending to home environments in a comprehensive way around basic needs and quality of life. It is the difference between saving the planet and saving the neighborhood. The former asks the kind of pretentious questions *Scientific American* did in its issue entitled "Managing the Planet"; the latter neither pretends such reach nor assumes such a posture. William C. Clark exemplifies the former: "Two central questions must be addressed: What kind of planet do we want? What kind of planet can we get?... How much species diversity should be maintained in the world? Should the size or the growth rate of the human population be curtailed...? How much climate change is acceptable?"[27]

In short, the questions raised by the notion of sustainable community are different from — and come from a different place than — those raised by the notion of sustainable development. Humbler questions from a humbler place, yet questions advocates hope will lead toward the fourth human revolution of mutually enhancing earth relations.

---

27. William C. Clark, "Managing Planet Earth," *Scientific American* 261 (September 1989): 48.

Message from Geneva

The only background knowledge the reader needs about the World Council of Churches (WCC), headquartered in Geneva, is that "the impulse toward the marginalized [has been] a consistent theme"[1] of this stream of the ecumenical movement.

Just as the WCC gave the world the phrase "the integrity of creation," so too the currency of "sustainability" and "the sustainable society" is from the council, though not "sustainable development" (a telling omission, as we shall see).[2] We must discuss the term "the sustainable society" in order to understand the WCC's reticence to embrace the term "sustainable development."

Shortly after the 1972 UN Stockholm meeting on the environment, and for a 1974 conference in Bucharest entitled "Science and Technology for Human Development: The Ambiguous Future — The Christian Hope," a WCC statement introduced "sustainability" as a term applying to human behavior and society, in contrast with its earlier reference to renewable natural resources only. (Prior use of "sustainability" referred to the sustained yield of forests and fisheries.) The 1974 statement gave sustainability human flesh:

> The goal must be a robust, sustainable society, where each individual can feel secure that his quality of life will be maintained or improved. We can already delineate some necessary characteristics of this enduring society. *First,* social stability cannot be obtained without an equitable distribution of what is in scarce supply and common opportunity to participate in social decisions. *Second,* a robust global society will not be sustainable unless the need for food is at any time well below the global capacity to supply it, and unless the emissions of pollutants are well below the capacity of the ecosystem to absorb them. *Third,* the new social organization will be sustainable only as long as the rate of use of non-renewable resources does not outrun the increase in resources made available through technological innovation. *Finally,* a sustainable society requires a level of human activity

---

1. This is the judgment of J. Philip Wogaman in his *Christian Ethics: An Historical Introduction* (Louisville: Westminster/John Knox Press, 1993), 267. In connection with this, Wogaman notes the related controversy and fractiousness regarding John Courtney Murray's view that civilization is people "locked together in argument." By that measure, Wogaman says, "the ecumenical movement has always been quite civilized!" (268).

2. A letter from a former staff person, Wesley Granberg-Michaelson, to David Wellman, author of the paper "Defining Sustainable Development," includes this: "You may be interested to know that in the restructure of the WCC, we named the staff team in Unit III dealing with these related concerns [of sustainable development] 'Economy, Ecology, and Sustainable Society,' partly to avoid the ambiguity over the definition of the term 'sustainable development'" (letter of June 23, 1995; used with permission).

which is not adversely influenced by the never ending, large and frequent natural variations in global climate.[3]

The next year, at the WCC's Nairobi assembly, delegates adopted a new program: toward a "just, participatory, and sustainable society." Sustainability here is a norm for society. Yet there is firm insistence that sheer sustainability is of itself not enough. Or, rather, that "sustainable" should always be defined in ways that give equal normative emphasis to "just" and "participatory." Justice and participation were in fact the first characteristics listed in the 1974 statement (see above). While sustainability is ecological, it includes transformation of present societal structures, systems, and practices so as to render them sustainable in their own right. "If the life of the world is to be sustained and renewed,... it will have to be with a new sort of science and technology governed by a new sort of economics and politics," Charles Birch told the assembly.[4] "The rich must live more simply," he added, "that the poor may simply live."[5]

Sustainability thus encompasses not only the environment but, in our terms, "earth and its distress," its human distress included. Moreover, this scope has an explicit ethical focus. Moral dimensions, including human rights, are *basic* qualifiers and determinants of socioeconomic choices. They do not first appear after development engines are in place and running full tilt. Nairobi thus meant to emphasize, and did, that "a sustainable society which is unjust can hardly be worth sustaining. A just society that is unsustainable is self-defeating."[6] Insistence on this scope and moral focus carries through like a strong pedal line in subsequent WCC programs, including those called "Justice, Peace, and the Integrity of Creation"; "The Ecumenical Decade of the Churches in Solidarity with Women"; and "Theology of Life."

The ethical focus necessitated a certain kind of analysis — a critique of reigning socioeconomic and political practices. Had the WCC considered sustainability an "ecological" term only, members might have concluded, for example, that soil erosion and exhaustion undermining the agricultural base of a society render it "unsustainable." Farming marginalized lands intensively and neglecting adequate soil conservation would be critical causes of unsustainability.

While obviously important, these reasons do not push the analysis far enough. What causes the farming of marginalized lands? is the prior question. Marginalizing peasant farmers — driving them from better land so the land can be used to grow export crops for global agribusiness — may turn up the plausible answer. Marginalized peoples make for marginalized lands, and vice versa. These are socioeconomic and political causes of ecological unsustainability. And they raise or expose moral issues about the ordering of society. They follow from WCC questions about society as "just, participatory, and sustainable."

---

3. Cited from *Study Encounter* 69, vol. 10, no. 4 (1974): 2.

4. Cited by Dieter T. Hessel, "Sustainability in Social Ethical Perspective" (paper for MIT working session, February 13–14, 1994), 1.

5. Also in ibid.

6. Cited in ibid.

If sustainability has been a WCC theme since the 1970s, development has even longer tenure. But "development" initially referred to the appropriate technology and community-development movements of the 1960s and 1970s, themselves born of dissatisfaction with the dominant form (that of Truman and Rostow). E. F. Schumacher, himself an active friend of the WCC, was an influential source, not least because of his classic of 1973, *Small Is Beautiful.* Yet Schumacher and the Church and Society Unit of the WCC were only reporting the experience that development as mass economic growth invariably faced environmental bottlenecks and did not yield sustainable local communities with power to control or use their own native resources well.

From the very beginning, there was an anticapitalist, anti-Western tone in much of the WCC's approach to development. In the Cold War context, development of such ilk was invariably construed by the WCC's many critics as leftist, if not Marxist. This cost the council and affiliated development organizations dearly.[7] Yet the fundamental point was not socialist at all. It was that capitalist economics prosper most when labor, technology, and capital are fluid; they are thus driven to international integration at high levels; and this takes place at the expense of local communities, their resources, culture, and ways of life. The criticism was of capitalism as a "creative destruction"[8] process in which destruction and benefits fall out very unevenly. And in a resource-scarce world, where, to remember Boutros Boutros-Ghali at the Earth Summit, "progress is not necessarily compatible with life" and "the time of the finite world has come in which we are under house arrest,"[9] the final result is that rich and poor end up in mortal competition for a dwindling base. In this competition, the increasing poor lose.

The WCC critique was not an economic one only, however. Global development as the "self-evident organizing principle of modern society" bore a materialism and secularism that undermined and destroyed traditional ways and worldviews.[10] Mass-oriented consumption and communication undercut diversity and its value as a creational good. Technical rationality and consumption as a way of life were not advances and progress at all, but forces that disdained local culture, work, technology, lifestyle, religion, philosophy, mores, and social institutions. Ancient religious and cultural practices were rendered quaint, and age-old questions of ultimate meaning and purpose were not even considered significant matters of attention for "development." Spiritual vacuousness and

---

7. In the case of the WCC, the costs came via both internal controversy among member churches and through broad media exposure and condemnation in such as *The Reader's Digest,* a segment on *60 Minutes,* and Ernest Lefever's *Amsterdam to Nairobi: The World Council of Churches and the Third World* (Washington, D.C.: Ethics and Public Policy Center, Georgetown University, 1979).

8. The reference is to Joseph Schumpeter's argument in *Capitalism, Socialism, and Democracy,* pt. 2.

9. As cited by Wesley Granberg-Michaelson, *Redeeming the Creation, the Rio Earth Summit: Challenge to the Churches* (Geneva: WCC Publications, 1992), 6–7.

10. Bruce Rich, *Mortgaging the Earth: The World Bank, Environmental Impoverishment, and the Crisis of Development* (Boston: Beacon Press), as cited by William Stevens, "Developing Ourselves to Death," *New York Times Book Review,* March 6, 1994, 29.

homelessness set in, as did estrangement from the sacred and alienation from the earth. Loss of identity and cultural schizophrenia were common experiences, as were generational chasms in societies undergoing compressed change. No place felt like *oikos*. Dislocation and disenchantment replaced "ecolocation."[11]

Such criticisms were voiced at meeting after meeting. Perhaps the sharpest of all came from Third World women and indigenous peoples who had been on the receiving end of a deadly correlation far too long: the farther "development" penetrated, the more their culture *and* their environment were despoiled, together. The very last thing they desired was to render all this "sustainable"!

Another criticism to the fore in WCC circles *was* an economic one. Sustainable development as a modified version of earlier development did nothing to change what many called the propensity of the global economy to create both internal and external "colonies." Nature, women, and the poor were just such colonies in many locales. This economic colonization was the mechanism for what we earlier called "appropriated carrying capacity," but seen now from the side of those from whom the appropriation took place. The charge they brought was that ongoing growth and prosperity for the well-off was sustained by externalizing the costs to communities elsewhere. If full economic, environmental, and social costs had instead been *internalized within* the centers of benefit (wealthier nations and wealthy sectors of poor nations), there would have been a quick end to claims about the glory and beauty of growth and development. In short, this was the economics of borrowing as reported by those who were squeezed, even crushed, rather than sustained.

Little of this would change, the criticism went on (sometimes ad nauseum), so long as basic changes in trade, debt, and aid were not forthcoming and so long as sustainable development policies were marginal to the basic economic policy-making decisions of government, business, and finance. It should be added that this criticism has in fact been borne out in efforts since. The global gatherings at Rio (UN Summit on Environment and Development), Cairo (Summit on Population and Development), and Copenhagen (Summit on Social Development) are testimony.[12]

In short, "sustainability," whether attached to "development" or "society," has come to mean something very different in WCC and other NGO circles from its prevailing notion in UN, business, finance, and Northern nation-state government circles. It means, *not* global economic growth qualified by environmental sensitivity, but local and regional communities that are economically viable, so-

---

11. The terms "dislocation" and "ecolocation" are Daniel Spencer's. See Spencer's "The Liberation of Gaia," *Union Seminary Quarterly Review* 47, nos. 1–2 (1993): 91–102. I use "ecolocation" differently from Spencer, however, in keeping with the earlier discussion of *oikos* as a vision. My use is normative, Spencer's descriptive.

12. To use Rio as illustrative: only two billion dollars over three years has been pledged for objectives that government heads meeting in Rio agreed will cost some six hundred billion annually. Significant here is the additional fact that national ministers of economy and development are not even involved in the UN process. Rather, ministers of environment are sent, without the backing of financial and business interests as channeled through governmental or private representation. See "Panel Finds Lag in Saving Environment," *New York Times*, May 29, 1994, 13.

cially equitable, and environmentally renewable. Earth itself in these locales is part of the communities' very sense of being. It is their habitat, their patch of *oikos*, "down home."

The search, therefore, is for local and regional self-reliance and economic and environmental sustainability, as well as the global institutions needed to serve these on a contracting planet. The approach usually focuses upon basic needs, with poverty eradication as a major goal. Since poverty eradication and meeting basic needs are also goals of the Brundtland Commission and similar bodies, the difference of approach is not always readily apparent; differences are masked by a common vocabulary. Sustainable-development advocates talk of "community development," "empowerment," and "grassroots participation" as readily as sustainable-community advocates, and they often do so sincerely and without cynicism. Yet the differences are quite dramatic. They surface with the argument of sustainable-community advocates that a *sufficiency* revolution (redistributed power, access, and resources within and across societies) is every bit as important as the *eco-efficiency* one (more environmentally benign technologies and production). And the two belong together. "Healthy" eco-efficient business isn't enough. In fact, under conditions of fierce competition, healthy, eco-efficient business might generate higher unemployment and send manufacturing down the same track as agriculture. (What was once the largest U.S. employer — agriculture — now feeds the United States and more with only 3 percent of the workforce.)[13] In a small world doubling its population, eco-efficiency and other technical solutions are often socially insufficient. Yet sustainability requires equity, and the first step toward equity is to transfer resources from nonessential to essential consumption, from wasteful wants to basic needs.

Differently said, our two issues — sustainability and power in society — are interrelated dimensions of a common matrix in the WCC view. To borrow words from elsewhere, the debt to nature cannot be paid person-by-person in recycled bottles or ecologically sound habits, but "only in the ancient coin of social justice."[14]

COMMUNITY SUSTAINABILITY

We can illustrate all this with two documents, both of them conscientious responses to earth issues. One is a WCC document prepared by representatives from around the world entitled *Accelerated Climate Change: Sign of Peril, Test of Faith*.[15] In addition to illustrating the community alternative to sustainable de-

---

13. This is not to argue that U.S. agriculture is a paragon of eco-efficiency. It is not! This is only to argue that businesses seeking eco-efficiency under capitalist conditions will, by way of efficient technology, likely shed employees in the pursuit of productivity rather than generate humanly labor-intensive work.

14. Barry Commoner as cited by Wolfgang Sachs, introduction to *Global Ecology*, ed. Sachs (London: Zed Books, 1993), xvii.

15. World Council of Churches, *Accelerated Climate Change: Sign of Peril, Test of Faith* (Geneva: WCC Publications).

velopment, it reflects the WCC's integrity-of-creation theme and its ethic. The other document is a report of the secretary-general of the United Nations, *An Agenda for Development*. This is a statement not of the secretary-general's personal views but of his mandated task to "gather the widest range of views about the topic of development... from all Member States, as well as the agencies and programmes of the United Nations system"[16] and offer them to the general assembly for policy deliberation. Secretary-General Boutros Boutros-Ghali's report came two years after the Earth Summit of 1992 and may be said to represent widespread views among international representatives of nations and transnational organizations. Comparing and contrasting these documents will highlight the divergent approaches to sustainability.

The introduction to *Accelerated Climate Change* moves quickly to the ethical issues. "Accelerated climate change represents not only a threat to life but also an inescapable issue of justice. It throws into sharp relief the unjust balance of wealth, resources and economic power between the rich and poor that characterizes the world today."[17] Argument is then marshaled to support this. Global warming itself mirrors the injustice of economic relationships between the North and the South. The root of the global-warming problem is excess consumption in the North, which also deprives the South of resources. The appropriate balance between consumption and available resources and an equitable use of these resources over the globe lead to a distinction — between the "luxury emissions of the rich" and the "survival emissions of the poor." The industrialized nations need to achieve drastic reductions of emissions, while emissions in the South may legitimately increase as development strives to meet the basic needs of people.[18] In short, the setting of emission targets cannot be dissociated from issues of social and economic justice.

Nor can the social-ethical issues be dissociated from justice for (the rest of) nature. One conclusion is that "in this crisis, compounded in the violation of intricate natural systems and the disruptive exploitation of vulnerable human communities, nature has become co-victim with the poor. Earth and people will be liberated to thrive together, or not at all."[19]

The document in this first section includes both the presentation of scientific evidence and the theological rationale for addressing climate change and the ecocrisis. It finishes with implications for ethics.

Seeing the degradation of natural systems, we know that "*we must cherish the whole of creation, not for our sake alone, but for its sake and for God's sake*, for God made and loves all."[20]

Seeing the marginalization and deprivation of many brothers and sisters, we know that "*justice today requires sufficiency without excess, and that we must*

---

16. *An Agenda for Development: Report of the Secretary-General to the Forty-eighth Session of the General Assembly*, 3.
17. WCC, *Accelerated Climate Change*, 7.
18. Ibid., 12.
19. Ibid., 13.
20. Ibid., 14.

build communities that enable all members to participate in obtaining and enjoying sufficient and sustainable sustenance from nature."[21]

Seeing the mounting peril that looms ahead, we know that love and justice are transgenerational and that we must protect this earthly habitat so that it will sustain the lives of our children's children's children, together with other life, into the indefinite future.[22]

Seeing the mounting tensions and conflicts among ethnic, religious, and racial groups in various regions of the world, and the evidence of reluctance by Northern governments to attend to the suffering of the South, we know that "*we must not allow either the immensity or the uncertainty pertaining to climate change and other problems to erode further the solidarity binding humans to one another and to other life.*"[23]

Moral commitments clarified, the WCC paper proceeds with analysis of "current political socio-economic realities." It affirms eco-efficiency, yet not without a keen sense for the kind of economy of which it is a part. "Using resources more efficiently in a world of ever-expanding material production and consumption only delays the inevitable. While producing more efficient automobiles is an important step, all reductions in petroleum consumption will eventually be cancelled out if the total number of automobiles continues to rise."[24] The eventual conclusion is that "[w]e are pushed to go beyond efficiency, renewable energy technologies and individual lifestyle changes to the formidable task of reconceiving and transforming the economic system. This task belongs within the context of a renewed vision of community."[25]

Since a renewed vision of community is key for this approach to sustainability, the document turns to it next, in a chapter entitled "Building Community." The heart of it is a set of five "ethical claims" (lettered *a.* to *e.*) brought to bear "on economics." The introductory paragraph is reminiscent of the earlier discussion of *oikos:*

> "Economy" in its Greek-root meaning is simply the ordering of the household for the sustenance of its members. Economy signifies the arrangements that people make in a household — or a community — or a community of communities — to draw sustenance from nature to meet their needs and wants. In accordance with Gen. 2:15, it signifies the work that humans do to "till the garden" but also to "keep" it so that the tilling may continue. The ordering of work for the sake of life belongs to God's purpose for human beings and the whole creation. As such, it falls under God's will and intention that love-as-justice be expressed in the structures, policies, and practices of economics.[26]

21. Ibid.
22. Ibid.
23. Ibid.
24. Ibid., 23.
25. Ibid., 24.
26. Ibid., 26.

Ethical claims for economics then follow. For the sake of illustration, we pare already succinct discussions even more.

a. Ecological sustainability. Economics must come to terms with the ecological realities that make endless material growth unsustainable. The claim of sustainability, however, must transcend anthropocentric self-interest. Without a sense of awe and reverence before the goodness and grandeur of the whole created order, humans will not likely preserve their habitat.

b. Sufficiency of sustenance. A majority of the people in the South and a significant minority in the North need more access to energy, work, and land in order to obtain sufficient sustenance. There is an urgent need for a sense of "enough" based on a less-materialistic conception of the good life and responsibility on the part of the comfortable and the wealthy to lighten their demand upon earth's bounty so that it may provide enough for all.

c. Community through work. This requires economics to regard people not simply as individuals but as persons-in-community. For enterprises to contribute constructively to genuine community, they have to exhibit concern for the well-being of their personnel, the environment, and their consumers as an intrinsic part of their economic mandate. All economic arrangements should be conducive to community well-being.

d. Participation by all. This means both the right and the responsibility to share in work for individual and family livelihood and the common good of the community. This makes unacceptable the prevalent situation today in which millions of people have been made superfluous nonparticipants in economic activity. Human dignity requires that economics become more democratic, that people rightfully participate in decisions pertaining to their work and the kind of development that is best for their community.

e. Respect for diversity. While the need for community is universal, communities are far from identical in their definitions of the common good or in their needs and wishes pertaining to development. A global economy must respect a wide variety of local, national, and regional communities.[27]

From the ethical claims for economic life, the document moves to proposals for economic re-visioning and action. Again we curtail even further an already crisp presentation.

a. Shift from unneeded production to work that needs to be done. There are massive employment possibilities in work that would address major unmet needs, including housing, health care, public transportation, environmental restoration, and recreational, cultural, and aesthetic opportunities that would enhance life for everyone. Markets have an essential role here, but this is a challenge that market economics by itself cannot meet, since reliance upon unlimited economic growth is no longer tenable. New policies, involving governments and NGOs at many levels from the local community on up, are needed.

b. Make development increasingly community-based, focused on essential needs and the sustainable, equitable use of natural resources. Alternative kinds of devel-

---

27. Ibid.

opment are being tried in many places, in both South and North. These entail local community organization and development, with an emphasis on participation by all members of the community in determining development goals, methods, and technologies. The intention is to apply local resources more directly to local needs, to maintain or restore community support systems, and to reduce dependence upon outside forces and unstable commodity markets. This increases the capacity of communities to take care of their own most essential needs while it focuses trade on those items most relevant to meeting those needs. Larger economic units or entities — regions, nations, international agencies — would increasingly construe their role as enabling their subcomponents to be as self-reliant as realistically possible as well as maximally conserving of resources.

c. Organize and mobilize democratically to curb excessive power and empower all people. Political mobilization is critical. Strong citizens' coalitions to monitor public policy and business practices are the best antidote to the corruption and the domination by special interests that now violate the common good.[28]

The conclusion of the chapter called "Building Community" is posed as a question and a challenge to the WCC member churches:

> The transformation for which we call is already beginning in many places around the world as people struggle and organize to achieve their aspirations. What will the churches do to support localities, nations, and the world in moving towards economic and political arrangements that accommodate reduced consumption, promote new development models and nurture community? The answer will be a test of faith.[29]

ECONOMIC ACTIVITY IN THE LONG TERM

We turn next to the UN's *Agenda for Development.* "The dimensions of development" are listed in the table of contents itself.

A. Peace as the Foundation
B. The Economy as the Engine of Progress
C. The Environment as a Basis for Sustainability
D. Justice as a Pillar of Society
E. Democracy as Good Governance[30]

Most helpful for comparison are items B, C, and D.

Section B, "The Economy as the Engine of Progress," begins with this. "Economic growth is the engine of development as a whole. Without economic growth, there can be no sustained increase in household or government consumption, in private or public capital formation, in health, welfare and security

---

28. Ibid., 26–27.
29. Ibid., 27.
30. *Agenda for Development*, table of contents.

levels."[31] The supporting paragraphs echo the Brundtland Commission Report and the Business Council for Sustainable Development:

> Accelerating the rate of economic growth is a condition for expanding the resource base and hence for economic, technological and social transformation. While economic growth does not ensure that benefits will be equitably distributed or that the physical environment will be protected, without economic growth the material resources for tackling environmental degradation will not exist, nor will it be possible to pursue social programmes effectively in the long term. The advantage of economic growth is that it increases the range of human choice.
>
> It is not sufficient, however, to pursue economic growth for its own sake. It is important that growth be sustained and sustainable. Growth should promote full employment and poverty reduction, and should seek improved patterns of income distribution through greater equality of opportunity.[32]

Of more than passing interest is the secretary-general's statement that in the quest for sustained and sustainable economic growth, "governments can no longer be assumed to be paramount economic agents." Nonetheless, they retain the responsibility "to provide a regulatory framework for the effective operation of a competitive market system."[33]

"The Environment as a Basis for Sustainability," section C, leads to the clearest differences with the WCC document and *oikos* community development. The differences can be seen by examining the emphases of the UN document (the separation of environment and humanity with the latter as managers; the view of nature and its subordination to economic ends; the uncritical affirmation of progress; and the wholly anthropocentric view of the existence of otherkind) in a few key paragraphs:

> Development and environment are not separate concepts, nor can one be successfully addressed without reference to the other. The environment is a resource for development.... Successful development requires policies that incorporate environmental considerations.
>
> Preserving the availability and rationalizing the use of the earth's natural resources are among the most compelling issues that individuals, societies and States must face. A country's natural resources are often its most easily accessible and exploitable development assets. How well these natural resources are managed and protected has a significant impact on development and on a society's potential for progress.[34]

---

31. Ibid., 9.
32. Ibid., 10.
33. Ibid.
34. Ibid.

The fundamentally anthropocentric and utilitarian nature of development in this scheme is clear in the next paragraph: "Environmental degradation reduces both the quality and the quantity of many resources used directly by people." An enumerated listing of various kinds of degradation then follows. All are judged by their outcome for humans only. There is a sentence in another section about "the intrinsic worth and value of nature." But even it is raised in the context of the "tangible and material aspects of the environment" for social welfare and the pleasures of nature as an aesthetic good for society.[35]

"Community involvement in all development efforts" is a subject mentioned only in passing as well, though its importance is underscored, and the "pioneering initiatives" of community groups are applauded. The language is again telling, however. "By making local inhabitants incentive partners rather than simply collateral beneficiaries, these programmes have broken new ground." The programs of these "incentive partners" have led to "greater community cooperation in preserving tourist assets and higher rural incomes. They are examples from which many others can learn and benefit." "Sustainable tourism strategies" are singled out as a concrete means for "preserving the natural environment."[36]

All this makes clear that what needs to be sustained is "economic activity in the long term." Sustainable development means economic activity as qualified by environmental considerations in such a way as to secure sustainable economic growth. The secretary-general is also clear that "at present, economic policy coordination among the major economies centres on the Group of Seven industrialized countries" (United States, Japan, Britain, France, Germany, Italy, and Canada) but that it should be "more broad-based." Surprisingly, he accepts a human construct as fixed natural law itself and, as a consequence, shifts the locus for change elsewhere: "The laws of economics cannot be changed, but their social consequences can be eased."[37] Presumably, broadening the base of policy coordination would aid this easement.

Other sections of the report stress the importance of *social* stability. "Poverty, resource degradation and conflict" are becoming "an all too familiar triangle" that raises the specter of waves of refugees and resource-based wars. This underscores the need for "justice as a pillar of society." The link of justice to security is then made on the terms of what has gone before: namely, sound investment. "While investment in physical capital is an important aspect of stimulating economic growth, investment in human development is an investment in long-term competitiveness and a necessary component of stable and sustainable progress.... A stable economy and a stable political order cannot be built in an unstable society. A strong social fabric is a prerequisite to sustainability."[38]

At this point the UN's picture of sustainable development is similar to the World Bank's. The fiftieth anniversary of the World Bank in 1994 brought stinging criticism and a broad review. In response the anniversary occasioned

35. Ibid., 14–15.
36. Ibid., 15.
37. Ibid., 13, 12, 20.
38. Ibid., 17, 19.

a restatement of the bank's vision. In this statement, entitled *Embracing the Future*, bank president Lewis Preston enumerated five basic challenges: promoting broad-based economic growth that particularly benefits the poor; stimulating private sectors so developing countries can be more competitive in the global economy; reorienting governments to become more efficient at complementing the private sector; protecting the environment; and investing more in human-development programs supporting education, nutrition, family planning, and the role of women.[39]

Perhaps this suffices to contrast the approaches as well as note important agreement that "a strong social fabric is a prerequisite to sustainability." All that remains is to cite further arguments from the WCC, some NGOs, and some peoples' movements against the dominant notion of sustainable development and its realization by various transnational organizations. The latter group doesn't here include the United Nations but does include the World Bank, many transnational corporations, and certain NGOs.

BAD NUMBERS, INVERTED POLICY

Differences are greatest on two matters. The economics of the reigning approach to sustainable development is bad economics, in the WCC view. And its rendering of the good society and sustainability is "too economistic."

"Too economistic" means that social well-being is measured by rising or falling gross domestic product (GDP). Of course for millions of destitute people, to *have* more is required in order to *be* more. Championing socio-economic equity and bending the market to address basic needs have thus been unrelenting themes of the WCC and many NGOs in community-development work. But once people are above the poverty line, there is very little correlation of happiness and well-being with increased consumption and rising incomes. Satisfactions in life relate more closely to the quality of family life and friendships, work, leisure, and spiritual richness. None of these is well measured by the GDP. In fact, in societies with the highest levels of consumption, where the basic choice of serving God or mammon has been faced squarely and decided in favor of the latter hands down, there seems to be inordinate psychological and spiritual emptiness. (Which commerce then plies in order to sell more goods and services for greater self-fulfillment!)

But development as economic growth is not only too economistic — it is bad economics as well. Basically the criticism of the WCC, many NGOs, and peoples' movements is this: domestic and global wealth is generated without eradicating poverty or making local communities more viable and sustainable.

The argument, put somewhat differently, is that the high-level efforts of the Brundtland Commission, the Business Council for Sustainable Development, and *Agenda 21* cannot attain their own good ends because the means fail them.

39. "World Bank, at 50, Vows to Do Better," *New York Times*, July 24, 1994, A4.

The numbers just don't work. Official sustainable development counts on environmentally safe economic growth to alleviate poverty, rather than attempting the much more politically difficult path of socioeconomic redistribution. To this end Brundtland targets a 3 percent annual growth in income. But economists at the World Bank, in a working paper entitled *Environmentally Sustainable Economic Development: Building on Brundtland,* show the following results:

[A]n annual 3 percent global rise in per capita income translates initially into annual per capita income increments of $633 for USA; $3.6 for Ethiopia; $5.4 for Bangladesh; $7.5 for Nigeria; $10.8 for China and $10.5 for India. By the end of ten years, such growth will have raised Ethiopia's per capita income by $41.0....[40]

Former World Bank president Robert McNamara had earlier observed that "even if the growth rate of the poor countries doubled, only seven would close the gap with the rich nations in 100 years. Only another nine would reach our level [that is, that of the United States] in 1000 years."[41]

Conclusion? Growth does not eliminate poverty.

Worldwatch Institute offers other, longer-term corroborating evidence. *State of the World 1990,* in its overview entitled "The Illusion of Progress," notes that on average the additions to global economic output during *each* of the last four decades matched economic growth from the beginning of civilization to 1950! Yet during the same decades the ranks of the destitute soared. The World Bank puts the number of the "bottom billion" at 1.2 billion in 1990 and says these people scrape together a dollar a day or less per capita. The global rich number about the same as the destitute — 1+ billion — and include all middle strata as well as the upper strata of Northern nations. The 3.6 billion between are the "managing" poor. They do more than exist but do not live well or without fear. The ranks of both the destitute and the managing poor are almost certain to grow in coming decades, even with 3 percent annual economic growth. Too, these same four decades of unprecedented economic growth saw environmental destruction grow alarmingly. Since midcentury the world has lost nearly one-fifth of cropland topsoil, a fifth of the tropical forests, and thousands of plant and animal species. The rending of the social fabric has proceeded apace during this boom time as well.[42]

The numbers get worse when growth is tracked by "developed" and "developing" sectors. Inequality between and within societies accompanies high growth rates. The share of total world production in developing countries a century ago

40. Cited by David Korten, "Sustainable Development," *World Policy Journal* (winter 1991–92): 167, from Robert Goodland, Herman Daly, and Salah El Serafy, eds., *Environmentally Sustainable Economic Development: Building on Brundtland,* working paper (Washington, D.C.: Environment Department, World Bank, July 1991), 6.

41. McNamara as cited by Paul Hawken, *The Ecology of Commerce: A Declaration of Sustainability* (San Francisco: Harper Business, 1993), 135. Hawken does not give the source.

42. Lester R. Brown, "The Illusion of Progress," in *State of the World 1990* (New York: Norton, 1990), 3–4.

was 44 percent. By 1950 it was 17 percent! It recovered to 20 to 21 percent by 1980, still less than half what it was the century earlier. By this time (1980), however, the developing world's population share was about 75 to 80 percent and counting.[43] This means that 20 to 25 percent of the world's population, the advanced industrial countries, contributes 80 percent of world production. Conclusion? "Catching-up development" is not working.

All this is graphed by population and consumption levels in the accompanying figure (see fig. 1). The chronological graph tracks world population, the pie chart reflects world consumption.

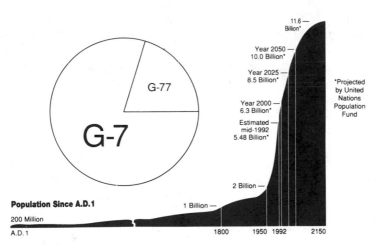

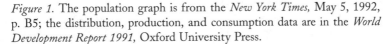

Figure 1. The population graph is from the *New York Times,* May 5, 1992, p. B5; the distribution, production, and consumption data are in the *World Development Report 1991,* Oxford University Press.

By way of explanation, the G-7 nations are the large nations of North America, Europe, and Japan listed earlier. The G-77 nations were once only 77 but now include some 128 lesser developed countries, most of them in the South. G-7 nations have about one-fifth of the world's population but produce and consume four-fifths of all goods and services. Thus the G-77 countries, with four-fifths of the population, produce and consume one-fifth. By way of comparison, then, every person added to the population of the North represents, in terms of production and consumption, twenty persons in the South. Or, in Holmes Rolston's analogy, "for every dollar of economic growth per person in the South, twenty dollars accrue to each individual in the North."[44] That is, ex-

43. Shridath Ramphal, *Our Country, the Planet: Forging a Partnership for Survival* (Washington, D.C.: Island Press, 1992), 147–48.

44. The graph and this description are taken from Holmes Rolston III, "People, Population, Prosperity, and Peace," in *Ethics and Agenda 21: Moral Implications of a Global Consensus,* ed. Noel J. Brown and Pierre Quiblier (New York: United Nations Publications, 1994), 35–36.

panding economic pies bring greater rewards to the already rich and increase the absolute gap between rich and poor.

Nonetheless, Brundtland's recommendation is a five to ten times increase in world output, as stimulated by the North, in order to effect poverty reduction everywhere. This must be done without increasing environmental degradation, in fact reversing much of it. If one adopts the formula sometimes used to calculate environmental impact, $I = PCT$, where I = environmental impact, P = population, C = consumption per capita, and T = the environmental intensity of consumption (i.e., all the changes in technology, factor inputs, and the composition of the GDP), and if one strives for eco-efficiency as the second industrial revolution, then a growth rate of 3 percent annually for the projected world population would mean the following: the environmental impact of each unit of consumption would need to fall by 93 percent over the next fifty years to meet the rather conservative definition of sustainability that Brundtland, the Business Council, and *Agenda 21* have adopted.[45] Since the consumption rates are much higher in the North than South, and about 75 percent of global environmental degradation has Northern causes, both the greatest efficiencies and the greatest falloffs in consumption need to happen in the North.

Bearing such projections in mind, Friends of the Earth U.S. adapted a consumption model proposed for the Netherlands and used it to answer this question: "If the earth's resources are consumed at a sustainable rate and distributed evenly among the people who will be alive in the year 2010, how much would consumption levels need to change?" For the United States, the results played out as follows. If global warming is to be held to less than 0.1 degrees centigrade per decade, U.S. per capita carbon dioxide emissions must drop from 19.5 tons today to 6.6 tons by the year 2010 and 1.7 tons by 2030. On the order of 60 percent drops will be needed in the use of timber and eating of seafood. And the consumption of natural resources overall will require reductions of 60 to 90 percent![46] Yet the UN Earth Summit (Rio), Population Summit (Cairo), and Social Development Summit (Copenhagen) treaties were virtually devoid of targets in consumption levels, an omission that marks *Agenda 21* as well. Further, although heads of state at Rio agreed that $600 billion would be required annually for cleaning up the planet and allowing poorer countries to develop their economies in an environmentally sustainable way, rich countries have so far provided only $2 billion over three years to the fund that was set up, the Global Environmental Facility.[47]

Two comments almost write themselves in response to what crunching the numbers means for the reigning model of sustainable development.

The first is that persons often regarded as consummate, worldly realists — businesspeople, economists, politicians, and diplomats — expose themselves as

45. Paul Ekins, "Making Development Sustainable," in *Global Ecology*, 92–93.

46. Friends of the Earth U.S., "Sustainability and U.S. Resource Consumption," as reported in *ECO* (NGO newsletter at the second session of the Commission on Sustainable Development) 6 (May 23, 1994): 2.

47. "Panel Finds Lag in Saving Environment," 13.

pure utopians in the pejorative sense; they are dreamers with no place to stand on the solid ground of sound evidence. Schmidheiny's earlier admission that, for the moment, merging economic growth and environmental sustainability is a dream turns out to be even more ecomystical than he himself suspected. His hope, and that of sustainable development on the going model, is hallucinatory.

The second comment is that the approach itself is backward, given its own objectives. Its argument is that governments and societies need not limit economic growth so long as they stabilize use of natural resources. They will in fact need to increase growth dramatically (Brundtland) while achieving stabilization through the second industrial revolution of eco-efficiency. But why not, as per Sharachchandra Lélé, turn the matter around and suggest that if economic growth is not correlated with environmental sustainability, then there is no reason to have economic growth as an objective of sustainable development? If the reply (a good one) is that economic growth is absolutely necessary to remove poverty, then the question is: Why has it not done so, especially during boom decades when environmental constraints were little imposed? The question is: *If* economic growth of itself leads neither to environmental sustainability nor removal of poverty, then why is reviving growth the first-listed objective of the Brundtland Commission Report and the object of the first substantial chapter of *Agenda 21*? Why not consider the converse and explore "whether successful implementation of policies for poverty removal, long-term employment generation, environmental restoration and rural development will lead to growth in GDP, and . . . to increases in investment, employment and income generation"?[48] This would seem especially promising in the South, where it is most needed.

CONCLUSION

Two approaches to sustainability and its meaning are before us. One begins with economies and their growth, the other with communities and theirs. To recall Hildyard's terms, sustainability for sustainable development belongs to those for whom the environment (and sustainability) is "around their economies." Sustainability for sustainable community belongs to those for whom the environment (and sustainability) is "around their homes" (*oikos*).[49]

48. Lélé, "Sustainable Development: A Critical Review," *World Development* 19, no. 6 (1991): 614.

49. I am keenly aware of the deficiencies of using a two-term typology: development qualifiers and development dissenters, sustainable development and sustainable community. There are important groupings within each, and the loyalties and work of many individuals and organizations intersect both. Sustainable development camps, for example, parallel virtually all the variants on capitalism, from something close to eco-attentive laissez-faire to welfare-state capitalism. Yet all of them, as indicated above, move into sustainability from within global capitalist economic frameworks and considerations, that is, economic growth on a free-market model. Development dissenters are even more varied, in part because most all of them begin with and focus on local issues. Thus Love Canal toxins and health issues create one kind of movement in resistance against government and industry in New York State, mining and logging and cattle ranching in tropical forest lands of indigenous peoples in the Amazon region another, and sewage treatment, a huge city bus terminal, and

For sustainable-development advocates, "sustainable" seems essentially a qualifier of the prevailing development path and largely assumes both its layout and its economistic ethic.[50] The issue its critics pose so sharply is whether sustainable development is any more than an environmentally qualified continuation of the trek of "Sweet Betsy from Pike," rather than another path altogether, as Brundtland and most delegates to the Earth Summit hoped. Is sustainability here not "the ecologization of world society" as "the latest version of Western development dynamics,"[51] a kind of "green globalism"?

Differently said, Brundtland, *Agenda 21,* and the Business Council for Sustainable Development — the development qualifiers — all hope to bend an international global economic order in ways that make "poverty removal, long-term employment generation, environmental restoration and rural development" (Lélé's list) the prime determinants of economic polity, policy, and practice. If they were, Brundtland and the others must know, the basic dynamics of the economic system itself would be fundamentally transformed. By contrast, the WCC and other members of a growing international civil society are skeptical in the extreme. In their judgment the global economy has no intention of going this direction. Should it happen to change course, it will be because its own ills require harsh treatment undertaken to prevent its own demise. In any case, sustainable development of this sort (read: unsustainable maldevelopment) is not what a restless and fragmented but vibrant international civil society is hoping and working for. The development dissenters are hoping and working for the breakup of both traditional economic domination *and* green globalism, in favor of an ecumenism focused on viable communities and sustainable livelihoods within them.

There is yet a third "take" on sustainability, however. "Development qualifiers" and "development dissenters" alike draw upon it. This is the take of the scientists, especially ecologists. Intelligent decisions about sustainability would certainly seem to rely upon sound scientific data and the good counsel of the empirically informed. So what are scientists saying? How does sustainability look under the microscope and in the field? What does nature, via these students, say about sustainability?

lines of idling garbage trucks for New York's Upper West Side off-loading onto barges, all nearby a grade school in Harlem, yet another. Nonetheless, this diversity shares the general characteristics of community approaches to sustainability outlined above.

50. This is Dieter Hessel's important point from "Sustainability in Social Ethical Perspective," 2.

51. The phrases are Guy Beney's in " 'Gaia': The Globalitarian Temptation," in *Global Ecology,* 181.

Fish Stories

Sustainability assumes predictability. We need to know how ecosystems behave and what follows from changes prompted by human intervention. We certainly need to know what sustainable yields would be for cultivated lands, forests, or fisheries. Some reasonable level of reliability and, from it, predictability is necessary for sustainable development and/or sustainable community.

We have a problem. The problem, it turns out, is a serious outbreak of agnosticism among ecologists. They are less confident than they once were about knowing with any precision the indices of stress and the points of environmental collapse. They are less confident as well about marking nature's bounty and its harvest capacity and limits. Moreover, just when "ecosystem" becomes a common notion that makes the evening news, supermarket ads, and even books in religious ethics, some scientists quit using it. Why? The short answer is that sometime in the 1970s nature in the purview of science changed rather markedly, with broad ramifications for the human uses of nature. The long answer takes a little time but must be pursued. One way to do so is to see what has happened recently to fish.

NET LOSSES: TRUE FISH STORIES

Articles in the *New York Times* over a couple years are sober testimony to the consequences of changed nature and a new scientific agnosticism. The world fish-catch, like world food production in general, quadrupled from 1950 to 1990. In 1989, the catch began to decline, and after further decline seventy nations met in 1993 to face the fact that the world's edible fish stock had thinned "to levels unimaginable only a few years ago."[1] The UN Food and Agriculture Organization (FAO) was still confident in 1991 that global fish-catches would increase. But by 1995 it, too, was saying that "the ocean's most valuable commercial species are fished to capacity" and an estimated 70 percent of global fish stocks are "depleted" or "almost depleted."[2] Russia reported that its stocks of pollack had virtually collapsed. Canada's cod industry did go belly up (with a loss of twenty thousand jobs),[3] as did Peru's anchovy fisheries and much of California's sardine commerce. (Indeed, marine life in general off southern California

---

1. "Fishing Countries Split on Harvests," *New York Times*, August 1, 1993, K9.
2. "Introduction," *The Ecologist* 25, 2/3 (May/June 1995): 42.
3. Ibid. Other sources, cited below, say thirty-three thousand. The difference seems to be that the latter includes workers in fish-plants and others tied to the industry, not only fishing crews.

155

is in precipitous decline, probably because a critical part of the food chain, zoo-plankton, has tumbled by 80 percent.)[4] Overall, the productivity in all but two of the fifteen major fishing regions of the world has fallen, and in four of the hardest hit areas — the northwest, west-central, and southeast Atlantic and the east-central Pacific — the decline has exceeded 30 percent. Only Indian Ocean fisheries were still increasing their total output as of 1995. *Worldwatch*'s conclusion is that *all* the world's major fishing grounds are presently at or beyond their sustainable limits, as best we can judge that.[5]

The reasons are many. Of most immediate interest is one signaled in the title of a *Times* article: "Biologists Fear Sustainable Yield Is Unsustainable Idea." "Science is probably incapable of predicting safe levels of resource exploitation," the article says, and goes on to suggest that policy-makers had best abandon "the pretense of scientific certainty."[6]

The culprit, it turns out, is nature. Boom, bust, and flux may be the norm. Scientists trying to find the cause for occasional drastic fluctuations in natural populations have concluded there probably isn't any single one. In a study of the life-cycle of the Dungeness crab, knowing how the populations changed over a few generations, "or even a few hundred," allowed no insights about either past or future behavior. In an effort to model the world they found and then project results, scientists discovered total population numbers could remain stable for thousands of generations, then without warning suddenly boom or crash, *without* notable changes in the environment as the trigger. (The model allowed the normal changes in Dungeness crab behavior but no significant changes in the surrounding environment.) Their conclusion: "nature is more unpredictable and unstable — and difficult to study — than researchers had guessed." Reflecting an implicit criticism of past ecology, Peter Turchin remarked: "The important message is that transient behavior becomes the end point, so to speak."[7] (Until recently ecologists had largely assumed, with Eugene Odum, that erratic change is *ab*normal or part of a phase in a system on its way to equilibrium as the end state.)

In the case of fish, the resulting mood of caution was accompanied with embarrassment. It turns out that scientists had advised the California sardine industry, just as they had worked closely with the Canadians on cod. They had done their level best to tally the "maximum sustainable yield," that is, the number of fish that could be taken without endangering the resource. "The net result of putting trust in scientists [in the Canadian case] was to put 20,000 people out of work," Carl Walters, one of the scientists, commented.[8]

---

4. See "Marine Life Disappearing Off California," *New York Times,* March 5, 1995, A23.

5. Peter Weber, *Net Loss: Fish, Jobs, and the Marine Environment,* Worldwatch paper 120 (Washington, D.C.: Worldwatch, 1994), 13.

6. "Biologists Fear Sustainable Yield Is Unsustainable Idea," *New York Times,* April 20, 1993, C4.

7. "Boom and Bust May Be the Norm in Nature, Study Suggests," *New York Times,* March 15, 1994, C4.

8. "Biologists Fear Sustainable Yield Is Unsustainable Idea," C4.

In reviewing where they had gone wrong, the scientists concluded that they had badly misjudged the range of fluctuations and did not in retrospect see how they might better gauge them in the future. If the fish count was taken when environmental conditions seemed good, and projections were made accordingly, the populations would be high. But when the conditions turned bad — as often they did rather suddenly and with little forewarning — good fishing would yield a disastrous result for future fish stocks. (Keep in mind the effective technology of huge nets and the precision guidance systems of sophisticated information technologies. Good catches can be had even with low stocks. Fishing has become, like agriculture, a fine-tuned, technically sophisticated, highly competitive industry. Commercial fish are not "caught" one by one. They are "harvested" in grand sweeps from factory ships.)

There seem, in short, to be peaks and troughs in fish populations for many reasons, and only some of the reasons are known. So Ray Hilborn, one of the scientists whose work is reported, recommends that something like the following be done, for safe measure. The number of Alaska salmon needed to spawn in order to maintain a healthy population apparently can range anywhere from one million to fifteen million, depending on whether the population cycles are at peak or in a trough. "Sustainable yield" is not an impossibility under such conditions, but it does mean that fisheries need to leave enough fish whether the year is a good one or a bad one. And they cannot know with assurance in advance which it might be. All in all, it requires taking considerably fewer fish than fishing would yield in a good year.

But it means more. It means that a wide range of uncertainty is part and parcel of the scientific picture and therefore of sustainable development or sustainable community. What complicates matters is that sustainability itself can no longer be conceived "as some permanent, steady-state condition. Rather, the normal fluctuations of nature make the sustainable use of biological resources 'a moving target.'"[9] Contrary to its own connotations, then, sustainability is inherently dynamic and significantly unpredictable. This does not render science useless or sustainability an utter illusion. But it does mean that reliable notions such as carrying capacity, essential to sustainable development and society, are not reliable in a stable sense at all. Carrying capacity changes. It, too, is "a moving target."

So what is the scientists' advice? "Rely on scientists to recognize problems but not to remedy them. Distrust claims of sustainability, since past resource exploitation has rarely been sustainable. And confront uncertainty honestly." "As long as you recognize your limitations," scientist Donald Ludwig advises, "you can begin to cope with things."[10]

Yet another finding bears hard on sustainability. There is so much change and fluctuation in nature as impacted by humans everywhere now that "the rate of learning might not be faster than the rate of change." Consequently, Hilborn

---

9. Ibid.
10. Ibid.

says, "Over time we might not get more certain." Which is only to say that knowledge necessarily *trails* dynamic changes. There goes refined predictability! "It's a very scary thought," Hilborn adds.[11]

Sobering as that may be, it is not the most important finding. That is another point emphasized by Hilborn. "Most fishery problems are the result of not understanding how the fishermen behave rather than how the fish behave."[12] Commercial fishers will fish for scientists, too, it seems, and exploit their differences in favor of numbers that best serve a competitive industry. Gibson's law that for every Ph.D. there is an equal and opposite Ph.D. evidently holds for scientists as well; and fishing lobbies will find the ones they need.

Still, that is only the surface expression of the human factor. Doubts about sustainability have as much to do with the way human beings organize society, including fishing, as they do with nature's inherent restlessness. How we chase fish reveals as much about our environment as the fish's.

Immediate competition among those in the fish industry is obviously one of the most powerful forces at work in declining fish take. But other economic factors risk fish as well. There is the buildup of infrastructure and markets that occurs when fishing and commerce are good. When they take a dive, the pressure is still on to protect the large capital investment and market share by finding the fish. And as mentioned, the technology is at hand to fish even depleted stocks effectively. So when the U.S. government closed Georges Bank in October 1994, the unemployed fishers of this traditionally rich fishing ground off New England did not respond by training for other work. "We've put a lot of time into learning how to catch fish and where to sell them," one of them commented, adding, "so we've got to go elsewhere."[13] Add to this the backing of governments representing both important domestic industries and the habits and demands of the citizenry at large. (Worldwide, fish provide more than half of all the animal protein humans consume.)[14] So the human factor pushing unsustainability grows, whether the fish do or not. It gets more intense as fish get fewer.

Hilborn's "human factor" favoring unsustainability was underscored in two separate incidents. In August 1993, Russia, for the sake of future sustainable yield, halted pollack fishing within its two-hundred-mile-wide territorial zone. It called on Japan, Poland, South Korea, and China to do the same for three years in the Okhotsk, an important fishing ground that straddles the Russian zone and international waters. Only Japan agreed.[15]

In May 1994, Canada granted itself the legal authority to seize foreign fishing vessels outside its two-hundred-mile limit in order to stop the depletion of cod, flounder, and redfish in the Grand Banks of Newfoundland. Although

---

11. Ibid.
12. Ibid.
13. "Fishing Fleets Are Battling over Territory along Coast," *New York Times,* March 5, 1995, 25.
14. "Fishing Fleet Trawling Seas That Yield Many Fewer Fish," *New York Times,* March 7, 1994, B7.
15. "Fishing Countries Split on Harvest," K4.

the law was popular at home, Canada was excoriated abroad for extending its authority into international waters, thus threatening to unravel the Law of the Sea treaty. "Canada is acting purely because it is driven to act," the architect of the law, Brian Tobin, replied. "We have conducted diplomacy down to the last few pounds of fish. We either continue to talk about it, or we try to save those last few pounds and use them as a basis for restoring the stock."[16] For its own sake, Canada could not let foreign fleets exhaust the future supply of the Grand Banks's stock.

Both these examples expose the difficulty of solving problems of the global commons via nation-state jurisdiction. More than that, all these dynamics together show that everything bends in the same direction — exploitation of the natural economy by the human one. Everything also seems to reach for one conclusion: the problem is not so much that nature needs to change or be changed, but that humans, as the most dynamic element of nature, do.

This is perhaps as convenient a point as any to inject the necessary history of ecosystems promised at the outset of the chapter. While this history switches examples from fisheries to forestry, it shows from whence our new doubts about nature arise and what consequences for changed human response are in order. That in hand, we will return to the fish stories.

CHANGED NATURE

The eighteenth and nineteenth centuries saw extensive deforestation in Germany. Alarmed at this, Germany — a nation that was then teacher to the world — turned to scientists. The question was whether scientific management of the forests was possible in such a way that harvests would match rather than overshoot growth. German scientists responded with an emphatic *Jawohl!* Not unimportant was that in this era "nature" was viewed in Newtonian terms; that is, it was viewed as controlled by a set of predictable cycles that could be clearly tracked and, where necessary, influenced through human intervention — the same picture of nature that led to agriculture as a full-fledged industry in the next century. German forests (in this case) would follow steady, reliable cycles subject to human management. Thus was born the notion of "sustained yield" and, with it, one version of "sustainable growth."[17]

Early in the twentieth century this management philosophy of sustainable growth found its public advocate in Gifford Pinchot. Trained in German forestry, Pinchot became the first chief forester of the U.S. Department of Agriculture and highly influential in one of the trademark ventures of Theodore

---

16. "Canada Acts to Cut Fishing by Foreigners," *New York Times*, May 22, 1994, L9.

17. This is a foreshortened version of the story told by Donald Worster in *The Wealth of Nature* (New York: Oxford University Press, 1993), 144–46. Worster includes the rather fascinating speculation that sustainability was "cultural" good news and thus well received in a Germany beset by instability and chaos. It promised ordering economic and social conditions on the basis of something as objective, secure, and "apolitical" as science.

Roosevelt's presidency — conservation. Because Roosevelt himself was a world leader here, Pinchot's enterprises became models for efforts elsewhere.

Pinchot was not the lone advocate, nor the first. His ideological predecessor was Bernhard Fernow, who also brought German forest management to the United States. Fernow and Pinchot shared a common philosophy, quite remarkable for its day. It spoke of both sustainability and development:

> The first great fact about conservation is that it stands for development, ... [not just] husbanding of natural resources for future generations ... but the use of natural resources now existing on this continent for the benefit of the people who live here now. ... In the second place conservation stands for the prevention of waste. ... [Third,] natural resources must be developed and preserved for the benefit of the many, and not merely for the few. ... Conservation means the greatest good for the greatest number for the longest time.[18]

Conservation management thus meant stewarded development of natural resources for the majority over time. At the same time, Fernow and Pinchot were utterly utilitarian in their view of nature and firmly committed to "economic progress" as "the primary goal of social life."[19] They had no argument with steadily increasing production as a good or with nature as a commodity to be harvested for the public welfare, elements retained in all later notions of sustainable development. They only insisted that the state provide professional management so that the resource base would be carefully maintained for present and future generations. Sustainable development thus had a clear meaning; it was a synonym for the scientific management of resources and the use of natural capital for socioeconomic good.

It was also a faith, a faith in human intelligence to understand, predict, and direct nature. Environmental historian Donald Worster even goes so far as to say that the only new element in the Brundtland Commission Report and other documents on sustainable development three-quarters of a century after Pinchot and Fernow is that their ideas have been extended "*to the entire globe.*"[20] "Now it is Planet Earth, not merely a beech forest, that is to be managed by trained minds, an eco-technocratic elite."[21] Whether Worster is entirely fair to Brundtland or not, he is certainly correct that for millions, especially Northerners, sustainable development means the long-term management of resources in an economy committed to growth and dependent upon scientific expertise to signal potential use as well as danger and limits.

---

18. Cited from J. Ronald Engel, "Sustainable Development: A New Global Ethic?" in *The Egg: An Eco-Justice Quarterly* 12, no. 2 (spring 1992): 4. The words are Pinchot's.

19. This is Worster's distillation of their common stance in *Wealth of Nature*, 145.

20. Ibid., 146.

21. Ibid. A sophisticated version of this proposal is Norman Myers, ed., *Gaia: An Atlas of Planet Management* (Garden City, N.Y.: Anchor Books, 1984).

This picture of nature in place, intact, and subject to scientific inquiry for the purpose of managed resource policy not only continued late into the twentieth century. It was taken up in modified ways into the dominant notion of the century's new science — ecology. In 1953, Eugene Odum published perhaps the most influential work of ecology ever written, *The Fundamentals of Ecology*. The earth, he and his pathfinders taught, is organized into interlocking ecosystems. Ecosystems exhibit a steady tendency toward order marked by a shift from competition in the early stages to increasing interrelatedness and cooperation. In this process they pass through a succession of traceable stages, and when they are mature, ecosystems achieve an equilibrium that Odum labeled "homeostasis." Homeostatic dynamics continue until decisive damage breaks the basic patterns. The chief culprits here, Odum thought, are human beings, at least those moderns who have viewed nature as, in his words, "a supply depot" rather than "the *oikos* — the home — in which we must live."[22]

Odum portrays ecosystem life as dynamic rather than static. Despite assumptions shared with Pinchot and Fernow, he and his disciples wandered far from the mechanistic models of Newtonian science and economics and joined the vast majority of scientists since Darwin who understand that nature has a long and lively history of dynamic, evolutionary change. Nonetheless, Odum and his generation of ecologists generally lined up behind the conviction that living nature, if not unduly interfered with by human beings, eventually attains an *equilibrium* where *production is at a steady rate*.[23] Furthermore, and whatever the deviation from Newton, this more recent presentation of nature, adjusted for evolutionary dynamics, is continuous with wider cultural legacies that reach into deep-seated premodern understandings that remain with us: the biblical view of creation's orderliness and dependability, the Greek sense of a rational cosmic *logos*, the medieval Christian metaphysics of harmony and order assembled as the great chain of being, and eighteenth-century rationalism itself. An orderly and predictable nature has thus had and still has a considerable hold in our heads; it belongs to our "think withs" as we "think about" things. This continues in popular ecology and in reigning notions of development. In fact it has *created* the popular notion of "sustainability" itself. Sustainability is basically "a matter of accommodating the human economy to [nature's] constancy and orderliness."[24] And scientists, we believe, can tell us where the dangers of human interference lie as well as reveal earth's capacities as a stock of resources and a sink for waste disposal.

Something happened to all this beginning in 1973. William Drury and Ian Nisbet published a major study on temperate forests. Unfortunately, almost nothing substantiated Odum and the going manuals on ecology. Drury and Nisbet's evidence argued that change goes on continuously, without reaching a point of equilibrium or stability. There isn't any noticeable progressive increase in the diversification of species or some greater cohesiveness among

22. Eugene Odum, "Strategy of Ecosystem Development," *Science* 164 (April 18, 1969): 266, as cited by Worster, *Wealth of Nature*, 160.
23. Worster, *Wealth of Nature*, 149.
24. Ibid.

plant and animal populations, or even notable success in regulating the environment all together. Rather, nature is *"fundamentally"* erratic, discontinuous, and unpredictable, . . . full of seemingly random events that elude our models of how things are supposed to work. As a result, the unexpected keeps hitting us in the face."[25] The scientists' conclusion is that "[t]he world is more complex than we ever imagined . . . and indeed, some would add, ever can imagine."[26]

Very strong orderliness at some levels (the sun's behavior, for example) may have no counterpart at other levels (local plant and animal populations). Nature is a many-splendored thing, we were certain. It turns out to be simply many-splendored and not *a* thing at all. More precisely, while there is a many-sided, complex integrity to creation, its patterns are far more dynamic and perplexing than either our pictures of nature or our imaginations themselves seem able to accommodate. And the more powerful our capacities to describe become, chiefly through more sophisticated technologies, methods, and information bases, the more complex, dynamic, and unstable nature renders itself.

In fact, the patterns may be more unstable than our psyches permit. Andrew Knoll's opening sentence in *Natural History*'s special issue on evolution is this: "We live on an ever changing planet, where stability — as much as humans might yearn for it — has no place."[27] Or, in the picturesque language of Jonathan Weiner, drawing from his minutely detailed studies of bird evolution: "Life is always poised for flight. From a distance, it looks still, . . . but up close it is flitting this way and that, as if displaying to the world at every moment its perpetual readiness to take off in any of a thousand directions."[28] The nature of nature, it seems, is change. Stasis has no tenure. The universe itself "is a restless adventure,"[29] the cosmos a pilgrim on a journey, with nature bent on more peregrinations than one of its expressions — namely, us — can ever know.[30] Creative turbulence, with no enduring equilibrium, appears to be nature's way.[31] In Diane Ackerman's nice terms: "Think of a niche and life will fill it, think of a shape and life will explore it, think of a drama and life will stage it."[32]

HUMANLY ORGANIZED UNSUSTAINABILITY

Nature as intricately interrelated, but too unstable to sustain sustainable development itself as most circles conceive that, brings us back to the fish stories.

---

25. These are Worster's words on Drury and Nisbet, not direct quotation. In ibid., 162–63, 168.

26. Again, this is Worster's summary of his study of the scientists' discussions, in ibid., 168.

27. Andrew Knoll, "Life's Expanding Realm," *Natural History* (June 1994): 14.

28. Jonathan Weiner, *The Beak of the Finch: A Story of Evolution in Our Time* (New York: Knopf, 1994), as cited in Douglas H. Chadwick, "Evolution, Right before Their Eyes," *New York Times Book Review*, May 22, 1994, 7:7; no page number given for the citation from Weiner.

29. John F. Haught, *The Promise of Nature* (New York: Paulist, 1993), 61.

30. An unusually good piece of graffiti in the bathroom at the North Country Co-op, Minneapolis, is worth including here: "And nothing that remains but doth suffer a sea of change into something rich and strange."

31. Berry, "The Meadow across the Creek" (unpublished manuscript), 188.

32. Diane Ackerman, "Worlds within Worlds," *New York Times*, December 4, 1995, op-ed page.

This time the attention is not to what is happening to nature and the fish, how-ever, but society and people. The question is what the fallout for sustainability is when our notions of nature and its predictability are upended.

Bumping up against natural limits and unpredictability about their location is, it turns out, costly in several senses. The first costs are literal. In 1992 the Canadian government declared a moratorium on cod fishing. The ban contin-ues indefinitely. Initially, $1.5 billion will be spent between 1992 and 1997 for a retraining program for thirty-three thousand unemployed Atlantic fishermen and fish-plant workers.[33] In a parallel case, Secretary Bruce Babbitt of the U.S. Department of the Interior in 1993 took the lead in negotiating a buyout in the amount of $400,000 for each of two Atlantic salmon seasons, 1993 and 1994. The $800,000 was paid to the Greenland Hunters and Fishermen's Organiza-tion, whose members forwent their right to catch 235 tons of salmon per year, itself down sharply from 957 tons earlier. The 235 tons was still too high in 1995, and the fishing crews simply had to stop fishing, period, to let the stocks revive. In the case of the Georges Bank shutdown, the U.S. Department of Commerce put aside $2 million in a pilot program to buy back boats from strug-gling New England fishers[34] while the State of Massachusetts asked the federal government to declare the state's fishing industry a "natural disaster" eligible for federal relief, admitting that while "a management plan to address overfishing" is clearly needed, "we haven't been able to put a finger on environmental factors."[35]

The U.S. and other national governments have often paid farmers and others not to produce. But the reason has been *overproduction*. Overproduction messes up markets. Moratoria on fishing are something quite different, quite new, and likely a portent of things to come. These industries are paid because of *depleted and endangered supplies*. How far sustainable development under these conditions is from the Truman-Rostow criterion for achieving it — that is, sustainable eco-nomic growth as measured by unlimited rising production and consumption — further boggles the already boggled mind. It may seem less distant from the Fer-now and Pinchot vision of sustained growth and sustainable yield at the hands of trained ecomanagers, yet predictability about yield is far more complicated than ever they imagined, and nature's limits in a saturated world evidently closer.

There are other costs. At first less tangible and pressing, they are no less real. For Canada, the immediate toll was unemployment in hundreds of coastal villages. The sinking feeling is that the correlative damage is a way of life gone, a way of being in the world. "Our culture is tied to fishing. Without fishing there is no culture," remarked Cecil Stockley, a teacher in the elementary school of Twillingate, Newfoundland, one of the hardest-hit villages. (A novel by Pulitzer author E. Annie Proulx, *The Shipping News*, speaks of these communities as "a joinery of lives all worked together.")[36] "I'm sure the town is going to survive," said Mayor Cooper, referring to Twillingate. "But in what capacity, that's not at

---

33. "Cod Are Almost Gone and a Culture Could Follow," *New York Times*, May 14, 1994, A3.
34. "Fishing Fleets Are Battling over Territory along Coast," 25.
35. "Massachusetts Seeks to Help Fish Industry," *New York Times*, March 22, 1995, A13.
36. Cited in "Cod Are Almost Gone and a Culture Could Follow," A3.

all clear. I hope not as a museum."[37] Brian Tobin, Canada's minister of fisheries and oceans cited earlier, warned in a UN speech that the entire Grand Banks, and not only cod stocks there, "may become a marine desert" if "tough measures" aren't taken internationally. "This is an ecological disaster," he said, but "it is also a societal calamity. What do you do if your life, your family and your community are all linked to the fishery, but there are no fish?"[38]

Yet even unraveled ways of life are not all the fish stories tell. The stories have other dimensions, submerged in the political economy of overfishing. A discussion in *The Ecologist* surfaces further reasons for humanly organized unsustainability.

"Quite simply, there are too many people chasing too few fish," author James McGoodwin is quoted as saying about the "deforestation of the deep."[39] Lawrence Juda, chair of the Department of Marine Affairs, University of Rhode Island, uses exactly the same words.[40] But are McGoodwin and Juda and this commonsense analysis right? Do the numbers ("too many people" / "too few fish") generate the real-life equation? Are there too many boats out there with too many people fishing from them, as well as too few fish for all those lines? Are the systemic reasons for overfishing revealed by this analysis? Naming the problem "too many boats" or "too many people" lumps all fishing interests together and obscures differences. It doesn't answer the question, for example, of *whose* boats are too numerous. Is it the three million canoes, skiffs, and workboats that "provide a living for about 20 million fishers and their families," most of them in the South?[41] Or is it the few thousand highly capitalized factory ships of the industrial fishing corporations whose disproportionately large catches are destined as much for fishmeal (animal feed) as for human consumption? And can the lines of the twenty million fishers and their catches be compared, from earth's point of view and its fish, with the trawler fleets' nets, scooping up shoals of fish, without discrimination, in sweep after sweep?[42]

What about "too few fish"? Too few fish for whom, and for what? Too few for the industrial fishing fleets? Too few for the millions dependent on the twenty million nonindustrial fishers? (The latter have seen their per capita consumption of fish decline in recent years.) Too few to supply the market in the North, where levels of consumption of fish have increased and are three times those in the South and where some, such as Japan, have doubled fish consumption since 1950?[43] These are definitions of scarcity ("too few fish") generated not by the number of the fish in the ocean but by socioeconomic expectations and dynamics.

37. Ibid.

38. "U.N. Envoys Fear New Cod Wars as Fish Dwindle," *New York Times*, March 20, 1994, K6.

39. From the introduction by the editors of *Overfishing: Causes and Consequences*, a special double feature of *The Ecologist* 25, 2/3 (May/June 1995): 42.

40. "Fish Don't Observe the 200-Mile Limit," *New York Times*, April 8, 1995, 22.

41. Introduction to *The Ecologist*, 43.

42. Ibid.

43. Ibid., 44.

If the causes, rather than the fish, are chased further, layer after layer of the Big Economy and its globalizing ways are lifted. We need not uncover each in turn, only list some of the factors. In addition to the patterns of consumption that fuel the trade in fish we have just noted, there is the use now of highly mobile companies that "pulse-fish" the world's waters. They do so in much the same way companies play world markets — that is, move in where the fishing is good, then pull out when it turns bad. The effect is to plunder, then leave, without responsibility for preservation or restocking. Too, the factor we noted earlier — harvesting fish in quantities sufficient to pay for highly capitalized boats and gear — is a steady pressure on supply and marketing. To it is added the pressure on governments to subsidize national fisheries and combat the indebtedness, sometimes bankruptcy, some ailing fisheries face. (Malaysia recently cut its support for inshore fishers and expanded its subsidy for its industrialized deepwater fleet.) In short, the number of boats *and* people is being cut, while the size and technical efficiency of the remaining boats increase. This is not "too many people chasing too few fish." It is the unsustainable practice of a limited number of certain kinds of boats harvesting too many fish for certain markets. The reasons have to do with fishing as an industry in the global economy, not with the needs of peoples and communities traditionally tied to fishing or even their mode of fishing.[44] From a moral point of view, then, "too many people chasing too few fish" is a mystification of the injustice of current global economic and environmental dynamics. Wealth is generated at one pole, and poverty and deprivation of communities and resources are generated at another. If our earlier analysis is correct (recall the Gowdy discussion),[45] then the sustainability of societies and communities is threatened by these quintessentially socio-ethical factors. The "great instabilities of injustice, unpeace, and the disintegration of creation" (Charles Birch) are at play. "Too many people chasing too many fish" obscures the actual dynamics.

CONCLUSIONS

What, finally, shall we say about sustainability in light of all this? Three things, for starters.

1. Sustainability will have to find its way with uncertain sciences trying to keep up with changing and unfinished nature. Nature is, unfortunately for planners, not a fixed, designed system but an evolving one. Ecosystems do comprise "wholes." But they do not have clear thresholds so much as loose gradients. They are communities of communities — water chemistries, soil types, light and atmospheric conditions, complex plant and animal lives — that interface and interact in such a way that "there is rarely a point where you can say with certainty

---

44. Ibid., 43.
45. See the chapter "Sweet Betsy and Her Avalanche," above.

that one stops and the next starts."[46] At what point, for example, does the soil stop affecting the air, and vice versa? While nature certainly does have absorptive capacities and limits, as well as regenerative capabilities and limits, we cannot use notions of fixed thresholds and settled systems as a measure. Of course, there are limits and stabilities we can count on. Trees grow to a certain height and then stop. Organisms thrive on certain kinds and levels of nutrients. Humans need oxygen in precise balance with nitrogen, argon, and carbon dioxide. Nature knows about "enough" and "the right amount" in regular patterns. Nonetheless, fixed thresholds and settled systems of interacting communities don't exist in a way stable enough for confident human prediction over the long term.

2. Sustainability not only is as complicated as human society and nature but is as complicated as society and nature knit and knotted together in ways large and small, with smaller margins for error now than those we assumed and lived by when the planet was large and flat. Increasing human influence is a part of biospheric dynamics themselves on a contracting planet.

3. While major technical and scientific changes are needed, the even more crucial and difficult changes are human. (The habits of some humans more than others, to be sure.)

These call for comment.

Regarding (1) and (3), the key idea of a normal "yield" from the natural economy is rendered ambiguous, as is carrying capacity. We are not even certain what "normal" itself means when applied to ecosystems and nature as a whole. It certainly means less assurance of results than "normal" does on the wash cycle! Ecosystems are not as harmonious as once we considered them. They are not "the orderly flow of energy through the food chain,... in timeless equilibrium, no struggle or imperfection or failure."[47] Rather, biology textbooks now present ecosystems and nature in toto as "unfinished and flawed."[48] (Remind you of anyone else you know?) But if we are uncertain what normal means, how do we "develop" in a sustainable way, *except and unless* we do so with a view to far humbler powers of prediction, far wider tolerances of error, and far less aggressive treatment of earth? To play the margins close is to compound social and political conflict as well as tease tragedy. Yet playing the margins close is inevitable in our world, given steep population climbs, pressing needs, deeply ingrained, wasteful ways of life, and dominant economic dynamics.

Regarding (2) and (3), the scientists cited earlier reluctantly became psychologists and advised that human attributes such as "the hunger for wealth and the tendency to disregard long-term consequences" be considered "*as part of* the natural system to be studied and managed."[49] How fish behave is only one element of nature in the matter of sustainability. How fishers and consumers and voters and peoples and their nations behave is another, and at least as critical.

---

46. William Ashworth, *The Economy of Nature: Rethinking the Connections between Ecology and Economics* (New York: Houghton Mifflin, 1995), 211.
47. This is Worster's summary of the earlier presentation, in *Wealth of Nature*, 39.
48. Ibid.
49. "Biologists Fear Sustainable Yield Is Unsustainable Idea," C4; emphasis added.

All the human attributes, including ignorance, the propensity for self-deception and group aggrandizement, and the compulsions of both affluence and survival, are *part of* the totality *of nature* to be taken into account, since human culture is the single most fateful force for earth as a whole.

Said differently, there is no outside place to stand and no "scratch" from which to begin. All is within. We had best abandon the Western notion that we can "develop" as though we are external to a nature we understand and as though we can then alter it in accord with our hopes, dreams, and appetites. Rather, all the human factors *are part of* that comprehensive nature that must be taken into account.

To state the significance in yet another way: not just knowledge (*scientia*) but wisdom (*sapientia*) and the Psalmist's contrite heart and humble spirit are requirements of sustainable community itself. Issues of sustainability are as much dispositional and ethical as they are technical.

In short, sustainability is, like Alice's experiences in Wonderland, far "curiouser" than we thought. We had enough to contend with: ecologically empty economic theory and practices, appropriate scale and a miserably feeble conceptual apparatus for thinking about it, totally upended culture and nature because of the last few hundred years of aggressive commerce and conquest, exploding population numbers despite declining birthrates, and the need for safety nets for all species, including humans, on a planet under assault from its own vaunted "masters" and "stewards." We did not need an uncertain science, much less uncertain nature. But we have both. Uncertain science chasing uncertain nature.

So can we say anything? Can we describe sustainability as a positive set of conditions? Are there characteristics and norms, including moral norms, by which to identify sustainable development or sustainable community and distinguish it from the unsustainable? If we draw from foregoing chapters in toto, the doubts and fights included, can we sketch elements that might prove a meeting ground for people deeply at odds with one another yet stuck with one another on the same boat, under house arrest on that Ark of Life, the Good Ship Oikoumenē?

While details for policy direction and the supporting ethic are postponed until part 3, a significant summing up can be offered before then.

Conclusions

Sustainability, we said, is the capacity of social and natural systems to survive and thrive together, indefinitely. What does that require?

- Sustainability means thinking sideways and around corners, not just up and down in hierarchies, or forward and back in our usual sense of time and history. Too few of us are good at thinking in rounded ways and proceeding on the assumption that when we try to pluck something from the rest it is already hitched to all else. Thinking inside multiple connections does not mean we cannot act, only that the subject of thought is the entanglements of relevant wholes rather than pieces and parts. Tinkertoy thinking does not for sustainability make. Whole-systems perspectives are required.

- Sustainability thinking means latitude for both error and adaptability. Vice President Al Gore writes early on in *Earth in the Balance* that "ecology is the study of balance, and some of the same principles that govern the healthy balance of elements in the global environment also apply to the healthy balance of forces making up our political system. In my view, however, our system is on the verge of losing its equilibrium."[1] "Balance" and "equilibrium" are, we have seen, deceptive metaphors for perceiving and conceiving nature or society, much less nature and society together. Gore is closer to the point when he says that "civilization is not a frozen image; it is in constant motion."[2] This includes nature as part of, and encompassing, civilization. Even Aldo Leopold might have been closer to what we now understand as earth's ways had he substituted "dynamics" for "stability" in his ethical baseline: a thing is right when it serves the "integrity, [dynamics,] and beauty" of a biotic community and wrong when it does otherwise. But the larger point is that the dynamism of ecosocial communities requires room for changes that allow adaptability without catastrophe and an elasticity tolerable for the (uncertain) evolution of most creatures. Calibrating balances and equilibrium finely is not wise.

This point is served by returning to *the* big rub, the Big Economy with the Great Economy, and there exegete three images: cowboy economics, spaceship earth, and day care. Cowboy economics is an image of Kenneth Boulding's, himself a renegade economist who once panned his profession by remarking that

---

1. Al Gore Jr., *Earth in the Balance: Ecology and the Human Spirit* (Boston: Houghton Mifflin, 1992), 11.
2. Ibid.

anyone who believes we can have infinite growth on a finite planet (the cowboy illusion) is either a madman or an economist! Boulding's alternative borrowed Adlai Stevenson's image, spaceship earth. Earth is a closed system and we're the crew, said Boulding. We live in a small capsule where everything circulates in feedback loops, and we must recycle it all, without remainder. We'll achieve sustainability when we order everything as one must in a closed system. Spaceship earth is thus Boulding's effort to dramatize the radical interdependence that makes sustainability a requirement for continued existence. In short, everything goes somewhere; nothing goes away; so deal with it as if the consequences of your actions are going to hang around indefinitely.[3]

On one level, the contrast of cowboy economics with spaceship earth could hardly be more dramatic. Overall scale is negligible for cowboys; but for spaceship earth the scale of economy to ecosystem is total. They are one and the same. The human economy grows, so to speak, to match the biosphere.

Yet as a contrast to cowboy economics, does spaceship earth adequately picture sustainability? For long-haul sustainability (the only kind that qualifies!) it does not, despite its capacity to mirror ethical responsibility by showing the reach and impact of our actions. The failure is the one intimated: there is little margin for error. Indeed, a very large part of *un*sustainability now is precisely that the human economy pushes the unforgiving boundaries of nature's economy.

Ecologists and economists picture this encroachment with an important index. It measures the percentage of human appropriation of the total world product of photosynthesis. Photosynthesis is key because green growing things hold the trump card. They can exist well enough without us; we cannot exist at all without them. "Net primary product" (NPP) is the ecological-economic concept here. NPP is "the amount of solar energy captured in photosynthesis by primary producers, *less* the energy used in their own growth and production."[4] NPP is thus the basic food source for everything on earth that is not capable of photosynthesis.

Ecological economists calculated in 1986 that humans were then appropriating 25 percent of potential NPP. This included both aquatic and terrestrial production. Since humans use relatively little ocean product, the more significant figure is the human appropriation of 40 percent of terrestrial NPP. This means only somewhat more than one doubling is arithmetically possible before total human capture of net primary production! Since humans cannot live without ecosystems, made up of innumerable other species, the doubling cannot happen. Of course the effort can be made to increase total photosynthesis. More green space on earth can be created. But while theoretically possible, increased green

---

3. Kenneth Boulding, "The Economics of the Coming Spaceship Earth," in *Environmental Quality in a Growing Economy,* ed. Henry Jarrett (Baltimore: Johns Hopkins University Press, 1968), 3–14.

4. Herman Daly, "The Economist's Response to Ecological Issues," in *Sustainable Growth: A Contradiction in Terms?* report of the Visser't Hooft Memorial Consultation (Geneva: WCC Publications, 1993), 43.

space runs counter to economic growth, which has steadily increased the "take" of NPP by humans. It runs counter to increasing the standards of living for more people (read: higher consumption) and counter to meeting the basic needs of increasing numbers of people overall.

The point is that spaceship earth is nearly as faulty an image as cowboy earth. It renders the planet one vast human-controlled machine in which the human life-system and earth's coincide and collide. With virtually no room for major breakdown, there are greater chances of it. Even minor flaws, like faulty O-rings, can be major ones for spaceships. This is yet another reminder of Liebig's law of the minimum: carrying capacity is not ultimately determined by general conditions (i.e., most things working most of the time). It is determined by the single crucial factor in short supply in a given, interrelated ecosystem.

There are other fault lines for spaceship earth. All mind and spirit are arrogated to human beings. (The rest of) nature is mindless, soulless, and shorn of any true life of its own. It is thus bereft of moral and religious consideration and claim. Too, "singular man" (the crew) faces "singular nature" (the ship),[5] a notion that runs full grain against the diversity that generates evolving life and permits the adaptability essential to sustainability. In short, the threat to sustainability is not overgrazing, as in the cowboy economy, but the illusions of planetary management by a species not known for fine-tuned perfection, yet in charge of earth's primary production. When human appropriation of nature's economy pushes the natural boundaries hard, as it does more and more at human hands, sustainability is rendered more and more precarious. A spaceship is a bad guide indeed. So is its mind-set, mirrored in a *Scientific American* article:

> It is as a global species that we are transforming the planet. It is only as a global species — pooling our resources, coordinating our actions and sharing what the planet has to offer — that we may have any prospect for managing the planet's transformations along the pathways of sustainable development. Self-conscious, intelligent management of the earth is one of the great challenges facing humanity as it approaches the twenty-first century.[6]

Great challenges we have! But day-care earth is a better image than spaceship earth. After all, it is not so much the planet that has gone haywire. A used-lemon-of-a-spaceship is not the basic problem. We are. Wrongheaded notions, human numbers, greed, arrogance, ignorance, prejudice, and stupidity are what most need to be taken into account. And they are not much given to cure by spaceship management. Thus David Orr suggests that the needed orientation is "more akin to child-proofing a day-care center than to piloting spaceship earth."[7]

---

5. The terms are from Raymond Williams, *Keywords: A Vocabulary of Culture and Society* (New York: Oxford University Press, 1983), 188–89, 220–21.

6. William C. Clark, "Managing Planet Earth," *Scientific American* 261 (September 1989): 47.

7. Cited from Daly, "Economist's Response," 45.

In day care, space is secured in which the child is both free *and* protected from the excesses of its own raw and raucous freedom. The electrical circuits cannot be chewed; dangerous objects are placed out of reach; the schedule allows for the unexpected and unpredictable. Using our earlier lingo, the human scale of appropriation is kept light enough to leave the Plimsoll line above water. We don't, can't, and shouldn't control all the actions of the child; nor do we, can we, or should we the actions of all ecosystems and one another. The human percentage of NPP is kept low enough for the independent life of those very systems our own lives depend upon. Which is to say that like day care, we remove grave dangers and give wide margin for errors and accidents that will not, *because* there is wide margin, be fatal for most of the kids, their caretakers, and their environment. Theologically stated: day care gives sin and error room; spaceship doesn't. And the habits of sin and error are quite deleterious enough without being pushed to do even "better"!

Of course, like all analogies, day care is limited. Some dimensions don't work. A book urging extended moral responsibility (like this one) does not desire to cast most persons in the role of children and all responsible care in the hands of a separate, gnostic elite. But for the present point — sustainability's requirement of dynamic adaptability, with margin for inevitable error — day care works better than cowboy economics, which is still in the saddle from the point of view of global economic practices, and spaceship earth, which is the going management theory in the mode of the information revolution.

Sustainability is, in view of all this, a process with a beginning but no end; and in considerable measure, it is a social construct. It requires, absolutely, recognizing and respecting ecological integrity. It also requires a human vision for nature's duration, on terms hospitable to us and millions of other creatures. It is thus a matter of human imagination and dreaming as well as concrete technologies, tasks, and policies. Among other things, this means that all those questions of participation in the play of human power are crucial to sustainability: who gets to sit at the table, whose knowledge and experience count, who casts what kind of votes, who represents future generations and those of the present generation who cannot speak for themselves, whose values and character carry the day. These questions are vital because "sustainable" modifies "environment" and "society" together. Both environment and society, together, are dynamic processes without foreseen ends, processes depending heavily upon human constructs of both heart and mind, indeed upon some modicum of justice and peace, together with sober respect for the integrity of creation.[8]

Such processes as these, and questions of power and its play in society and nature together, mean the ascendancy of ethics in matters of sustainability. So we finish by lifting the implicit moral norms from the discussion to this point, as norms of and for sustainability.

---

8. This paragraph is indebted to the paper of Stephen Viederman, "The Economics of Sustainability: Challenges," 3–4 (the paper is available from the Jessie Smith Noyes Foundation, New York City). Used with permission.

Participation as the optimal inclusion of all involved voices in society's decisions and in obtaining and enjoying the benefits of society and nature, together with sharing their burdens. All primary "stakeholders" have opportunity to articulate their needs, propose and consult on projected solutions, and be part of considered implementation.

Sufficiency as the commitment to meet the basic material needs of all life possible. This means sufficiency for both human and otherkind's populations. For humans it means careful organization of the economics of borrowing and sharing, with both floors and ceilings for consumption. A sufficiency revolution must accompany the eco-efficiency one called for by the United Nations and business. Eco-efficiency is a first step into sustainability. Eco-sufficiency is a basic requirement.

Equity as basic fairness is the cousin of sufficiency. Both distributive and procedural justice are needed as means to ensure global equity among nations and societies, biotic equity among species, equity across generations stretching into the future, and equity between women and men. Such equity is necessary to assure that sufficiency levels are attained and sustained.

Accountability as the sense and structuring of responsibility toward one another and earth itself, carried out in ways that prize openness or "transparency." (Accountability is sometimes considered a dimension of participation, but here the character of that participation is made clear. People know who and how decisions are made, and there are structures and procedures for holding decision makers accountable.)

Material simplicity and spiritual richness as markers of a quality of life that includes bread for all (sufficiency) but is more than bread alone. Negatively stated, major disparities of wealth and poverty generate instability and unsustainability. So do soul-barren cultures.

Responsibility on a scale that people can handle. This is an argument for actions commensurate with workable community. It is also an argument for technologies whose consequences are more apparent rather than less, are smaller in their range of impact than larger, and are subject to alteration and correction without vast disruption.[9]

Subsidiarity as the means of participation most attuned to earth and the scale proper to *responsibility*. Subsidiarity states that problems should be resolved at the closest level at which decisions can be taken and implemented effectively. Negatively stated, one should not withdraw from persons and their communities and commit to larger entities that which they can accomplish by their own enterprise and means. Larger and higher bodies should not perform and

---

9. This listing of norms is a composite drawing from WCC discussions as well as the Presbyterian Church (U.S.A.) draft of "Hope for a Global Future: Toward Just and Sustainable Human Development" (report of the Task Force on Sustainable Development, Reformed Faith, and U.S. International Economic Policy, August 8, 1995).

provide that which can be performed and provided by smaller and subordinate bodies.[10]

Adding these moral requirements to preceding discussions, we might summarize the whole by recognizing that sustainability necessarily involves the total earth-human process. It encompasses the requirements of sustainable environments, sustainable societies, and sustainable livelihoods, economies, and ways of life. While the full range of human activities is thus engaged, together with a certain set of moral values, virtues, and obligations, *all must function as a phase of earth economics and earth community.* Any millennial drive of human power married to the conviction that we somehow transcend the natural community of earth and need not worry overmuch about otherkind's claims can only be deadly for all.[11] Earth community is basic.

With a view to the human task, all this points to whole-systems approaches that address overconsumption, population numbers, and inequality simultaneously. These are interlinked, and ways must be found to bring human society into balance with the earth and its economy — household by household, locale by locale, region by region.[12] If we turn to the planet's three socio-ecological classes, the 1.1 billion desperate, the 3.3 billion managing poor, and 1.1 billion overconsumers, sustainability means huge shifts. Roughly 80 percent of environmental damage is caused by the overconsumers and lives organized around cars and planes, high-caloried meat diets, prepackaged and disposable products, single-family dwellings, and wasteful ways. This must shift in a direction more in keeping with the 3.3 billion who occupy more modest, but adequate, quarters, whose diet is a healthy one of grains and vegetables and some meat, whose travel is via public transport and bicycles, and who use unpackaged, local goods while practicing extensive recycling. Above all, the lot of the 1.1 billion absolutely deprived must be addressed, with their nutritionally inadequate diets, contaminated water, rudimentary shelter, scraps for clothing, endemic disease, and daily insecurity.[13]

Such is only the barest orientation, but perhaps this suffices for the moment. Part 3 will add detail to this. And perhaps the entire discussion thus far suffices as an Earth Scan, an initial reckoning of where we are and how we got here. The next undertaking is to work out that which earth requires of us in the way of an alternative orientation to life, an orientation profound enough to constitute a faith to live by. Given the foregoing Earth Scan, then, what Earth Faith answers to it, for the sake of life? What manner of leaning into life bends us toward the fourth human revolution?

---

10. This paraphrases the wording in the encyclical of Pope Pius XI, *Quadragesimo anno,* 79, as cited by J. Philip Wogaman, *Christian Ethics: A Historical Introduction* (Louisville: Westminster/John Knox, 1993, 213.

11. The discussion echoes Thomas Berry's in "The Meadow across the Creek" (unpublished manuscript), 152–59.

12. David C. Korten, *When Corporations Rule the World* (San Francisco: Berret-Koehler; West Hartford, Conn.: Kumarian Press, 1996), 35–36.

13. All this is elaborated in some detail by Alan Durning, "Asking How Much Is Enough," in Lester R. Brown et al., *The State of the World, 1991* (New York: Norton, 1991), 153–69.

On a chance visit to the little village of Sdeh-Boker ("Herder's Field") in the Negev Desert, David Ben-Gurion, a founder and the first president of Israel, made a sudden, astonishing decision. A leader at the peak of his power, he resigned from government and took up residence in this desert village in order to do land reclamation. He lived out his days in Sdeh-Boker and is buried in the Negev. He also enunciated his credo in that reclaimed place:

> The energy contained in nature — in the earth and its waters, in the atom, in sunshine — will not avail us if we fail to activate the most precious vital energy: the moral-spiritual energy inherent in man; in the inner recesses of his being; in his mysterious, uncompromising, unfathomably, and divinely inspired soul.[1]

Part 2, "Earth Faith," is about renewable "moral-spiritual energy" in the "inner recesses" of humanity's "mysterious, uncompromising, unfathomably, and divinely inspired soul." Not moral-spiritual energy in the abstract, or as the worthy and seductive subject of itself, but as directed to an earth-honoring faith. The goal is to find our bearings anew, this time as an exuberant expression *of* nature that reclaims nature, ourselves included. Sustainability, another name for the future, requires it.

The reasons we need new bearings are essentially practical. The most basic is that earth cannot industrialize but once in the manner and on the scale it has. The bulging, throbbing modern world cannot be replicated and extended indefinitely. For one thing, costs cannot be met. Maintaining what the world already has is driving enough communities into poverty, bankruptcy, even destitution, to say nothing of what quadrupled or octupled infrastructure and consumption would exact. Nor are natural resources present in the abundance or availability they once were. Even allowing for human ingenuity and material substitutes yet undreamed of, a single factor such as the end of the petroleum era, the lack of further fertile lands for all manner of productive uses, unmeetable demand for freshwater, or substantially altered climate will yield huge resource-related problems. Then there is population, a world of first six, then eight, then ten billion within the lifetime of a child born today. Whatever else these numbing numbers mean, they are a sure multiplier of all other problems, from destitution, poverty, unemployment, and refugee agonies to overconsumption, resource depletion, and destruction of habitat. Not least, psychic energy is largely spent.

---

1. Cited in Daniel Hillel, *Out of the Earth: Civilization and the Life of the Soil* (Berkeley: University of California Press, 1991), 8.

The bright side of the agricultural, industrial, and information revolutions was their lure and drive. Now, facing the disintegrative downside of modernity on a pitched scale, a palpable global fatigue wears us away. We seem less willing to make hard demands upon ourselves and sacrifices for future generations.[2] One of the requisites for moral-spiritual energy such as Ben-Gurion's, then, is a different sense of the universe and our place in its processes, another way of existence with symbols powerful enough to guide the cumulative powers we exercise; in short, a different earth faith issuing in an ethic appropriate to the altered world. An earth faith narrative, or several, is needed, together with other senses of who we are and what we ought to do in matters both grand and pedestrian. This practical mythic vision is what part 2 strives toward, ambling among dreams that might guide the action.

What is rejected is the Promethean self and world forged and formed by the ethos and ethic of Western messianism, the ways of those E. F. Schumacher once tagged "the forward stampede people." Sought are the ethos and ethic appropriate to healed and transformed "homecomers."

There is a turn to the religious here. The reasons are many. Not least is something deep down that links culture to cultus (worship). Both derive from *colere*, meaning to attend to, respect, worship, take care of, till, and *culti*vate. Evidently some primal sense, a religious sense, is convinced that tending nature, soul, and society must all be done together, all at once. But in the final analysis, the basic reason for the turn to the religious is simple realism. Humans are by and large incorrigibly religious, and have been ever since that warm night aeons ago when the first baby was lifted on strong dark arms toward a full moon in the awesome presence of waiting gods. Whatever the wishes of the cultured despisers of religion, as a species we yearn to see things whole and sacred. We insist on telling a cosmic narrative and locating ourselves somewhere in it. Something primordial wants assurance that we are supported by the same powers that brought earth into being and threw sun and stars into orbit. And something wants to name with the name of God "the fathomless mystery that surrounds the burning mystery of our own lives."[3]

This is a report, not an endorsement. Religion is not ipso facto "a good thing." Religions and other faiths (chauvinist nationalisms, for example) have been, can be, and are demonic as well as redemptive. The holy destroys as well as saves. Lucifer is dressed out as an angel of light, and what Goethe's Mephistopheles calls "the cruel thirst for worship" can wreak unspeakable horror. The world within, the world of the spirit, is no sure guide for the world without. It, too, can deceive and defeat. This only underlines, once again, the ascendancy of ethics. Religious impulses are as subject to measure by a moral plumb line as anything else that beats in the human breast and issues from human hands. Ben-Gurion's "moral-spiritual energy" of this "mysterious" and "divinely inspired

---

2. This discussion of reasons is taken from Thomas Berry, "The Meadow across the Creek" (unpublished manuscript), 171–72. Used with permission.

3. Elizabeth A. Johnson, *She Who Is: The Mystery of God in Feminist Theological Discourse* (New York: Crossroad, 1993), 10.

[human] soul" must be judged by its outcomes. "By their fruits you will know them" is the wording from one of Ben-Gurion's ancestors.

Still, while Earth Faith is the subject now, and is incurably religious and moral, the stimulus is less the nature of human nature than the prospect of intensified unsustainability. These chapters follow, then, from the analysis of part 1. Assuming that analysis, these chapters take up the matter of getting our bearings anew so that earth is not damned by our presence but rejoices in it.

That can be done in numerous ways. The one chosen here is to highlight symbols that might key a whole pattern, a pattern that makes for a viable, sustainable way of life. Such symbols can communicate a mystique, a mystique that discloses how the ultimate mysteries of existence are manifest in the universe. These symbols ought not to disparage intellectual and scientific quests, but aid the imagination's quest for the highest levels of comprehensive knowledge. And they ought to foster critical competence, not least the moral discernment and judgment best befitting care of an embattled earth.[4]

Symbols are chosen because all people "symbolize." We see and speak and live through symbols. People do not stir their whole being and die for rational scientific truths. Something filled with the senses and carrying nature in forms that grasp us is needed. Something from without that resonates with worlds within. Symbols do that. They do that not least because nature itself — stars, sun, water, animals, plants, other people — lives within us as emotional patterns and as archetypes our ancestors' ancestors also knew.

Of course, symbols, too, are subject to ethical assessment. They, too, must be judged. But the point for the moment is the one made by Michael Thomas. What he says of art is true of symbols more broadly: namely, that they "take over when observed and documented fact hits the wall, and existential or metaphysical truth still lies somewhere on the other side."[5] We can document earth's distress without end and only hit the wall. In fact, we have done that. Beyond our Earth Scan, what is needed are forms of insight and illumination that take us to the other side and show the way, forms that bridge from the expressible to the ineffable, forms that both specify and suggest. Ethical reflection's own initial task may in fact be no more or less than connecting human experience and empirical reality to those symbols that make discernment possible in the first place, including the discernment of salient facts. No symbols, no discernment.

Hope stirs in the chapters that follow. On one level, it is protesting hope in the manner of St. Augustine: "Hope has two lovely daughters, anger and courage. Anger, so that what must not be, shall not be. Courage, so that what must be, shall be." Yet hope's source is not in anger and courage, and its end is not protest. Nor does it arise from historical optimism and sure confidence about earth's next journey. On the contrary, *Global 2000 Revisited* is probably

---

4. This paragraph draws from Brian Swimme and Thomas Berry, *The Universe Story: From the Primordial Flaring Forth to the Ecozoic Era* (San Francisco: HarperCollins, 1992) 4, 223, 229.

5. Michael M. Thomas, "'Depressing' Schindler?" *The New York Observer* 8, no. 10 (March 14, 1994): 1.

right. Knowing what to do and making sound choices are particularly difficult now because

(a) there is some scientific and economic uncertainty about the severity of the difficulties ahead, (b) it is difficult to believe that such major unprecedented change can be occurring, (c) it is generally thought to be easier to adapt to whatever comes than to make changes in advance of necessity, (d) there is widespread lack of awareness of what is happening, (e) the steps which must be taken are extremely difficult, and (f) we lack a set of common moral values on which to base collective action.[6]

"Most difficult, however," the paragraph concludes, "is to accept that our concept of progress has failed."[7]

Hope here, then, is not official optimism, and it is not the confidence of the agricultural, industrial, and informational revolutions. On the contrary, we are all unsteady and uncertain in an age of grave threats to existence itself. T. S. Eliot's questions are searching and formidable: "Where is the life we have lost in living? Where is the wisdom we have lost in knowledge? Where is the knowledge we have lost in information?"[8] Rather, hope wells up from within the same irrepressible nature that renders us religious. Just as nature itself seems to abhor a vacuum, so human history in nature resists dead-ends and demands a resurrection. This doesn't mean progress. But it does mean charged creativity of mind, heart, and hand. It is creativity in urgent search of life. It is Ben-Gurion's inspired soul on fire. In religious tongue, it is hope against the death of hope itself and faith in a God (or Reality) "who gives life to the dead and calls into existence the things that do not [yet] exist" (Rom. 4:17). For the moment, that is enough.

---

6. Gerald O. Barney with Jane Blewett and Kristen R. Barney, *Global 2000 Revisited: What Shall We Do?* (Arlington, Va.: Millennium Institute, 1993), 4. *Global 2000 Revisited* was prepared as the "critical issues of the 21st century" document for the 1993 Parliament of the World Religions in Chicago on the occasion of its centennial.

7. Ibid.

8. Cited from Charles Birch and John B. Cobb Jr., *The Liberation of Life* (Denton, Tex.: Environmental Ethics Book, 1990), 7.

Midges and Cosmologies

Those soaring marble inspirations, the Lincoln and Jefferson Memorials in Washington, D.C., are eroding. Sex, the food chain, and the habits of the National Park Service are slowly doing them in.

Each evening at sunset, small insects called midges congregate at the memorials to mate. At about the same hour, automated floodlights, triggered by dusk, wash the marble in yellow. Excited midges cannot resist. Hormones in drive, they crash by the thousands against the walls and columns. Spiders, in turn, are drawn to the instant protein. They soon fill crevices here and there with themselves, their webs, midge bits, and droppings.

In no time the word is out, and birds chase the food chain into spider heaven. Then they defecate with abandon in and on the temples. The Park Service is left with undigested midge bodies, spider feces, and bird debris. Perhaps worst of all, the detergents required to clean the mess slowly eat the marble as well, abetting the gradual erosion.

Fortunately, a solution exists. Wait an hour after sunset before tripping the light switch. The midges go elsewhere. So do the spiders and birds. The temples are clean and serene, ready for tourists and cameras. Simple.

There *are* technical solutions to ecosocial problems. They must be vigorously pursued. We will need all the earth-friendly technologies we can muster. At the same time, and several planes deeper, appropriate technology may not even be conceived, much less pursued, if reigning worldviews (or "cosmologies") are haywire. Technologies express cultures. Ways of doing things reflect ways of seeing things.

Consider the following:

If you put God outside and set him vis-a-vis his creation and if you have the idea that you are created in his image, you will logically and naturally see yourself as outside and against the things around you. And as you arrogate all mind to yourself, you will see the world around you as mindless and therefore not entitled to moral or ethical consideration. The environment will seem to be yours to exploit. Your survival unit will be you and your folks or conspecifics against the environment of other social units, other races and the brutes and vegetables.

If this is your estimate of your relation to Nature and *you have an advanced technology*, your likelihood of survival will be that of a snowball in hell. You will die either of the toxic by-products of your own hate, or simply of over-population and overgrazing. The raw materials of the world

181

are finite. If I am right, the whole of our thinking about what we are and what other people are has got to be restructured.[1]

These are not a theologian's, philosopher's, or ecologist's sentences, but an anthropologist's. They were not written in 1992, but 1972, by Gregory Bateson. They speculate — and warn — that a particular religious worldview, and a way of life in keeping with it, when armed with the powers of modern technology, will do us in. The intriguing matter is Bateson's assumed connections. He assumes the reality and efficacy of a cosmology and its ethic: "the whole way of thinking about what we are and what other people are." And he assumes that earth's present distress is a foundational challenge to the reigning cosmology. Its consequences expose this cosmology and ethic as death-dealing. Bateson concludes that an alternative worldview must be put in place.

Bateson is right. Others concur but express it differently. "Nothing less than the current logic of world civilization"[2] runs counter to the well-being of the earth, Vice President Gore writes. This logic drives not only the massive achievements of the industrial and postindustrial world but earth's distress. In searching for the cause Gore says that "the more deeply I search for the roots of the global environmental crisis, the more I am convinced that it is an outer manifestation of an inner crisis that is, for lack of a better word, spiritual."[3]

The spiritual crisis rests in the alienated way in which we conceive ourselves apart from nature. "We have misunderstood who we are, how we relate to our place within creation, and why our very existence assigns us a duty of moral alertness to the consequences of what we do." Gore ends his book with his own statement of Christian faith as the reason for the hope that is in him and as the ultimate beliefs that buoy up his own part in the collective action "to change the very foundation of our civilization." "Faith," he writes, "is the primary force that enables us to choose meaning and direction and then hold to it despite all the buffeting chaos in life." In brief, Gore seems to mean by "spiritual" what others mean by "worldview," "cosmology," and "ethics": namely, "the collection of values and assumptions that determine our basic understanding of how we fit into the universe."[4] Here are Bateson's assumptions argued twenty years later.

As subjects of debate, cosmology, ethics, religion, faith, Ben-Gurion's "moral-spiritual energy," and moral responsibility turn up with increasing frequency as earth's distress deepens. Their down-to-earth role may be judged a largely negative one, as Bateson's citation implies or as Lynn White Jr. contended about Western Christianity in a famous essay (Christianity "bears a huge burden

---

1. Gregory Bateson, *Steps to an Ecology of Mind* (New York: Random House, 1972), 472. I am grateful to Anne Primavesi for the citation from Bateson, together with her discussion. See her *From Apocalypse to Genesis: Ecology, Feminism, and Christianity* (Minneapolis: Fortress, 1991), 24–43.
2. Al Gore Jr., *Earth in the Balance: Ecology and the Human Spirit* (New York: Houghton Mifflin, 1992), 269.
3. Ibid., 12.
4. Ibid., 258, 14, 368, 12.

of guilt" for the ecocrisis).[5] Nonetheless, religious cosmologies are consistently called upon to muster a response commensurate with the expanding challenge. It is as though ecocrisis consciousness and religious consciousness are, in our time, rivers that, though arising in different terrain, converge to cut a common channel.

The evidence is there. In 1972, the same year as Bateson's book and John Cobb's *Is It Too Late? A Theology of Ecology,* Jørgen Randers finished his work on the MIT report *The Limits to Growth.*[6] Reflecting upon the massive changes the report implied, Randers concluded in another publication of the same year: "Probably only religion has the moral force to bring about [the necessary] change."[7] Mihajilo Mesarovic and Eduard Pestal followed in 1974 with *Mankind at the Turning Point.* It argued that "drastic changes in the *norm* stratum . . . are necessary in order to solve energy, food, and other crises."[8] Neither Randers nor Mesarovic and Pestal considered themselves musical or literate in religion and ethics and did not, at the outset, anticipate the implications of their studies.

More recently, in January 1990, thirty-four internationally renowned scientists led by Carl Sagan and Hans Bethe issued "An Open Letter to the Religious Community." After detailing horrific environmental deterioration — "what in religious language is sometimes called 'Crimes Against Creation'" — the letter goes on:

> Problems of such magnitude, and solutions demanding so broad a perspective must be recognized from the outset as having a religious as well as a scientific dimension. Mindful of our common responsibility, we scientists — many of us long engaged in combatting the environmental crisis — urgently appeal to the world religious community to commit, in word and deed, and as boldly as required, to preserve the environment of the Earth.[9]

The letter even says what scientists, qua scientists, have been reluctant to say: "Efforts to safeguard and cherish the environment need to be infused with a vision of the sacred."[10]

---

5. See Lynn White Jr., "The Historical Roots of Our Ecologic Crisis," *Science* 155 (March 10, 1967): 1203–7.

6. Donella H. Meadows et al., *The Limits to Growth: A Report on the Club of Rome's Project on the Predicament of Mankind* (New York: Universe Books, 1972).

7. Jørgen Randers, "Ethical Limitations and Human Responsibilities," in *To Create a Different Future: Religious Hope and Technological Planning,* ed. Kenneth Vaux (New York: Friendship Press, 1972), 32.

8. Mihajilo Mesarovic and Eduard Pestal, *Mankind at the Turning Point* (New York: Dutton, 1974), 54; emphasis added.

9. From "An Open Letter to the Religious Community," 3. The letter is available from the National Religious Partnership for the Environment, 1047 Amsterdam Ave., New York, NY 10025.

10. Ibid. The open letter sparked a response that took the form of the Joint Appeal in Religion and Science. From the joint appeal itself came the Summit on Environment in June 1991, a gathering of religious leaders and scientists. The following was included in the summit statement written by religious leaders: "Much would tempt us to deny or push aside this global environment crisis and refuse even to consider the fundamental changes of human behavior required to address it. *But we religious leaders accept a prophetic responsibility to make known the full dimensions of this*

Why this religious appeal to jump-start a sustained response to the challenge of the ecosocial crisis? The reason is certainly not self-evident, especially for the religiously nonobservant. Why the turn to worldviews, faith, basic values, and moral-spiritual energies?

The crude but real reason we engage religious cosmologies and moralities is that we seem unable to do otherwise. We are, as a species, storytellers who refuse to stop short of the cosmic story itself, despite (or because of?) its pretentious dimensions. One might think, says John Shea, that faced with impenetrable mystery, we would all be more modest. Not so. Rather, when confronted with what is transcendent to every human capacity to fully fathom, we quest for more. Clifford Geertz in fact finds that the quest for meaning becomes most acute when human limits are broached. When we cannot adequately explain events, or when we cannot bear the evil that defeats us and the suffering that overwhelms us, we do not become more submissive so much as more imaginative, even if desperately so. "The effort is not to deny the undeniable — that there are unexplained events, that life hurts, or that rain falls upon the unjust," says Geertz, "but to deny that justice is a mirage. The instinct to meaning will not be denied."[11]

This habit of telling grandiose meaning-stories is as ancient as the oldest human records and apparently unique to *Homo sapiens*. We wonder about origins and worry about destiny, probably because we are both self-aware and death-aware. We ponder, but not only ponder; we organize our ponderings into grand narratives that become part and parcel of a way we live. We pray, but not only pray, as primal utterance; we think about prayer and the kind of communication it might be. We conceive, but not only conceive; we draw the imposing maps we call "cosmologies."

All the questions can be posed as a child does: Where did I come from? What am I doing here? What will happen when I die? Will the sun burn up? Or they can be asked as people who have paid lots of tuition do: What is an adequate comprehensive picture of the origin and evolution of the cosmos? What is the creation story, and destination story, of the universe? Where do we fit in the natural order? What way of life honors our appointed place in a creation that both includes and surpasses us? What manner of living does earth now require of us?

The cultural forms of our cosmologies and ethical systems vary. But invariably they are laced with religious symbols, myths, rites, and reflection. A

---

challenge. . . . Furthermore, we believe a consensus now exists, at the highest level of leadership across a significant spectrum of religious traditions, that the cause of environmental integrity and justice must occupy a position of utmost priority for people of faith." Out of the joint appeal emerged the National Religious Partnership for the Environment, launched in Washington, D.C., with the presence and support of Vice President Gore and headquartered at the Cathedral of St. John the Divine in New York City, itself the initiator of the joint appeal and the Summit on Environment.

11. Clifford Geertz, "Religion as a Cultural System," in *The Religious Situation 1968*, ed. Donald R. Cutler (Boston: Beacon, 1968), 663, as cited in the discussion of John Shea, *Stories of God* (Chicago: Thomas More, 1978), 43.

straightforward, exacting empirical cosmology is also possible, and in our time this steady work of good science is far-reaching, charting worlds too small for our sense experience and regions and realities too expansive for our imagination. Yet age-old religious and moral traditions hang on tenaciously, for better and worse. Why? While religious cosmologies cannot do what good science does, they hold distinct advantages science does not. Because their language is mythic, they can speak to the end and purpose of existence without abstracting them from experience. Indeed, they speak mythically from and to our deepest experience. Furthermore, religious cosmologies characteristically trade in ultimates — ultimate origins, destiny, meaning, value. Science can venture close to these but cannot leap, as science, into the final interpretations of faith.

Too, religious cosmologies tend to rest at the foundation of cultural structures as well as their pinnacle and show their many faces in the most common forms of art and language as well as the most arcane and elite. Religious cosmologies also provide not only meaning power but survival power and promise deliverance, healing, and well-being. Not least, they engender hope, the very nerve of moral action itself.

In sum, religious cosmologies display our thirst for exhibiting in symbols, and explaining in ritual, rite, and reflection, nothing less than the totality of things! Yet how we are set in the grand scheme of things and what we make of it is almost as diverse as life itself and part of that diversity. Charlene Spretnak provides a glimpse at but a few of the traditions that offer cosmic riches. The nature of the serene mind-self in Buddhism, she writes, cuts the cord of the continuous cravings of distraction, greed, and anxiety and the perverse encroachments of fear, indifference, rage, and ill will. It simultaneously seeks profound unity in the communion of all things and in daily practices of nonharming action. Native American spirituality, like that of other indigenous peoples, knows everything in our life to be kin to us, everything the wondrous yield of a common cosmic birth and a common, sustaining earth. The universe is not a collection of objects to be used. It is a vast community of living subjects who, as part of nature itself, are each on intimate terms with all the rest. We live best — indeed we only live at all — when we hearken to the wisdom of nature's ways and let the mystery of the spheres envelop us. Goddess spirituality is yet another religious cosmology and set of practices. It knows that the cosmic body, the sacred whole that is in and around us, is present and experienced in our own bodies. So we rightly honor and embrace both earth's body and our personal bodies as sacred. The divine is not in some distant seat of power but in life immediately at hand. To live aright is to "come to our senses." Islam, Judaism, and Christianity, like these other traditions, are internally diverse and ongoingly creative. Yet "the peoples of the Book" hewed a bold line from the beginning, a certain focus and concentration on community and social justice as a God-given vocation. A way of life is to be lived that emphasizes righteousness and moral responsibility. This is responsibility for the steward-

ship of all life and especially the "other" as neighbor.[12] We are all children of
God who share in the common good, in response to life as a gracious gift
from God.

Common themes that trail across religious faiths could be underscored as
well: a global vision of human unity, of care for the earth, of a generosity toward
all creatures great and small, of compassion for all that suffers, and of seeing all
things as precious, even holy.[13] Yet whatever the common threads and differing
emphases in varied traditions, religious cosmologies immodestly take on all of
life, bundled together. Moreover, they detail the manner of life that accords with
these grand explanations and their promises of healing and fulfillment. They offer a "way," a "way of life." So it is no coincidence at all that "ritual" is from *rita*,
the Sanskrit signifying "order" and the etymological source of both "rite" and
"right." Telling the arching cosmic story, learning the great narrative and giving
it ritual expression, is the "rite" that offers the "right" ordering of existence and
the guidance for living the "right" way. Ethics and cosmology are inextricable
and indissoluble.

Differently said, we simply will not allow our experiences to dissipate in
purely private sentiments. We are "inveterate seekers of the transcendent." We
will not let pass "the smiles of infants, the beauty of the mallard, the gentle
budding of the rose, the generous fecundity of the earth, the fire of heroism,
the ecstatic promise of intelligence,"[14] without finding in them encompassing
meanings and a sense of life's giftedness. Furthermore, we insist that the larger
meanings of life and their implications for behavior find public and institutional
forms. We insist they become the material of conscious socialization and precious inheritance. And we insist they make their way to God in prayer, song,
celebration, and mourning. We insist upon hymning the earth.

The point, however, is not only what cosmologies "do" and their intimacy
with moralities and ritual. The point is that we human creatures would not be
the species we are apart from these. We are incorrigibly cosmic storytellers, and
without cosmologies and their symbols we *literally would not know what to do.*
They are revelatory of Reality for us. It is no surprise, then, that when the
reigning cosmology and morality fail us or threaten to do us and others in —
Bateson's citation — the reply is not something like, "Be done with all such
supercilious pastimes! Take your poetry and go home!" The response instead is
to grope for a different or transformed narrative by which "the whole of our
thinking about what we are and what other people are [might] be restructured."

In sum: when the challenge is as far-reaching as earth's present distress,
human cosmologies and moralities will not only be reviewed, scrutinized,

---

12. Charlene Spretnak, *States of Grace: The Recovery of Meaning in the Postmodern Age* (San
Francisco: HarperSanFrancisco, 1991), passim.
13. See Joel Beversluis, project editor, *A Sourcebook for the Community of Religions* (Chicago:
Council for a Parliament of the World's Religions, 1993), passim.
14. Daniel C. Maguire, "Atheists for Jesus?" *The Christian Century* 110, no. 35 (December 8,
1993): 1228, 1229.

and radically judged; they will be invoked anew, drawn upon, changed, and employed. Something in us insists upon it. The theologian is right:

> In the critical moments we do ask about the ultimate causes and the ultimate judges and are led to see that our life in response to action upon us, our life in anticipation of response to our reactions, takes place within a society whose boundaries cannot be drawn in space, or time, or extent of interaction, short of a whole in which we live and move and have our being.[15]

Oh, yes, about midges. Turning the lights on an hour later does help. Midge body count is down 80 to 90 percent. But, it turns out, that's not enough. Lincoln and Jefferson still need a periodic washing down. So we will have to find a detergent that cleans grimy marble without eroding it.

Then we will have to turn to the auto and bus fumes, not to mention the steady stream of jet exhaust from approaches to National Airport just down the Potomac. What do we do about such pollution in a society that cannot imagine itself carless, busless, and jetless, and could not function in that mode, either? Too, the marble suffers from what the sweet line of wind and rain brings — acid. How do we keep sacred places from fizzing when the heavens open to bless life with the water of life? Shut down industry? No. Insist that it not burn fossil fuels? Maybe, but like Augustine's vow of chastity, not yet. And when we *do* have to make the big changes, and create very different technologies, how will we convince one another, and how will we make those changes with some sense of fairly shared benefits and burdens? Midge bashes are amenable to technical solutions. But many other matters, wrapped tightly around peoples' ways of life, are cosmological and ethical at the core. "Outer" human ecologies mirror "inner" ones.

---

15. H. Richard Niebuhr, *The Responsible Self* (New York: Harper and Row, 1963), 88.

The Vine Languishes,
the Merry-Hearted Sigh

"Education, I fear," Aldo Leopold once wrote, "is learning to see one thing by going blind to another."[1] So it is with our symbols, cosmologies, and ethics. In E. F. Schumacher's terms, the particular "think withs" we use when we "think about" things are both powerful and destructive for living with earth (not "on" it) in the modern world. They let us see and do some things while blinding us to others.

This brief chapter profiles some sophisticated theologies, chiefly Western and Protestant, of the past few decades. Their significance here rests in their enlightened and critical reflection of deep-seated cultural perspectives and practices. They represent some of the best thinking of modern times. Nothing is gained in the direction of more adequate symbols, cosmologies, and ethics by choosing the easiest and most obvious targets of reproach, as some environmentalist writing does in matters of religion. Something *is* gained, however, if the flaws of already nuanced and influential thought are exposed in the quest for something more adequate to earth's distress. We need to understand why so many of us still "instinctively" think and act the way we do. As Lord Keynes said of the influence of long-dead economists, we may unwittingly still be in the grip of some dead philosophers and theologians.

So just as the preceding chapter showed the inevitability of religion, worldviews, and moral systems in response to earth's agony and ecstasy, this one demonstrates the connection of some recent and particular inner ecologies to outer ones. When this task is finished, we can turn to the constructive enterprise of alternative perspectives.

FAULT LINES

For all their power as articulations of faith amidst several historical crises, canonical Protestant theologies from the 1930s to the 1970s were miserably deficient as cosmologies. They located human beings in the cosmos in ways that alienated us from the rest of nature and set the living substance of nature's infinite variety over against us. Nature was submissive objects at the disposal of creative subjects, human beings. Reigning theology thereby adjusted to modernity and its assumptions: that life is no longer fated, that nature exists for us, that we are the artisans of a world of our own making, that human power and purpose can

---

1. Aldo Leopold, *A Sand County Almanac* (New York: Ballantine, 1970 [1949]), 168.

shape nature, society, and psyche in indeterminate ways. The meaning of life, read in terms of this split humanity/nature relationship, was stripped down to a creative struggle between the human mind and the rest of the natural order.

Francis Bacon's stance, articulated early on in the modern era, has had real staying power as the assumed humanity/nature connection, especially in Protestant theologies. The power of some citizens in a country to subject others and the power of some countries to dominate other countries are crass and vulgar, said Bacon. The supreme human vocation, by contrast, is "to establish and extend the power and dominion of the human race itself over the universe."[2] (Evidently an ambitious man must subjugate something, Langdon Winner comments wryly, and nature and the universe, unlike human beings, will not mind subjugation.)[3] Of course, these theologies knew sin's capacity for corruption in relationships of unequal power. Their very sense of sin was the inordinate reach of overweening pride, and they excelled in unmasking its varied forms. They also set the face of theology toward social justice and excoriated the crass extensions of power. Yet they never truly argued with modernity's "disenchantment of the world," including its leeching of the sacred from nature so that a cosmic community of a million living subjects became little more than a ready collection of user-friendly objects. Granted, nature enjoyed high aesthetic status and stirred hearts as the stuff of poetry, psalm, sermon, and song, or as a place of retreat from the throbbing, phobic world. But even this was nature in the unquestioned service of human need, albeit as an appreciated enclave rather than an ensemble of raw resources. True dignity, intrinsic worth, existence valued for its own native sake, and moral limits on use were thus restricted to that one species among millions that was created in "the image of God." *Imago dei* set us apart from the rest of nature as free agents who act upon it in responsibility before God. *Imago dei* in fact situated us on the great chain of being *between* nature and God, hitched to (inescapable) finitude below us and to spirit as transcendent freedom above (the human quality most reflective of the divine). Anxiously bridging heaven and earth, finitude and freedom, nature and spirit, humanity had the vocation of stewardship, to be sure. But it was stewardship in the mode of mastery, control, and good management, all determined by an anthropocentrism of interests. Protestant theologies never stated this cosmology as bluntly as Bacon did. Nonetheless, the effective call was, like dominant powers in the wider culture, "to establish and extend the power and dominion of the human race itself over the universe."

Existentialist theology from Kierkegaard through Bultmann, for example, located the truly human in the realm of freedom to decide, freedom to fashion one's own world and make meaning along with everything else. This freedom was explicitly contrasted with a deterministic, even mechanistic, view of nature. Nature and freedom were on different, unrelated planes, the former the mute

2. Francis Bacon, *Novum Organum*, in *Selected Writings*, ed. Hugh G. Dick (New York: Modern Library, 1955), 537.
3. Langdon Winner, *Autonomous Technology* (Cambridge, Mass.: MIT Press, 1977), 23.

stage for the drama of the latter. Nature and history were divorced, and all that humanly counted was the making of history. Religious activity itself was the value-creating, meaning-making, symbol-producing work of these largely abstracted human beings. This work of symbols, myths, and ritual produced culture. Culture joined history. But nature joined neither, except culture and history as what humans did *to* nature to give it meaning and render it serviceable. The heart of the Christian ethic in this theology — neighbor-love and justice — thus never included five to ten million other species of God's fecund imagination. (A question from the wings: If God's love knows no bounds and embraces the interests of all creatures, why would a species created as *imago dei* and instructed to love as God does draw the line in the sand in front of its own feet?)

Luminaries of no less magnitude than Karl Barth and Reinhold Niebuhr joined Bultmann in this line. This may explain why so few systematic theologians are interested in the ecocrisis or know what to do with it theologically. What Gertrude Stein said of Oakland, we can say about much recent Protestant cosmology: "There's no there, there."

Paul Tillich was enough of a nature romantic that he did not quite fit neoorthodox and liberal Protestant ranks. Yet the true exceptions were the "process" theologians, led by John Cobb. The exception is telling. This school has vaulted from the minor leagues to the majors largely *because* nature also abhors a theological vacuum, and other theologies failed to meet the need.

Of course, nature-irrelevant theologies did not begin with twentieth-century Protestants. Even the references to Bacon and Enlightenment science do not go deep enough chronologically. For the Protestant Reformers, too, faith's concentration was on the knowledge of God and of self (or of "man") in such a way as to push aside much consideration of the world itself as a configuration of nature. Human subjectivity and individual human consciousness became *the* place of encounter with God. Reinforced by the Cartesian notion of the self and by the science of the Enlightenment, the desacralization of nature thus yielded a nature that was set over against the knowing self in a fateful split of active subject and passive object.

Were this not enough, the Reformers' theological preoccupation with human subjectivity as the place of encountering God already had its generative theologian in the immensely influential St. Augustine. Augustine, despite his moving praise for all creation and ourselves as part of it (our bodies are "the earth we carry," he says), encouraged human estrangement from earth through Neoplatonic cosmology. For him the ascetic ascent of the soul from earth and body to higher spirit and heaven works hand in hand with the sense of world cataclysm and world alienation that attended the decline of the Roman Empire and of classical civilization itself.

What is uppermost here is not the thought of stellar theologians, however. Their numbers wouldn't stir the heart of any serious census-taker anywhere in the world. The important matter is that *commonplace* thinking and practice in both church and society shared this faith, even if on simpler terms. It is a faith that acts as though mental and historical events occur outside nature rather than within it, and a faith that sets nature over against us and renders it religiously and morally mute. Like religion, then, justice and peace, too, draw the circle around human welfare only. The rest of life drops from sight and is pulled back onto the moral horizon only when it surfaces for human well-being. A faith like this, when armed with powerful, ravenous human structures, habits, and technologies begets earth's distress. Isaiah (sixth century B.C.E.) is suddenly poignant:

> The earth lies polluted
>   under its inhabitants;
> for they have transgressed laws,
>   violated the statutes,
>   broken the everlasting covenant.
> Therefore a curse devours the earth,
>   and its inhabitants suffer for their guilt;
> therefore the inhabitants of the earth dwindled,
>   and few people are left.
> The wine dries up,
>   the vine languishes,
>   all the merry-hearted sigh....
> The city of chaos is broken down,
>   every house is shut up so that no one can enter.
> There is an outcry in the streets for lack of wine;
>   all joy has reached its eventide;
>   the gladness of the earth is banished.
>
> (Isa. 24:5–7, 10–11, NRSV)

In short, neither existentialism, neoorthodoxy, liberalism, common church practice, nor society at large in the North Atlantic world has a cosmology worthy of the name in many influential circles. The cosmology they do have disconnects what belongs together. Languishing vines (or thriving ones), serious merrymaking, and earthly covenants with God are all necessarily of a piece. But these cosmologies have not seen it so.

There is an addendum that is far more than an addendum. Bacon again states starkly what is only somewhat buffered elsewhere in prevailing Western philosophy and theology of recent centuries. The new science and technology of the Enlightenment require a "virile" mind cleansed of feminine scruples. Then it will be possible to establish "a chaste and lawful marriage between Mind and Nature" that is capable of "leading Nature with all her children to bind her to [man's] service and make her [his] slave."[4] Here the parallel treatment

---

4. See the discussion of Evelyn Fox Keller, *Reflections on Gender and Science* (New Haven, Conn.: Yale University Press, 1985), 36–37, from which this citation of Bacon is taken. See also the extended discussion by Sandra Harding, *Whose Science? Whose Knowledge? Thinking from Women's Lives* (Ithaca, N.Y.: Cornell University Press, 1991).

of women and nature by men over centuries of patriarchy reinforces male-minded domination of nature in the emerging science, technology, and culture of modernity.[5]

DIRECTIONS

What do we do? Where do we go from here? These are the questions for the next chapters, not only the next pages. But they need to be linked to religious dimensions of cosmologies that have been offered thus far, in the interests of more adequate response than the ones just sketched. The guide is John Haught.

Being religious always entails some degree of a certain "homelessness," Haught contends. Living faith is never a perfect fit for any given present reality. In some significant measure, things can and ought to be other than they are. Yet this homelessness and restlessness are the pilgrim's. They are part and parcel of a journey in quest of true home, of *oikos* realized, of justice done, of things "on earth as in heaven," of final peace and rest. In fact, nothing less than the cosmos itself, with its grandeur and adventure, is the proper horizon of our humanity and its abode. Stingier dimensions don't accommodate. "Nothing less than inexhaustible mystery can be the appropriate abode for the human spirit."[6]

What does accommodate spirit, in addition to the wide reaches of the cosmos, are the wide reaches of religion. Religion, to cite Haught again, is an orientation "toward the inexhaustible, enlivening and liberating depth of reality that we may call by the name 'mystery,'"[7] mystery that transcends weight and measure. Religions certainly do not all say the same things. But they do seem to share this orientation toward an encompassing mystery and ask that our lives correspond to what is portrayed as Ultimate Reality.

Religions also largely share the four marks Haught discusses. One is sacramentalism. Religion is sacramental in that concrete symbols derived from ordinary experience, and especially from nature, are used to enter and reveal the extraordinary world of mystery and the experience of the extraordinary in this world. There is an "otherness" about reality and something about it that resides in the depths of, or perhaps beyond, the empirically described world, just as there is a sensed "oneness" or unity that embraces all things. Yet all this can only be represented as dimensions of sacred mystery with objects, persons, or events we know firsthand. Natural phenomena especially — "the luminosity of bright sunshine, the freshness of wind and air, the purifying power of clear water, the fertility of soil and life" — symbolize the way Ultimate Mystery seems to

5. One of the early expositions of this parallel treatment in religion, theology and culture was Rosemary Radford Ruether's *New Woman, New Earth* (New York: Seabury, 1975).

6. John F. Haught, "Religious and Cosmic Homelessness: Some Environmental Implications," in *Liberating Life: Contemporary Approaches to Ecological Theology,* ed. Charles Birch, William Eakin, and Jay B. McDaniel (Maryknoll, N.Y.: Orbis Books, 1990), 161.

7. John Haught, *The Promise of Nature* (New York: Paulist, 1993), 73.

us and affects us.[8] An implication is that any religion that loses touch with the natural world will grow indifferent toward it and be susceptible to sanctioning abuse of it.

Haught's second mark of religion is mysticism: the sense of the "other" and the transcendent, as well as the "oneness" or unity just discussed. The mystical means that nature, ourselves included, is not all there is but points to all *that* is and participates in it. A mystical ecstatic state can, of course, take the form of escapism of differing sorts and *contemptus mundi*, a disdain for the world and even hatred of the natural as the prison of the spirit. Not least for this reason, mysticism should be nurtured by intimate touch with nature and the sacramental, Haught argues. In any case, mysticism is that dimension of religious life that makes transparent the infinite depth and mystery of the Reality that surrounds us.

Reverential silence also belongs to religious experience. Classically termed the "apophatic" quality of religious experience, it means a necessary reserve. Religious experience as reverential silence knows it cannot adequately represent inexhaustible mystery and reality and so of necessity falls quiet before them. No symbols or sacraments are thus fully adequate to the Reality they praise. All are subject to a necessary iconoclasm if they are not to be vehicles of idolatries large and small.

Differently said, an endless multiplicity of images is needed to portray Ultimate Reality and the reverence and reserve due it. "Every creature is a word of God and a book about God," writes the medieval mystic Meister Eckhart.[9] Haught also cites Aquinas: "God's goodness could not be adequately represented by one creature alone. [God] produced many and diverse creatures so that what was wanting to one in the manifestation of divine goodness might be supplied by another."[10] Yet whatever the images, reserve and reverential silence are essential traits of common religious consciousness.

Action is the last of Haught's components. For Haught this is essentially ethical: action's end is the transformation of our lives and world so that the latter becomes a better place to be and live. Of course, the other components are "actions" of sorts, too. The sacramental enjoys the world in its beauty in God. The mystical relativizes all things before the Ultimate and happily loses the self in a greater Whole. The attitude of silence, like that of Sabbath observance, lets things simply "be" in the midst of its reverence for them. Yet these "actions" are not all, or enough, and religion typically calls for change as guided by moral consciousness. The ethical belongs to the religious not accidentally but essentially. Faithfulness, love, and trust, or whatever words describe religious bonds, carry moral claims.

For Haught, all four dimensions are required for "the integrity of religion." His is thus a normative, and not only descriptive, account. In this normative

---

8. Ibid., 76.
9. Ibid., 81; no source in Eckhart's writings is given.
10. Thomas Aquinas, *Summa Theologia*, I, 48, ad 2, as cited by Haught, *Promise of Nature*, 81.

(or recommended) account, religion is essentially a questing entry into the realm of cosmic mystery. Its heart is trust, its orientation mystery, its style adventure, its marks the sacramental, the mystical, the reserved, the active.[11] Its wisdom knows and has always known that human beings can truly live only when they are situated in a greater context not at their command. Haught adds his judgment that earth's distress is made worse not *by* these components interrelated as religion but by the degradation of religion as the sacramental, mystical, silent, and active. In his judgment, the integrity of creation is served by the integrity of religion.[12] But that means, as well, that degraded religion degrades creation. Choose your religion and cosmology carefully.

Haught's rendition of a religious story that would let us inhabit Habitat Earth in a graceful way includes important remarks about symbols and human creativity. As a stunning emergent of earth itself, human creativity is an expression of cosmic creativity, which is an awesome expression of divine creativity. We ought thus to view our myths, metaphors, symbols, and even religions themselves as a flowering of the energies of earth and the universe itself. Our symbol-making, no less than a tropical rainforest, is a blooming of the universe.[13] We are not lost subjects standing alone on a dumb stage in vast, cold reaches, tossing alien projections into an incomprehensible emptiness. We are a living expression of a pilgrim universe as it continues on its adventurous way. We share divine purpose and power, that wondrous divinity resident "in, with, and under" all things finite (the phrase is Luther's).

All this is a song of celebration, a hymn to the universe we ourselves sing. But it should be clear that earth does not revolve around humankind. Nor does our relationship with the cosmos turn on the exclusive salvation of the human being, despite our preciousness as a member of the Community of Life.

What should also be clear from Haught's words is that while we do not simply create or conjure up symbols (rather, Reality presses itself upon us through such means), the need now is for those symbols that effect a "reenchantment of the world" that edges out the deadly cosmology of mindless and valueless nature worked over by ghostly human freedom in all too much of modernity. Such a reenchantment may draw from the ancient Jewish and Christian confession of the Spirit in which the Spirit is the energy and power of God present in all creation as its very animation, a Spirit that effects redemption of both languishing vines and merrymakers, that is, all creation. The Spirit's presence is not amidst, nor its work for, one species only.

Whether reenchantment pursues this path or others, religiously tinged cosmologies and ethics will be part of it in the ways described by Haught.

This brings us to the constructive task. The first order is a closer look at nature as the source of religious symbols. The quest overall is for an earth cosmology and ethic worthy of the name.

---

11. Haught, "Religious and Cosmic Homelessness," 175.
12. Haught, *Promise of Nature*, 86–87.
13. Haught, "Religious and Cosmic Homelessness," 166.

Trees of Life

Rabbi Heschel may have been more prophetic than he knew. Humankind "will not die for lack of information," he wrote in 1965, but "it may perish for lack of appreciation." Lack of appreciation — by which Heschel meant acknowledgment, respect, value, care — redounds to maltreatment of human and otherkind. "The complete manipulation of the world," he concluded, "results in the complete instrumentalization of the self."[1]

Such outcomes need not be. But if they are not to be, the reason will be the symbolic creativity of humans in quest of rebeginnings, as those rebeginnings disclose *meaning* and *direction* discovered in *nature itself*. Short of discovering life anew there, "complete instrumentalization" is the sure outcome.

The creative power of natural symbols is the subject of this chapter and the following two. Continuing the quest for a viable earth ethic, we turn to natural forms among us as rich symbols bearing moral substance.

The nature of this creative power will be discussed only briefly, however. Instead, the power of natural symbols will be illustrated. Examples teach better than discussion about examples.

SYMBOLS

There is an assumption here. The assumption is that "A Rock, A River, A Tree" (Angelou) can become "hierophanies" for us — disclosures of Ultimate Reality that show the way.

While it may come as a surprise to some moderns, even offend them, disclosures about reality and what it requires of us have been mediated by nature cast as symbols throughout most of human history. Human stories of origin and destiny, for example, have typically been charged with such symbols — a garden or a city, with a sacred mount, river, tree, and rock as well as primordial parents and radiant natural abundance, all bathed in benevolent sunshine and cooling rain.

These symbols carry more than purely aesthetic power. Images of beauty bear a moral vision and carry moral weight. They encourage some behaviors and restrain others. Subtle "oughts" and "ought nots" are part of the ethos created by such symbols and the universe they carry. Their orienting presence shapes moral imagination and human character itself. In short, symbols and stories answer questions of how we are to be in the world and treat it, whether or not we are always conscious of the power of these stories and symbols as we invoke them.

---

1. Abraham Heschel, *Who Is Man?* (Stanford, Calif.: Stanford University Press, 1965), 83.

But *that* nature has, for most of human history, disclosed ultimate power and provided guidance does not mean it does so now or will for future generations. In fact, it certainly will *not* if creatures large and small are only information and resources for the utilitarian construal Heschel feared, if reality is only "virtual." Yet the only way to find out whether nature still discloses substance for a viable earth faith is to test an example or two. The one chosen for this chapter and the next is a tree. The one thereafter is darkness.

THE OLD REDISCOVERED

Trees as a life-form are among the oldest living creatures, long predating humans as a species. Some trees alive today predate all "family trees" as well. Their trunks record every season of a millennium and more.

Just as old and prominent are trees as religious symbols of a sturdy, renewed, and upright way of life. Trees speak and tell stories, stories of life, resistance, death, and new life. Trees breathe of human imagination and history as well as their own.

The Hebrew Bible, for example, begins with the tree of life (*etz chaim*), set in the midst of the Eden of origins:

> Out of the ground the Lord God made to grow every tree that is pleasant to the sight and good for food, the tree of life also in the midst of the garden, and the tree of the knowledge of good and evil. (Gen. 2:9)

The Christian Bible, which also begins here, ends with the tree of life as well, this time not in the garden of origins but in the city of destiny: New Jerusalem.

> Then the angel showed me the river of the water of life, bright as crystal, flowing from the throne of God and of the Lamb through the middle of the street of the city. On either side of the river, is the tree of life with its twelve kinds of fruit, producing its fruit each month; and the leaves of the tree are for the healing of the nations. (Rev. 22:1–2)

But even these texts come late in the history of the tree of life as a religious symbol and, if limited to Jewish and Christian sources, are too parochial. In fact, with the possible exception of the sacred mountain, no religious symbol is more ecumenical than the tree of life. In East Africa, sacred trees are the meeting place with a powerful spirit and often the place of important community decisions. In Ethiopia, for example, indigenous trees considered to be the abode of spirits are called *adbar* trees. The *adbar* represents the dynamic force of the community in that place and is the source of stability and guidance for the community. So *adbar* trees are typically the place of community rituals. A bride leaving her parental home is sent on her way with the blessing: "May the *adbar* follow or accept you." Urban dwellers use their doorways or other openings for common traffic as places of *adbar*.[2] In Cameroon, West Africa, the Gbaya people use the leaves of the soré tree, exactly as Revelation 22 has it, "for the healing of the nations" (*ethnoi* in Greek, or "peoples"). Soré leaves are the medium for rites of reconciliation when harm has been done and community peace has been violated.[3] And if the reader took a flight from Duola, Cameroon, to Athens, Greece, and climbed the Acropolis — itself now slowly dissolving from pollution — she or he would find on that windy, barren plateau a tree, an ancient olive tree. Legend says the goddess Athena planted it. From that tree we have the olive branch as the universal symbol of peace. It is a tree of life, whose leaves are "for the healing of the nations."

Or recall how Noah knew the planet was not wholly destroyed, that life had somehow survived mass death. When the rain stopped, Noah released a raven in hopes it would return with a sign of land and life. The raven did not return. Waiting a week, Noah released a dove. The dove returned, but without a sign. On the second try, seven days later, the dove came back with an olive branch in its beak. Noah waited yet one more week and sent a third dove. She did not return. She was home. She had nested somewhere in a tree of life. It was safe to anchor and disembark. The covenant with creation was not irredeemably shattered. Things could begin anew. Rebeginnings are possible.

Somewhere in the caverns of the Smithsonian Museum are Señor López's carvings. Like native peoples of the Southwest United States and Mexicans astride the border, the eyes of Luiz López of New Mexico twinkle as he carves the tree of life. His tree, like his people's, is a colorfully crowded cottonwood with fruits, flowers, birds, and people — signs of life on every busy branch.

Señor López's is one of thousands of traditions regarding the tree of life that stretch back into the mists of history. The Iroquois Confederacy pictures planet earth held together with the gently encompassing roots of the tree of life. Without this tree, things fall apart. The Oglala Sioux have a vision in which a life-long search for the tree of life is the central quest. The tree is at the intersection of the Red Road, where life begins, and the Black Road, where life ends. There the tree must be watered in order to save Earth Mother. If the roots of the tree of salvation wither, Earth Mother will as well. Human beings are responsible for the tree's life, and thus earth's. (Native American Paula Gunn Allen continues in this same tradition with the very title of one of her writings: "The Woman I Love Is a Planet; the Planet I Love Is a Tree.")[4]

Trees have also long been the legendary abode of divinities and enlighten-

---

2. Tsehai Berhane-Selassie, "Ecology and Ethiopian Orthodox Theology," in *Ecotheology: Voices from South and North*, ed. David Hallman (Maryknoll, N.Y.: Orbis Books; and Geneva: WCC Publications, 1994), 161.

3. See Thomas G. Christensen, *An African Tree of Life* (Maryknoll, N.Y.: Orbis Books, 1990), for an extensive discussion of soré as "an African tree of life." See Christensen also for a good discussion of the place of symbols and rituals in religious culture.

4. See Paula Gunn Allen, "The Woman I Love Is a Planet; the Planet I Love Is a Tree," in *Reweaving the World: The Emergence of Ecofeminism*, ed., Irene Diamond and Gloria Feman Orenstein (San Francisco: Sierra Club Books, 1990), 52ff. Like most everyone writing on the cosmological and ethical dimensions of "eco" issues now, Allen moves among many traditions. Her

ment: the olives of Artemis, the myrtles of Aphrodite, the oaks of Zeus at Dadona, and the laurel of Apollo at Delphi. (The latter two are associated with the oracles of the gods.) In the Nordic world, the cosmic ash tree Yggdasil was said, like the Iroquois tree, to fasten the world about its roots. The god Odin hung nine days and nights from the great World Ash to attain enlightenment.

For Buddhists the Bodhi Tree is fixed at the World's Axis (*axis mundi*), and in its shelter Prince Siddhārtha attained enlightenment as the Buddha. The legend is that in gratitude for the shelter provided throughout his meditation, Lord Buddha looked at the Bodhi Tree a week without blinking. A physical impossibility, it nonetheless symbolizes the heart of Buddhism — reverence for life. Perhaps such a Buddhist teaching as "destroy the forest of your derailments [or passions], not the forest of trees," stems from this original experience at the foot of the tree.[5] In early Japanese religion, both Shinto and pre-Shinto, particular trees and forests were the dwelling place of the *kami* — the inspiring power of the sacred. *Kami* pervaded all nature, yet certain trees and groves were singled out as places where the spirit of the *kami* was intense and intimate. Places in the forest "where trees grow thick" typically became the sites of shrines.[6]

---

chapter closes with a quotation of Hebrew scripture, one in which Sophia is pictured as a tree. Sophia, or Wisdom, was present with God at the world's creation:

> I grew tall like a cedar in Lebanon,
> and like a cypress on the heights of Hermon.
> I grew tall like a palm tree in En-gedi,
> and like rosebushes in Jericho;
> like a fair olive tree in the field,
> and like a plane tree beside water I grew tall.
> Like cassia and camel's thorn I gave forth perfume,
> and like choice myrrh I spread my fragrance,
> like galbanum, onycha, and stacte,
> and like the odor of incense in the tent.
> Like the vine I bud forth delights,
> and my blossoms become glorious and abundant fruit.
>
> Come to me, you who desire me,
> and eat your fill of my fruits.
> For the memory of me is sweeter than honey,
> and the possession of me sweeter than the honeycomb.
> Those who eat of me will hunger for more,
> and those who drink of me will thirst for more.
> Whoever obeys me will not be put to shame,
> and those who work with me will not sin.
>
> All this is the book of the covenant of the Most High God,
> the law that Moses commanded us as an inheritance for
> the congregations of Jacob. (Ecclus. 24:13–24)

(I have added to Allen's selection the portions "Come to me..." and "All this is..." so as to make clear that part of the context is the covenant with Israel and creation.)

5. One account is offered by a Sri Lankan Buddhist, Sarath Kotagama, in "Practice What You Preach: A Lesson for Sustainable Living," from *Story Earth* (San Francisco: Mercury House, 1993), 115.

6. Michael Perlman, *The Power of Trees: The Reforesting of the Soul* (Dallas: Spring Publications, 1994), 96.

Was it the tree of life Luther referred to? Martin Luther believed he was living in the last days. Once asked, "If you thought tomorrow might bring the Day of Judgment, what would you do?" he replied, "I'd plant a tree."[7] A Jewish rendition goes like this: "If you have a sapling in your hand and they tell you that the Messiah has arrived, first plant the sapling and then go out to greet him."[8]

It was certainly the tree of life the abbess and mystic Hildegard of Bingen spoke of in the twelfth century as God's Spirit:

> The Spirit of God
> is a life that bestows life
> root of the world-tree
> and wind in its branches.
> She is glistening life
> alluring all praise
> all-awakening
> all-resurrecting.[9]

We find the tree of life prominent in ancient Judaism. Proverbs 3:18 depicts the Torah itself, the embodiment of divine instruction and the first emblem of Judaism, as a tree of life. It is even said that abiding by the words of Torah restores the tree of life lost in the primal act of disobedience in Eden. (In Genesis 3 cherubs with flaming swords guard Eden's tree of life from expelled and wayward humans.) The Proverbs verse is recited as part of sung prayer every time the Torah is returned to the ark after a Sabbath reading: "It is a tree of life to those who hold fast to it, and its precepts are right. Its paths are paths of pleasantness, and all its paths are peace. Return us to thee, Lord, and we shall return. Renew our days as of old."

Another ubiquitous emblem of Judaism, the seven-branched menorah, is also a tree. An adaptation of the ancient cosmic tree of the Sumerians, for whom the tree of life grew on the cosmic mountain, the menorah was likely a sign for Jews of God's inexhaustible nourishment of the world. The detailed call for buds on the branches of hand-fashioned gold menorahs in ancient texts (see Exod. 25:31–40) may indicate that the early menorah form was a living tree.[10]

The third emblem of Judaism, the burning bush of the giving of the law, is not quite a tree, but nearly so!

---

7. This story is often told about Luther, though it may well be apocryphal. What is certain is his use of the tree as metaphor for the Christian life in his "Lectures on Isaiah" and specifically his commentary on Isa. 61:3c: "They will be called oaks of righteousness, the planting of the Lord, to display his glory." See vol. 17 of *Luther's Works* (St. Louis: Concordia, 1972), 335–36.

8. Cited from *Avot de Rabbi Natan* B31 in materials for "While the Messiah Tarries: The Past, Present, and Future of Jewish Messianism," a program of the Franz Rosenzweig Lehrhaus at the Jewish Theological Seminary of America, New York, spring 1993.

9. From Barbara Newman, trans. and ed., *St. Hildegard of Bingen: Symphonia; a Critical Edition of the Symphonia armonie celestium revelationum* (Ithaca, N.Y.: Cornell University Press, 1988), 140–41.

10. Ismar Schorsch, "Trees for Life," *Union Seminary Quarterly Review* 45, nos. 3–4 (1991): 243–44.

Throughout Hebrew scripture the tree is a metaphor for fidelity and right-eousness as well as healing and new life. Passages are far too numerous even to list. Two are cited for illustrative purposes:

> The righteous shall flourish like a palm tree,
> and shall spread abroad like a cedar of Lebanon.
> Those who are planted in the house of the Lord
> shall flourish in the courts of our God;
> they shall still bear fruit in old age;
> they shall be green and succulent;
> that they may show how upright the Lord is,
> my rock, in whom there is no fault. (Ps. 92:11–14)

Note: here God is a rock, and the righteous are trees. Except for "house" and "courts," all the images of divinity, humanity, and life are directly from non-human nature. (Incidentally, Psalm 92 is the main psalm for the Sabbath service, recited in both morning and evening services.)

In Hosea the compassionate God speaks to unfaithful Israel in the tender words of chapter 14:

> I will heal their affliction,
> Generously will I take them back in love;
> I will be to Israel like dew;
> They shall blossom like the lily;
> striking root like a Lebanon tree.
> Their boughs shall spread out far,
> their beauty shall be like the olive tree's,
> and fragrance like that of Lebanon.
> Those who sit in their shade shall be revived:
> They shall bring to life new grain,
> They shall blossom like the vine;
> Their scent shall be like the wine of Lebanon.
>
> (Hos. 14:5–8)[11]

Here God is dew and Israel is flowers, trees, grain, and vines. Forms of nature disclose divinity and humanity. The psalmody of trees is new as well as old. The opening stanza of a typical eccentricity by e. e. cummings goes like this:

> I thank You God for most this amazing
> day; for the leaping greenly spirits of trees

---

11. This is a translation from Hebrew as it appears in "'She Is a Tree of Life': Trees and the Jewish Experience," available from the Melton Research Center for Jewish Education, the Jewish Theological Seminary, New York City.

and a blue true dream of sky; and for everything
which is natural which is infinite which is yes[12]

The work of Rabindranath Tagore, perhaps India's premier poet, includes
these lines:

I asked the tree,
Speak to me about God,
And it blossomed. . . .
Silence,
my soul,
these trees are prayers.[13]

We must let the evidence rest here. Enough has been tallied to make the
point: the tree of life is an old ecumenical symbol discovered anew time and
again across innumerable cultures. It is religiously and morally charged.

TREES OF RESISTANCE

Perhaps it was also the tree of life that was intimated at a peace gathering in
Los Alamos, New Mexico, on August 6, 1989 — Hiroshima Day. A Native
American dancer danced against the greatest powers of death ever assembled
by humans and prayed for life by giving each of us, in silence, a small green
branch. The branch seemed both impotent and powerful. It was a mere fraction
of a tree. But it was juxtaposed, as a strong sign of life, to the mushroom cloud
as the towering image of death.

The tree of life at Los Alamos[14] was also the tree of resistance, as it
has been in some communities for centuries. Like the communities of native
peoples, these communities have resisted Western-led globalization and power
or other aggressions from within and without. The Chipko movement in India is an example. It resists forestry interests intent on export for profit by way
of monoculture cropping, a practice that destroys the forest diversity that local dwellers use for sustaining their communities. In the 1970s Indian women
took to tree-hugging (*chipko* in Hindi) to stop loggers and save the source of
their fruits, roots, tubers, seeds, leaves, petals, and sepals, as well as their fire-
wood and the soil of their steep valleys. This, it turns out, is only a recent
instance of their tree-embracing resistance. Other campaigns in this area, also

---

12. Cited by Sallie McFague, *The Body of God: An Ecological Theology* (Minneapolis: Fortress, 1993), 196.

13. Cited by Chung Hyun Kyung in "Ecology, Feminism, and African and Asian Spirituality," in *Ecotheology*, 175, 178.

14. "Los Alamos" itself means "the cottonwoods" and is named for these prominent trees in the city. Some of the oldest are around the lodge and boys' school that became the center for development of the A-bomb.

led by women, have taken place over the past three hundred years against other invaders. Not unimportant, sacred trees are images of the cosmos in the cultures of these communities, and sacred forests and groves are sacred community spaces.[15]

A century before this latest round of Chipko resistance, Walking Buffalo of the Stoney Indian Nation lost land and people to the white man. He complained of the inability of white people to hear the voices of nature, including Indian voices. He says this about trees as bearers of knowledge:

> Do you know trees talk? Well they do. They talk to each other, and they'll talk to you if you listen. Trouble is, white people don't listen. They never learned to listen to the Indians, so I don't suppose they'll listen to other voices in nature. But I have learned a lot from trees; sometimes about the weather, sometimes about animals, sometimes about the Great Spirit.[16]

(While Walking Buffalo was speaking these words, white men were felling old growth redwoods more ancient than Christianity itself. Ecological imperialism and environmental apartheid were underway.)

Alenda Bermúdez, a Guatemalan poet, writes a resistance poem — "Guatemala, Your Blood" — in answer to Chilean Pablo Neruda's complaint: "[W]hy doesn't your poetry talk to us about dreams, leaves, the huge volcanos of your native land? Come look at the blood in the streets." Her poem, written in Spanish even though Spanish is the tongue of oppression, ends with these defiant lines:

> I reserve the right of the precisely exact Spanish word
> to name death and to name life
> as long as the blood holds itself suspended in our trees.[17]

Bermúdez is Mayan, so she speaks of "dreams, leaves, [and] the huge volcanos of [her] native land" in a Mayan way. In Mayan the word for "blood" is also the word for "sap." Both are the same life-force. "Who cuts the trees as he pleases cuts short his own life" is a Mayan adage.[18] Bermúdez will thus name

---

15. See Carolyn Merchant, *Radical Ecology: The Search for a Livable World* (New York: Routledge, 1992), 201–2, or the more extended discussion in Vandana Shiva's *Staying Alive: Women, Ecology and Development* (London: Zed Books, 1989), 67–77, 208–11.

16. Cited by Vine Deloria in *God Is Red: A Native View of Religion*, 2d ed. (Golden, Colo.: North American Press, 1992), 104.

17. Alenda Bermúdez, "Guatemala, Your Blood," in *Women on War*, ed. Daniela Gioseffi (New York: Simon and Schuster, 1988), 26–27.

18. From *National Geographic* 182, no. 5 (November 1992). The Mayan tree of life itself was the great *ceiba*, a silk-cotton tree that was believed to stand at the center of the earth, support the heavens, and symbolize life itself. As an interesting twist on the relationship of Mayans to their Spanish conquerors, Mayans readily took to the image of the cross of the conquistadors because their own tree of life was represented in stylized form as a leafy cross. This portrayal predates Columbus by at least eight centuries, as recorded on temple walls. Mayans thus transformed the symbol of an execution stand and of "civilizing" conquest into a living archetype from their own tradition (see "The Tree at the Center of the Earth," *New York Times Book Review*, May 12, 1991, 39).

life and name death as she knows them, against all attempts of others to name reality or write her history as they wish. She will do so "as long as the blood holds itself suspended in our trees." Here the tree of life is the tree of resistance as she knows these in her own body. Or, conversely, as the trees know her life and struggle in theirs.

Something of the same intimacy of blood and sap, life and resistance, is told in other literature in light and delightful ways. In Alice Walker's *The Color Purple*, Shug is explaining to Celie how she got rid of God as "the old white man" and found the Spirit and new life in so doing:

[Shug] say, My first step from the old white man was trees. Then air. Then birds. Then other people. But one day when I was sitting quiet and feeling like a motherless child, which I was, it come to me: that feeling of being part of everything, not separate at all. I knew that if I cut a tree, my arm would bleed. And I laughed and I cried and I run all around the house. I knew just what it was.[19]

Later Shug says: "Listen, God love everything you love — and a mess of stuff you don't. But more than anything else, God love admiration." Celie asks: "You saying God vain?" Shug then replies: "Naw... Not vain, just wanting to share a good thing. I think it pisses God off if you walk by the color purple in a field somewhere and don't notice it."[20]

Arms bleed when trees are cut. Everything is that much a part of everything else, something like using the same word for sap and blood. Yet trees are also "other" and the means of resisting and exorcising false gods (God as the old white man or conquistadors naming life and death for Mayans).

Trees also document the globalizing oppression that is resisted. Nettie, now a missionary in Africa, writes Celie:

Samuel and I had never really thought about war. Why [the white missionary said], the signs are all over Africa. India too, I expect. First, there's a road built to where you keep your goods. Then your trees are hauled off to make ships and captain's furniture. Then your land is planted with something you don't eat. Then you're forced to work it.[21]

On another note, one theory of the fall of Mayan civilization is that it occurred as the result of a self-inflicted ecological disaster brought about by the felling of too many trees. While other causes are included as well — population pressures on the agricultural system, warfare, changing trade routes, and droughts and other climatic factors — the felling of trees in support of monumental building projects may have led to soil erosion and the filling in of seasonal swamps important for gardens. Much like John Gowdy, the scholar offering this theory, Richard Hansen sees the chief motivation for deforestation in the desire of the ruling class to display and extend its wealth and power (see "Did Maya Doom Themselves by Felling Trees?" *New York Times*, April 11, 1995, C12).

19. Alice Walker, *The Color Purple* (New York: Pocket Books, 1982), 178.

20. Ibid.

21. Ibid., 204.

Resistance can take many forms. One is simply to assert identity and pride against forces that degrade and diminish. Thus Mari Evans uses the tree as "other" and as a way of talking of herself as a black woman:

> I
> am a black woman
> tall as a cypress
> strong
> beyond all definition still
> defying place
> and time
> and circumstance
>   assailed
>   impervious
>   indestructible
> Look
>   on me and be
> renewed.[22]

Sometimes a tree is literally planted as life's emblem, set against its deprivation. On South Africa's Robben Island there is an exercise yard in Block B of the feared prison. The yard's walls are high enough that no sunrise or sunset can be seen, much less the majesty of "the fairest cape in the world" and Table Mountain (Cape Town). Blue sky can be seen and the hot sun felt, but no other signs of life were present in the bare concrete yard — until prisoner Nelson Mandela, held there nineteen years, received permission to excavate a narrow strip of rocky land at one end of the yard, itself previously a landfill. Now there are two small trees, a grape arbor, a few flowers, and vegetables. And Mandela is postapartheid South Africa's first president.[23] "The sense of being the custodian of this small patch of earth offered a small taste of freedom," the ex-prisoner writes.[24]

A tree grows in Brooklyn, too. In its own way it is also a tree of life as a resistance tree. "Bed-Stuy" is a raw, inner-city New York neighborhood. But in its midst is the Magnolia Tree Earth Center, an environmental and cultural meeting place. An old *Magnolia grandiflora*, native not to Brooklyn but to the South, was planted in Bed-Stuy in the 1880s. It survived because it was sheltered through cold winters by surrounding brownstones and had enough street smarts to shape itself so as to take advantage both of shelter and light. When the block suffered social degradation in the 1960s and those who could, left, Hattie Carthan (neighbors call her "The Tree Lady") established the Magnolia Tree Earth Center around the *Magnolia grandiflora*. Her explanation is straightforward: "We've

---

22. Cited from Perlman, *Power of Trees,* 164.
23. This information from my visit to Robben Island, with the International Bonhoeffer Congress, January 8, 1996.
24. Nelson Mandela, *The Long Walk to Freedom: The Autobiography of Nelson Mandela* (London: Little, Brown and Company, 1994), 582–83.

already lost too many trees, houses and people — your community — you owe something to it. And I don't care to run."[25] The text of the photo exhibit at the Earth Center says it equally well. Hattie's tree "is a sign the community has survived," and like the tree, which "wasn't expected to live when it came here," the community, too, "is a survivor." The magnolia "says something about us as a community."[26]

A tree of life carries its community. A community protects its own life by protecting its trees. Trees mean life that has staying power. Life also seems to be gentler in neighborhoods adorned with old trees that arch in the air across the street.

One of the most fascinating episodes of trees of life as trees of resistance is in the biblical materials themselves, albeit as an episode still subject to scholarly speculation.

Hebrew monotheism, numerous critics have noted, can edge into abstraction and a depopulated universe because God is not seen, graven images are forbidden, and the world is left in the hands of God and human beings as virtually the only agents. Few other buffers, intermediaries, and intercessors inhabit the exposed world — a largely pastoral world — of the Israelites. Yet it seems the Israelites resisted a life devoid of nature symbols and rites associated with gods and goddesses. These other divine beings had embodied natural forces before their slow demise at the hands of monotheism. The agonizing monotheistic prophets, Hosea and Jeremiah, complain that in both northern Israel and Judah these "stiff-necked" and backsliding people are worshiping "on every lofty hill and under every green tree,"[27] sometimes "blowing kisses to calves" (Hosea). The reforms of Hezekiah and Josiah tried to destroy the complex of altar, tree, hill, and megalith, apparently with only mixed success. Tikva Frymer-Kensky speculates that the remaining nature symbols in Israelite worship may in fact have *served* the monotheism by enabling the people to sense the immanence of God, thus making it possible for them to continue to worship "the [otherwise] abstract and demanding YHWH."[28] Nonetheless, the taboo strongly associated with pagan nature religion was not thereby weakened. A later tradition will even say of Torah, itself the tree of life of wisdom, "He who forsakes Torah for the contemplation of trees forfeits his life."[29]

Israel was not, it should be said, simply "iconoclastic." The prohibition of graven images did not rule out concrete images of God's presence — the ark, the

---

25. Cited from Perlman, *Power of Trees*, 166.

26. Ibid., 167.

27. The phrase is W. L. Halladay's summary of forms of local worship at this time; see his "On Every Lofty Hill and Under Every Leafy Tree," *Vetus Testamentum* 11 (1961): 170–76, as cited by Tikva Frymer-Kensky, *In the Wake of the Goddesses: Women, Culture, and the Biblical Transformation of Pagan Myth* (New York: Free Press, 1992) 153. Frymer-Kensky says "green tree" rather than "leafy tree," however.

28. Frymer-Kensky, *In the Wake of the Goddesses*, 154.

29. From *Pirkei Avot* 3:9, as reported by Eilon Schwartz in his paper, "Judaism and Nature: Theological and Moral Issues to Consider while Renegotiating a Jewish Relationship to the Natural World," 2. The paper was prepared for the Theology for Earth Community Conference at Union Theological Seminary, New York City, October 1994.

tablets of the law, the bronze serpent, animals of sacrifice. But the image that interests us *is* a contested one — the *asherah,* a stylized tree or an actual one placed next to the altar. It stood there until Hezekiah's reform in the eighth century, when both it and the bronze serpent were bidden to go. It made a reappearance with King Manasseh but was then banished as "Canaanite" with Josiah's reform in the seventh century B.C.E. The "Canaanite" charge does thicken the plot, since Asherah is the name of a Canaanite goddess. (Jezebel, hardly an exemplary figure in the tradition, supported the prophets of Asherah.) Yet the battles are not between Yahweh and goddess worship, and never were. The battles are with the male Canaanite gods Baal and El. In any case, *asherah* at the Israelite altar is clearly associated with Yahweh worship, not Baal worship, even though Baal, too, attracted Israelites. Yet, says Frymer-Kensky, there may be an intriguing connection between Asherah and *asherah.* Asherah, the Canaanite goddess, is tolerated until the eighth century B.C.E. in Israel, although probably because she *doesn't* hold special power or do anything in particular (she is not the consort of Yahweh, for example). She is associated with trees and groves, however, and may in this way have represented the natural world to a pastoral people for whom trees certainly did hold a special place as symbols of fidelity, righteousness, and regeneration. (The last-named is perhaps most important in this case.) She might also, in the *asherah* at the altar, have symbolized earth reaching toward heaven. Yet the most intriguing items linking Asherah the goddess and *asherah* the tree at the altar are the many female figurines found in Israel from this period. Curiously, male figurines are practically nonexistent, and the female ones are not Canaanite. They have no divine headdress or any of the other symbols of Canaanite divinity. Scholars call them *dea nutrix* (the nourishing deity), however, because they are found in areas of cultic preparation of food, eating, and drinking. (These are always of great importance religiously and socially in the ancient Near East.) *Dea nutrix* may still be something of a misnomer, however, in that crowns, stars, or other divine insignia are wholly absent. Rather, the figurines have a pillarlike base and are solid figures in the round, with breasts, a molded head, and arms upraised, holding the breasts, or in some instances, with no arms at all. The pillarlike base, furthermore, isn't really a pillar. It is flared in the manner of a tree trunk, just as the rest of the pillar looks for all the world like a tree trunk. So Frymer-Kensky is moved to ask, "Could it be possible that the figurine is a kind of tree with breasts?"[30] An answer in the affirmative has as further evidence that other Near Eastern civilizations had exactly such breasted trees. Thus she concludes: "It seems more than likely that Asherah and/or the asherah is identified as the force of vegetation and nourishment"[31] and that both the stylized tree at the altar and the tree-based breasted figurines "brought the divine and natural worlds closer together"[32] without "worshipping" the sort of goddess that would rival the one

---

30. Frymer-Kensky, *In the Wake of the Goddesses,* 160.
31. Ibid.
32. Ibid. This figurine is used as the cover for Frymer-Kensky's book, even though the book is about much more than this in the rise of ethical monotheism.

true God, Yahweh. Differently said, the people may well have resisted the one invisible, transcendent God who, even though this God was the only source of "the blessings of breast and womb" (Gen. 49:25), left no medium of fertility and nourishment they could see. The tree-based breasted figure may thus have represented a primal force of life whose fecundity and regeneration were religiously absent in the symbols of this monotheism. Or rather, this force's symbolization *in a familiar and cherished form of nature was absent.* The lactating tree may have been just such a tree of life. It may have been the tenacious way people resisted the distancing of the God they worshiped as the God of life itself.

There is a lesson here, as there is with the other trees of this chapter. When nature is lost to the senses, God is as well. So is people's sense of identity and direction as well as moral-spiritual energy for the journey. This is unacceptable. People will protest, above ground or below, softly or loudly, at the altar or away from it. No heaven without earth. It's that simple.

Or Bare Ruined Choirs?

TREES OF DEATH

The unspeakable happens. Resistance is too little and too late. Sheer terror, horror, and murder prevail. Such was the Holocaust. Still, even here trees stand for life when no human words can be formed at all. Or rather, the tree is there when the only voice heard in Ramah is "wailing and loud lamentation" and Rachel cannot be consoled because "her children...are no more" (Matt. 2:18). At Yad Vashem, the Holocaust memorial in Jerusalem, the Avenue of the Righteous marks one of the signs of life in a landscape of evil and death. The Avenue of the Righteous remembers those who rescued Jews, saving them from the death camps. A bush or tree is planted for each rescuer, and the avenue is lined with them. Each is a tree of life from a time when life was wrenched away.[1]

The State of Israel has had its own agonies in the conflict between Jews and Palestinians. It is of more than passing notice, then, that the speech of Israeli prime minister Yitzhak Rabin at the signing of the Middle East Accords in Washington, D.C., included the following, addressed to PLO chairman Yasir Arafat and the Palestinians: "We have no desire for revenge. We harbor no hatred towards you. We, like you, are people — people who want to build a home. To plant a tree. To love — side by side with you. To live in dignity. In empathy. As human beings."[2] To love, build a home, and plant a tree — such are the essentials of peace. Here, too, "the leaves are for the healing of the nations" (Rev. 22:2).

The tree of life, then, is anything but a stranger to suffering and death.[3] Yet sometimes the tree is not life at all but simply the place of death, with no sign of redemption anywhere. The lynching tree has never been a tree of life, any more than the gallows. Persons who know suicidal depression also speak of seeing themselves in a dark wood, with no escape. Perhaps this is the same wood with which Dante introduces *The Inferno* ("I found myself in a dark wood").

---

1. One is for Oskar Schindler. The movie *Schindler's List* closes with note of this and says, "It grows there still."

2. "Statements by Leaders at the Signing of the Middle East Pact," *New York Times*, September 14, 1993, A12.

3. At the memorial service for the victims of the bombing of the federal building in Oklahoma City, President Clinton's comments included these: "Yesterday Hillary and I had the privilege to speak with some children of other Federal employees, children like those who were lost here. And one little girl said something we will never forget. She said, 'We should all plant a tree in memory of the children.' So this morning before we got on the plane to come here, at the White House, we planted that tree in honor of the children of Oklahoma. It was a dogwood with its wonderful spring flower and its deep enduring roots. It embodies the lesson of the Psalms that the life of a good person is like a tree whose leaf does not wither" (*New York Times*, April 24, 1995, B7).

Perhaps it is the same wood many remember as the terrain of childhood fears.[4] Trees lining cemeteries are often quiet and sturdy signs of life, but they may also guard the dead with those evocations of imminent mortality one sometimes feels in heavy woods.

In Toni Morrison's *Beloved*, Sethe flees slavery after a brutal beating. Amy, a young girl, finds her in the woods, pregnant. Amy unfastens Sethe's dress to look at her wounds. "Come here, Jesus," Amy says in quiet horror at what she sees. "It's a tree. Lu. A chokecherry tree. See, here's the trunk — it's red and split wide open, full of sap, and this here's the parting for the branches. You got a mighty lot of branches. Leaves, too, look like, and dern if these ain't blossoms. . . . Your back got a whole tree on it. In bloom. What God have in mind, I wonder. I had me some whippings, but I don't remember nothing like this."[5] Years later, Sethe begins to cry when Paul D, her lover, "rubbed his cheek on her back and learned that way her sorrow, the roots of it; its wide trunk and intricate branches." Sethe couldn't feel that Paul D was touching "every ridge and leaf of it with his mouth," however, "because her back skin had been dead for years."[6] The conventional association of trees with flourishing life is here reversed, Michael Perlman comments. Here a tree captures flourishing pain and oppression.[7]

Trees, then, are not unambiguous symbols of life. They do not convey only life and certainly not all life or life wholly redeemed. They are limited bearers of truth and bent carriers of guidance, as all creatures are — *and* as all symbols are. This is only a comment in passing, but an important one and a reason for the ancient prohibition against graven images and idolatry. No one symbol and no one creature can fully capture the true, the beautiful, or the good. All images, including the image of God, are limited, broken images.

Which is not to deny that there is also power in such symbols. Life itself is broken, and symbols that show its real and raw dimensions, including the harsh commingling of life, pain, joy, and death, speak their own kind of sure truth. Janie Crawford, in *Their Eyes Were Watching God*, imagines her life "like a great tree in leaf with the things suffered, things enjoyed, things done and undone. Dawn and doom was in the branches. 'Ah know exactly what Ah got to tell yuh,' she says, 'but it's hard to know where to start at.'"[8]

The problem is usually from our side, not the form of nature itself. It is we who normally trim the symbol to fit, round off the hard edges, remove the offenses, ambiguities, and dross, and force it to align with our expectations and perversities. A deformed tree with both "dawn and doom" in its branches cannot,

---

4. The suggested parallels are Michael Perlman's in *The Power of Trees: The Reforesting of the Soul* (Dallas: Spring Publications, 1994), 136.

5. Toni Morrison, *Beloved* (New York: Knopf, 1987), 79. For use of this and subsequent quotations from *Beloved*, permission granted by International Creative Management, Inc. Copyright ©1987 by Toni Morrison.

6. Ibid., 17–18.

7. This is Perlman's commentary (in *Power of Trees*, 135) on the passage from Morrison.

8. Zora Neale Hurston, *Their Eyes Were Watching God* (Urbana: University of Illinois Press, 1978 [1937]), 20.

we seem to insist, be a true tree of life. So instead of finding it "hard to know where to start at" and saying so, we prune the symbol until we *do* "know exactly what . . . to tell" — without nuance and ambiguity, and often without truth. The dreadful consequence of this habit, because we read life *through* our trimmed symbols, is that all perceived deformations, including those belonging to fellow human beings, become grounds for prejudice, neglect, exclusion, or worse. Harsh social "purity codes" then prevail and brittle orthodoxies reign, even though nature apart from us, from whence we got the symbols in the first place, knows few of either. Perhaps if we accepted the rest of flawed and many-sided nature more readily, we would accept our flawed and many-sided selves as well.

That said, it is yet another kind of testimony that trees don't seem to rest easy for long with the brute reality of death. They struggle against it. They tell us not to accept death as the final utterance and not to see life's real deformations as finally acceptable. They make every effort to claim life anew. This is one reason that Sethe, even with the tree of slave pain on her back, loved the woods. It was there, in that "green blessed place" where "the woods rang" and the assembled sang to her mother-in-law's preaching, that she found life. The whole scene is so full of the pathos of hard living and hard-fought joy that it must be included here.

The ex-slaves, "in the heat of every Saturday afternoon," would come to the clearing in the wood where "on a huge flat-sided rock, Baby Suggs [Sethe's mother-in-law] bowed her head and prayed silently."[9] Morrison writes:

> The company watched her from the trees. They knew she was ready when she put her stick down. Then she shouted, "Let the children come!" and they ran from the trees toward her.
>
> "Let your mothers hear you laugh," she told them, and the woods rang. The adults looked on and could not help smiling. Then "Let the grown men come," she shouted. They stepped out one by one from among the ringing trees.
>
> "Let your wives and your children see you dance," she told them, and groundlife shuddered under their feet.
>
> Finally she called the women to her. "Cry," she told them. "For the living and the dead. Just cry." And without covering their eyes the women let loose.[10]

---

9. Morrison, *Beloved*, 87.

10. Ibid., 88. This is only secondarily a fictionalized account. Morrison is drawing from slave experience. One of the slave accounts, that of Susan Rhodes, includes this: "We used to steal off to the woods and have church, like de Spirit moved us — sing and pray to our own liking and soul satisfaction — and we sure did have good meetings, honey — baptise in de river like God said. We had dem spirit-filled meetings at night on de bank of de river and God met us dere" (from James Mellon, ed., *Bullwhip Days: The Slaves Remember* [New York: Weidenfeld and Nicolson, 1988], 194–95). I am grateful to G. Whit Hutchison for bringing this passage to my attention in his doctoral dissertation, "The Bible and Slavery: A Test of Ethical Method" (Union Theological Seminary, New York, 1995), 308.

It was also in the woods that Sethe's daughter, Denver, often slept. "She smelled like bark in the day and leaves at night."[11] But this was not an unwelcome smell. (Perlman speculates that Sethe's "tree-daughter" is Morrison's appropriation of Janie Crawford's mother in Zora Neale Hurston's *Their Eyes Were Watching God.* She was named "Leafy," was born in the escape from slavery, and was "wrapped up in moss and fixed . . . good in a tree" until her mother was sure that "all of us slaves was free.")[12]

Yet the boldest attempt to wrest a tree of life from the tree of unspeakable death was one that rooted a new religion. "They put him to death by hanging him on a tree," Peter preached to the crowds, "but God raised Jesus on the third day and made him manifest" (Acts 10:39b–40a). Celsus, a second-century Roman philosopher, ridicules Christians because they call this dead tree of Golgotha, the Place of the Skull and a refuse dump outside Jerusalem's city gates, "the tree of life."[13] They also expect the "resurrection of the flesh through the wood," Celsus scoffs. "What drunken old woman, telling stories to lull a small child to sleep," he continues in disgust, "would not be ashamed of muttering such preposterous things?"[14] Celsus knows whereof he speaks. The cross was an execution form reserved for the lowest classes and the seditious. ("The rich buy judgment," a cynical folk saying had it, "the poor obtain the cross.")[15] Romans had, in the barbarism of the wars with the Jews, lit roadsides with Jews torched on crosses. "Thousands of Jews before Jesus, with Jesus, on both sides of Jesus, and after Jesus were put to death by crucifixion," Pinchas Lapide recalls.[16] So Celsus was only reflecting high society when he condescendingly said of Jesus that he had been "bound in the most ignominious fashion" and "executed in a shameful way."[17] Celsus could have cited the poem of Pseudo-Manetho: "Punished with limbs outstretched, they see the stake as their fate; they are fastened (and) nailed to it in the most bitter torment, evil food for birds and grim pickings for dogs."[18] Yet early Christians saw in the tree of execution the tree of life, just as they also saw in Jesus new life from a tree given up for dead. "A shoot shall come out from the stump of Jesse, and a branch shall grow out of his roots" (Isa. 11:1). This tree, fully severed at the base, would still give life.[19] And someday, the prophet dares to promise,

---

11. Morrison, *Beloved,* 19.

12. Perlman, *Power of Trees,* 135, citing Hurston, *Their Eyes Were Watching God,* 18.

13. In Christian art this was a habit for a long while, with the cross sometimes depicted as the tree of life of Eden. In medieval illuminated manuscripts, frescoes, and paintings, for example, the tree of knowledge and the tree of life are frequently portrayed as emblems of the Fall and of redemption, respectively. In some the tree of life, that is, redemption, is the cross shown in leaf and bearing fruit.

14. Cited from Origen's *Contra Celsus* 6.34 by Martin Hengel, *Crucifixion* (Philadelphia: Fortress, 1977), 8 n. 15.

15. Hengel, *Crucifixion,* 60.

16. Pinchas Lapide, *The Sermon on the Mount: An Orthodox Jewish Reading* (Maryknoll, N.Y.: Orbis Books, 1986), 101.

17. Hengel, *Crucifixion,* 7.

18. Translated and cited in ibid., 9.

19. In somewhat the same vein it might be noted that the word "Lent," referring to the season of the passion and the cross, is from the Latin for "spring."

you shall go out in joy,
and be led back in peace;
the mountains and the hills before you
shall burst into song,
and all of the trees of the field shall clap their hands.

(Isa. 55:12)

WHY TREES?

Why do we reach for trees as disclosures of meaning and a way of standing upright in the world? Why do they, as a form of nature, move us and help create a habitable world for the soul? What do trees do that we sense, in our innermost parts, is the reaching upward of life itself? What connects sap and blood?

Civilization, it is said, begins with the felling of the first tree and ends with the felling of the last. In some cases it seems literally so. Plato, in his *Critias*, writes of a dismal patch of ancient Greece and in the process describes erosion with almost scientific precision:

What now remains compared with what then existed is like the skeleton of a sick man, all the fat and soft earth having wasted away, and only the bare framework of the land being left.... [T]here are some mountains which now have nothing but food for bees, but they had trees not very long ago.... [T]here were many lofty trees of cultivated species and...boundless pasturage for flocks. Moreover, it was enriched by the yearly rains from Zeus, which were not lost to it, as now, by flowing from the bare land into the sea; but the soil it had was deep, and therein it received the water, storing it up in the retentive loamy soil, and...provided all the various districts with abundant supplies of spring-waters and streams, whereof the shrines still remain even now, at the spots where the fountains formerly existed.[20]

"Shrines" in the form of statues are also all that is left "where fountains formerly existed," this time on Easter Island, another instance of civilization literally falling with the last tree. When Europeans first visited the island in the eighteenth century, it was already completely devoid of trees except for some few at the bottom of the deepest volcanic crater. Pollen analysis shows Easter Island once had extensive woods, however. By the best evidence thus far, what happened is this. As the population grew, the people felled more and more trees. One use was to move the enormously heavy rock statues to ceremonial sites around the island by way of a moving track built by laying tree trunks alongside one another. Other trees were felled for more ordinary needs — plots

---

20. From Plato's *Critias* as cited by Clive Ponting, *A Green History of the World: The Environment and the Collapse of Great Civilizations* (New York: Penguin, 1993), 76–77.

for agriculture, fuel for cooking and heating, construction material for household goods and pole-and-thatch housing, canoes for fishing. By 1500, trees were scarce enough that people no longer built houses of poles but made reed huts or moved into natural or dug caves. Reed boats soon replaced canoes. This limited the fishing, however, as reed boats could not make the longer voyages canoes could. Fishing also waned because the nets made from the paper mulberry tree, used also to make cloth, could no longer be made. Soil erosion hastened as well in the heavy Pacific rains, so crops declined. About the only food source unaffected were chickens. As these became more valuable to a hungry population, they literally had to be defended! (Stone-built chicken coops remain on the island.) After 1600, the islanders, without trees and canoes and with a diminishing food supply, were trapped and fell into almost continuous warfare. Slavery became common, even among kin. There are even signs of cannibalism. With the environment ruined by deforestation, society collapsed. So quickly, in fact, that half the great statues were still at the Rano Raraku quarry when the Europeans arrived. The Europeans, standing on a treeless island in an otherwise lush zone of the earth, were mystified as to how the great carved rocks had "walked" around the island. Half had "walked" over trees while the other half no longer had any way of walking. They slept dead in the quarry.[21]

So it seems true: trees hold up not only the sky, as some native peoples' stories of origins claim, but civilization as well. Trees mean homes and shelter for humans and other animals, tools, furniture, fishing vessels and other boats, and in fact all the products that have been and can be made from trunks, branches, and leaves, this book and page included. Trees also mean food, both directly from trees and by way of protecting watersheds and holding and nourishing the soils that grow other foods. Trees mean fuel. They remain the basic energy supply for cooking in half the world's homes even today and heating in almost as many. Trees also mean medicines, thousands discovered, some yet to be discovered. Not least, trees are the habitat for the majority of the world's species, many of them known and vital to civilization, many still unknown and potentially vital. Finally, forests are still home to thousands of peoples.

Trees also entertain the journey of the spirit — of forest peoples and others. They are the subject of story, poetry, painting, and sculpture and the site and substance of things religious. They are, then, essential to both the basic needs of civilization and its flowering. Trees of life are life.

The dependence is profound. Trees *do* hold up the sky! More precisely, without them and other green plants, "sky" would not be. Every single breath every single human has ever taken or will take depends on trees and other green plants. Oxygen is the work of trees, not the nurse and hospital supply room. In fact, in earth's slow womb trees together with other plants created the very conditions that eventually made for life as we know it. The coevolution of plants, air, water, and animals depends on the first-listed of these as much as any. And trees, by absorbing carbon dioxide and producing oxygen, link air, water, and biota (living

---

21. This account is from ibid., 5–7.

things) together in the unity that continues to make for life. The Native American legend is thus quite exactly the case, even when the language is mythic: "The sky is held up by the trees. If the forest disappears the sky, which is the roof of the world, collapses. Nature and man then perish together."[22] All trees are literally trees of life.

Nor does any civilization have a substitute. The reason is utterly basic. Green plants, billions of trees included, are nature's only "autotrophs." Autotrophs by definition are the only creatures capable of creating living matter from inorganic raw materials. Trees do this by combining atmospheric carbon dioxide with water from the soil while converting solar radiation into energy through photosynthesis. No creatures other than plants know how to feast on the sun in this way. This process not only provides food for all the world's heterotrophs (you and me, among others) but releases the life breath of oxygen into the atmosphere where we can get at it. (Heterotrophs require complex organic compounds of nitrogen and carbon for metabolism.)[23]

Isaiah was right, then, in a way he never expected. All flesh *is* grass! (40:6). That is, all animals either live on plants and because of them or live on other animals who live on plants. The order is important. There can be no carnivores without there first being herbivores, and there can be no herbivores without plants. Our own flesh would literally not "be" without trees and other autotrophs. All flesh *is* grass. Incidentally, this is the reason green plants hold the trump card. Humans may have the greatest capacity for the cultural organization of nature, but green plants have all the chlorophyll. And in the battle between civilization and chlorophyll, chlorophyll will take the last trick and final round. (Chlorophyll and photosynthesis are the reason why increasing human consumption of the Net Photosynthesis Product [NPP], discussed in an earlier chapter, is alarming. In the closed cycle of utter dependency on trees and other green plants we are literally devouring ourselves when we appropriate more and more photosynthesizers for our own direct consumption. This material sustains all life. Running out of it by consuming its sources means running out of life.)

In short, while thousands of people have lived without love, none has lived without green plants.[24] They — *all* green plants — are trees of life. One can understand why people, especially people who have lived close to trees in undeniable dependence, have pictured the *axis mundi*, earth's axis itself, as a tree. Or have seen in the canopy of trees the cosmos in its infinity and the arch of heaven. Or have found in tree life the power of all life: of fertility and growth, of seasons of change, of survival against all odds, of beauty, and of death and regeneration.[25] The spirit's own body is often a tree.

---

22. Cited from Shridath Ramphal, *Our Country, the Planet: Forging a Partnership for Survival* (Washington, D.C.: Island Press, 1992), 65.

23. Daniel Hillel, *Out of the Earth: Civilization and the Life of the Soil* (Berkeley: University of California Press), 41.

24. With apologies to T. S. Eliot, who wrote in the same vein but said "water" rather than "green plants." Both are true.

25. Langdon Gilkey, *Nature, Reality, and the Sacred: The Nexus of Science and Religion* (Minneapolis: Fortress, 1993), 106–7.

It is something less than surprising, then, that when Charles Darwin finishes his presentation of theory in *On the Origin of Species*,[26] this particular image from nature comes to him. In his own summary:

> The affinities of all the beings of the same class have sometimes been represented as a great tree. I believe this simile largely speaks the truth. The green and budding twigs may represent existing species; and those produced during each former year may represent the long succession of extinct species. At each period of growth all the growing twigs have tried to branch out on all sides, and to overtop and kill the surrounding twigs and branches, in the same manner as species and groups of species have tried to over-master other species in the great battle for life. The limbs divided into great branches, and these into lesser and lesser branches, were themselves once, when the tree was small, budding twigs; and this connection of the former and present buds by ramifying branches may well represent the classification of all extinct and living species in groups subordinate to groups.[27]

In short, Darwin's symbol for the whole drama of evolving life, death included, is a great branching tree. Yet not simply "a" tree, great or not. His presentation ends thus:

> As buds give rise to growth to fresh buds, and these, if vigorous, branch out and overtop on all sides many a feebler branch, so by generation I believe it has been with the great Tree of Life, which fills with its dead and broken branches the crust of the earth, and covers the surface with its ever branching and beautiful ramifications.[28]

The "branching and beautiful ramifications" hold yet one last reason why trees generally, and the tree of life as nature symbolized par excellence, embrace human spirits and inspire them.

"The forests were God's first temples,"[29] wrote William Cullen Bryant. The author citing him goes on to say why we sense the sacred in this place:

> The forest is where the "roots" are, where life rises from the ground. Trees pierce the sky like cathedral spires. Light filters down as through stain glass. The forest canopy is lofty; much of it is over our heads.... [F]orests

---

26. The theory is presented in the first four chapters. The rest of the book is spent in refuting objections and marshaling evidence for the theory.

27. As cited by Stephen Jay Gould in "This View of Life," *Natural History* 12 (1992): 19. "This View of Life," the title of Gould's regular feature, is itself a phrase from Darwin's *On the Origin of Species*.

28. Cited by Gould, "This View," 19.

29. From William Cullen Bryant's "Forest Hymn," as cited by Holmes Rolston III, in "Wildlife and Wildlands: A Christian Perspective," in *After Nature's Revolt*, ed. Dieter Hessel (Minneapolis: Fortress, 1992), 129.

invite transcending the human world and experiencing a comprehensive, embracing realm.[30]

"The clearest way into the Universe is through a forest wilderness," Holmes Rolston III adds from John Muir.[31]

The farmer-poet Wendell Berry has his own way of sensing forests as God's temples. In this case, Sabbath is the subject, and Berry writes that all the body's weariness is left behind, as is the six days' labor, by climbing up through the field "to the high, original standing of the trees" with its "brotherhood of eye and leaf."[32]

What the "high, original standing of the trees" reveals is the sacred as "other" and the "other" as sacred. *Foris,* or "outside," is the likely Latin root of both "forests" and "foreign."[33] Of course, we may feel wholly at home in forests, at one and not estranged. But if so, it is communion with the other *as* other. Here nature is "Thou," an "other" who is not a projection of ourselves or a redundant meeting with our own species.[34] "Thou" remains "other." Jay B. McDaniel says it nicely: "It is not enough to feel 'at one' with others. We must also feel 'at two' with them."[35]

Some would name other religiously charged forms of nature — mountains, rivers, and their valleys, the stars of the night sky — as the first temples of God. The point is the same, however: like Maya Angelou's "A Rock" and "A River," "A Tree" is the way we experience the other *as* other and yet intimate, close to us, indeed somehow truly us as well, even as trees remain truly other.

In this way trees and other forms of nature are a glimpse also of God as Other and as Thou. In many religious traditions, God is named the Wholly Other (*totaliter aliter*), a description profound for some experiences and moments. The same traditions, while underscoring this quality of divinity, usually go on to claim the deepest intimacy with God as Other and as Thou. "Closer to us than we ourselves" is said when other words fail. The point is that trees have a grip on us because they, like the *asherah* at the altar of the Hebrews, join earth and heaven. They bring the far God nigh, and they do so in ways we know the other *as* genuinely other and not ourselves in disguise. In this sense trees may often be a better guide than fellow *Homo sapiens* for our own earth journey of the spirit.

There is something here of direct moral import. Trees help disclose that we are located in nature, amidst deep kinship and deep difference in the same moment. Nature is *in* us as *essential relationship*, and this relationship is not

---

30. Rolston, "Wildlife and Wildlands," 130.

31. Ibid., citing John Muir's *Our National Parks* (Boston: Houghton Mifflin, 1901), 331.

32. Wendell Berry, *Sabbaths* (Berkeley, Calif.: North Point, 1987), 88–89.

33. Perlman, *Power of Trees,* 5.

34. When Martin Buber exemplifies nature as "Thou," he chooses a tree to do so (see *I and Thou* [New York: Scribner's, 1955], 7ff.).

35. Jay B. McDaniel, *With Roots and Wings: Christianity in an Age of Ecology and Dialogue* (Maryknoll, N.Y.: Orbis Books, 1995), 158.

peripheral: the relationship with the other constitutes *our being itself.* Deeper appreciation of difference — the "otherness" of the other — thus goes hand in hand (or hand in branch!) with a sense of the underlying kinship of all things together. In a contracting world in which we are under house arrest (*oikos* again!), a world round and well-wrapped yet bursting with life, learning to live with the many as the "other" who both remains other and is us, is a requirement of survival itself. Such shamanic insight as this is often borne by trees.

Trees as bearers of the Spirit return us to the prophets. Isaiah, with characteristic prophetic passion for earth's redemption, offers his picture of the just, sustainable society. In passing — or more likely not in passing — "the days of a tree" gather up his dream:

> No more shall there be...
>> an infant that lives but a few days,
>> or an old person who does not live out a lifetime. . . .
> They shall build houses and inhabit them;
>> they shall plant vineyards and eat their fruit.
> They shall not build and another inhabit;
>> they shall not plant and another eat;
> for like the days of a tree shall the days of my people be,
>> and my chosen shall long enjoy the work of their hands.
> They shall not labor in vain,
>> or bear children for calamity;
> for they shall be the offspring blessed by the Lord —
>> and their descendants as well.
>
> (Isa. 65:20–23)

CLOSING

This chapter closes with the words of a child, though only after present reality and Shakespeare introduce them.

Present reality is pictured by Alan Durning with an imaginary time-lapse film. Play back the time since the neolithic revolution (10,000 — 8000 B.C.E.) at a millennium a minute. For seven minutes, the marbled blue planet looks like a still photograph. Forests cover one-third of the land. The agricultural revolution makes changes, but on a small enough scale they are hardly visible. Then at about seven and one-half minutes the islands and lands of the Aegean Sea begin losing their forests. At nine minutes, that is, one thousand years ago, scattered parts of Europe, Central America, China, and India show threadbare sections in the forest mantles. Six seconds from the film's end, a century ago, eastern North America shows the same. Still, forests cover 32 percent of earth's lands.

Suddenly, in the last three seconds — since 1950 — an explosion occurs. Forests vanish in Japan, the Philippines, mainland Southeast Asia, Central

America, the horn of Africa, eastern South America, western North America, the Indian subcontinent, West and Sub-Saharan Africa, the Amazon Basin, Papua New Guinea, Malaysian Borneo. Central Europe's forests are there, but poisoned. Incursions take place in Siberia and Canada's north. In only forty-five years, forests fall to 26 from 32 percent of earth's land cover. And only 12 percent of this is intact forest ecosystems. The rest is commercial timber and fragmented regrowth, aimed at lumber and fuel but without a view to what forest ecosystems do: provide habitat for perhaps half the life-forms on earth, hold much of the genetic information of four billion years of evolution, buffer the global climate against greenhouse warming, moderate local climates, prevent floods and droughts, filter the air, secure soil, nurture fisheries in rivers and lakes, keep pests in check.[36] The last three seconds of the film are a fast-forward into unsustainability. Increased economic prosperity for some has accompanied dramatically accelerated deforestation and is part of the reason for it. That prosperity rides on a steep expense to earth's health.

Shakespeare's images in Sonnet 73 are those of ghostly trees whose last leaves have fallen or soon will. These "bare, ruined choirs" capture the poet himself.

> That time of year thou mayst in me behold
> When yellow leaves, or none, or few, do hang
> Upon those boughs which shake against the cold,
> Bare, ruined choirs where late the sweet birds sang . . .[37]

Charles Anflick, of Bridgeport, Connecticut, grade five (age ten or so), also writes of "bare, ruined choirs." His mood would perhaps fit some elders in the late autumn of life; it is far from where any child should be. More importantly, this child is trying, like the biblical writer, to say that when the parents eat sour grapes the children inherit the sour taste, that the sins of the parents work their sure way on down to the fourth and fifth generations of those who follow. Charles's poem, however, is not, like Shakespeare's, about the last leaves. It's "The Last Tree":

> This scene is not pretty,
> Not pretty at all.
> No one really cares,
> No one, big or small.
> The rivers are dirty,
> And the buildings are tall.
> The last tree is dying,
> Dying, that's all.

---

36. Alan Thein Durning, *Saving the Forests: What Will It Take?* Worldwatch paper 117 (December 1993): 5–6. Durning's "film" is accurate overall. But it does omit some important details. The eastern United States, for example, has more forested land, not less, than two hundred years ago.
37. Cited by Perlman, *Power of Trees*, 85.

Fishes and birds
Can't live at all.
They're tearing down a playground
To make room for a mall.
The sun is crying raindrops,
Raindrops, big and small.
And the last tree is dying,
Dying, that's all.

The clouds are smoke,
And grey and black.
No one is cleaning up,
No one with a sack.
No one is cleaning the water,
The air, and the ground.
We cannot live,
And now we are bound.
For the last tree is dying,
Dying, that's all.[38]

"And the leaves of the trees are for the healing of the nations" (Rev. 22:2). That or Charles's last tree? Is it Isaiah's future — "all the trees of the field will clap their hands"? Or the Bard's — "bare, ruined choirs where late the sweet birds sang..."? Trees of life, or dead wood?

There is a true yearning to respond to
The singing River and the wise Rock.
So say the Asian, the Hispanic, the Jew
The African, the Native American, the Sioux,
The Catholic, the Muslim, the French, the Greek,
The Irish, the Rabbi, the Priest, the Sheik,
The Gay, the Straight, the Preacher,
The privileged, the homeless, the Teacher.
They hear. They all hear
The speaking of the Tree.[39]

Life cannot live without trees. But for powerful symbolizing animals like us, whether trees themselves live depends in part upon whether they are disclosures of life itself, whether they are religious symbols from which we learn earth ethics. The tree of life is a religious symbol with sturdy moral import. So, too, the rest of nature.

---

38. Charles Anflick, Hillel Academy, Bridgeport, Conn., as printed in *Environmental Sabbath Newsletter* (of the UN Environment Programme) 1, no. 2 (winter 1990); no page numbers given.
39. From Maya Angelou, "On the Pulse of Morning" (New York: Random House, 1993, n.p.).

The Gifts of Darkness

Children's ways can be arresting. You never see three-year-olds pull their chairs up to the kitchen table, grab a mug of milk, and reminisce about the good old days. But they do love to indulge in nonstop conversation. Often it is at bedtime when the parental directive is "go to sleep," and the child's agenda is some perplexity left over from the day just past or some anticipation of the one to come. I no longer remember the hoary details of those conversations. But I recall verbatim our son's words now and again as I stepped from bedside to door: "OK, Dad, you can turn on the dark now."

Because we get the symbols wrong, we get reality wrong. Light means life, we think, and darkness means foreboding, danger, death with all its authority. Yet since Hiroshima and Nagasaki, light has ghastly meanings, as it has since flames lit the night sky in the serene Polish countryside at Auschwitz. Lucifer, demonic prince of the netherworld, comes as an angel of light.

And even in New York, in love with lights as all great cities are, light all night is exhausting. Civilization means city ways, and city ways do their best to ban and banish darkness. Everything is throbbing energy twenty-four hours a day. Our adrenaline is pumped, but this is life without respite, peace, solitude, Sabbath. The child's instruction would never occur to us: "OK, Dad, you can turn on the dark now."

This is a meditation on the gifts of darkness in a city and world that will have none of it. I understand full well I am running counter to civilization itself and that something like a conspiracy against darkness and its gifts spans the globe. But there is a reason for the eccentric choice.

The reason has already been given in part: because we get the symbols wrong, we get reality wrong. So we need to look at the symbols anew and ask what they are doing with our lives. Because nature has been pictured as ever-regenerative, infinitely malleable, and virtually inexhaustible, for example, and we a powerful species apart, so much of that from and by which we live has been destroyed. Wrong pictures of reality can mean dreadful treatment of it.

Yet this reiteration of cosmology's import for ethics is only part of the reason for this chapter. The other is the search for more viable symbols themselves as guides for an adequate earth ethic. That search has already asked whether some old ecumenical symbols have new life; whether, in this case, the tree of life in its factual, religious, and moral dimensions offers substance. We might have done the same with the sacred mountain or the river. This chapter undertakes something quite different. There is no strong symbol here with a long and rich history across multiple cultures. On the contrary, "darkness," if a symbol at all, is a negative one: foreboding, danger, and the authority of death, as we noted. Yet the requirements of a viable cosmology and ethic for the next journey in earth's history are not timid, but radical. An earth ethic worthy of the name needs not only rediscovered treasures of the past but newfound gifts in the present. Some of those symbols will be quite fresh, arising from a different twist on familiar experience, a matter of returning home and seeing something for the first time, or a case of the routine and obvious turned in a way that gives rise to thought and new appreciation — like the child's, "OK, Dad, you can turn on the dark now."

So while this chapter, like the others, intends to learn from the forms and rhythms of nature in order to understand how we might more lightly live upon the earth as part of it, it is unlike previous chapters in that it chooses and stays with an unexpected source — darkness — to ask whether it might not be what it has only rarely been: a positive, saving symbol, a vital element of earth's own redemption, a needed way of seeing. If eyes can be trained to see darkness as a contributing symbol, perhaps other needed reorientations can happen as well. Without some such radical reorientations, much of earth's cause is likely lost.

The language here is meditative and evocative. This, too, is a language of ethics. Ethics is not only, or chiefly, moral exhortation and instruction in the manner of parents, teachers, priests, or prophets. Nor is its language only that of narrative — locating ourselves in the story that gives right direction and meaning. The language of ethics is not only rational analysis and argument, either — moving carefully from premise and evidence to conclusions about what should be done. Ethics, while all of these, is also intuition and evocation from experience, a lifting from what we know in our bones or sense in our hearts so that our discernment about what we are to be and do is less misty, clearer-eyed, and truer to the world to which we belong.

There is a way of being in the world, then, that takes its clues from our resonances with nature's rhythms as we reflect on nature itself in one of its moods, even when, perhaps *especially* when, that reflection is not in the form of argument, exhortation, or plot. Instead the mood is sheer receptivity, and the language is pure evocation. Without some such resonances and receptivity we will not likely be re-formed.

Consider this, then, one abbreviated instance of what needs to be done not only with darkness but with any number of other rhythms and forms in nature.

SOULS SEE IN THE DARK

A Quaker friend has a sticker on her refrigerator door something akin to the Psalmist's "Be still, and know that I am God." It reads: "In case of emergency, keep quiet." Life is something of an emergency now, and it's long since time to consider the gifts of stillness and darkness, augmented with a few candles now and then.

Remember Wendell Berry's old farmer who, unable to walk any farther, stopped at the spring and beheld "wonders...little wonders of a great wonder"?[1] Berry's preceding sentence is this: "And it comes to Mat once more, by stillness, he has passed across into the wild inward presence of the place."[2] By stillness, yes, but also by darkness. And by still darkness most of all.

Darkness cleans the soul, especially darkness in the hills, the woods, or the desert. Shadows rise there. We get this reality backward, too. Contrary to what we think, shadows don't fall at eventide, they rise. If you live in the city, go to the park in autumn and let the trees teach you as the sun falls. You'll find nature there "in a painting mood," as Aldo Leopold once said of a river at eventide.[3] First grace will creep silently across the grass, itself already refreshed at the thought of sleep and new dew. Then grace as shadow will, more gently than any other movement in all nature, slowly climb the tree trunk, close the door of the woodpecker's hole for awaited rest, and pause just long enough to let the blazing crown of red and gold leaves strut their stuff before the night and another hard day of big changes.[4] Shadows rise. What falls are the encumbrances of the soul.

Darkness is also dreaming time. We daydream, of course — on the subway, at the bus stop, in a meeting. But daydreaming runs a poor second. It is not the real thing. True dreams come by night, when whales dream and swallows and angels and the Spirit. This, too, is gracetime.

In this grace of dreamtime and darkness the soul is restored. If there were only light, the soul's restless dreams would soon flame out. Darkness restores the soul because in the night all things are reconciled. They all merge, they all join the same soft dark color, they cease their predation, they rest quietly in praise. In darkness, especially candlelit darkness, we all stop trying to outshout one another. Ever notice how voices grow hushed as evening silently advances, how we listen more carefully and far better with one another when we are still, together, in the dark? Often eventide is the only time of day when we even hear the Spirit, God's soft breath and gentle breeze, the sure presence of the divine. It's the time rage subsides and we can be one Body. Sometimes darkness is the only time we even notice sounds as silent as doves' feet or the pacifying whispers of serenity. Otherwise even hope is drowned out by rattling jackhammers and a million other noises of empires at work. But with the dark they cease; eternity steals a visit; and the soul recovers.

In fact it is the silent working of life itself that comes by night. In the growing darkness, nature and the Spirit's closeness are felt even on the skin and are palpable in the musty air of autumn. The breath of life the trees breathe out, I breathe in; the breath I breathe out, the soil breathes in; the breath the soil breathes out, the birds breathe in; the breath the birds breathe out, the water breathes in. Life is one. And this mighty presence, the truly mighty God, is

---

1. Wendell Berry, "The Boundary," in *Late Harvest: Rural American Writing*, ed. David R. Pichaske (New York: Paragon, 1992), 140.

2. Ibid.

3. Aldo Leopold, *A Sand County Almanac* (New York: Ballantine, 1970), 55.

4. Some of the images in this chapter as well as the idea to use darkness as a symbol are inspired by the chapter "The Gift of the Night," in Erazim Kohák's *The Embers and the Stars: A Philosophical Inquiry into the Moral Sense of Nature* (Chicago: University of Chicago Press, 1984), 29ff.

quietly praised in the darkness by the *ruach* of life (spirit, breath) no one even notices by day.[5]

Too, darkness teaches. It teaches true place. We are not masters of all we survey or even captains of our own tormented souls. We are simple little creatures alive with other simple little creatures in a cosmos too colossal to catch, except perhaps in the small twinklings of distant suns, a wrinkle or two in time, and whatever tiny portion of the unfathomable imagination of God we are able to borrow for a passing moment. Darkness teaches the utter giftedness, even strangeness, of life and comes to us without earning or requirement, status or striving, of our own. We are simply justified in God's reclaiming of creation, justified in the order of being itself, and saved with all the rest of it, together — the cottonwoods (*los alamos*), maples, jays, all creeping, crawly things, things visible and invisible. This is none other than the gift of God and the feast of life. It is best known when darkness reconciles us all in the beauty of blackness[6] and wee little candlelights.

I enjoy burgeoning cities. I love New York's energy and wild collection of peoples, and I'm glad to walk crowded sidewalks. But only when the gifts of darkness now and again surround and embrace, when darkness quells all compulsion to do more or to have more, when shadows and stillness let being simply, quietly, be. For we (some we's far more than others) have *made* life an emergency now. The planet reels from centuries of trying to make it light and white, fast, packaged, and marketable. So darkness saves, at least tries to save, all who have been wrecked by five hundred years of manic commerce and a half-century of development. Trees and darkness save, as in Langston Hughes's "Dream Variation":

> To whirl and to dance
> Till the white day is done.

---

5. Stephanie Kaza describes the practices at the Green Gulch Zen Center in northern California. Zen meditation is silent rather than verbally guided and, as is the case at this Zen center, is often held in the dark or semidark. To cite one instance, students rise at 5:00 A.M. for a walk through the darkness to the candlelit hall for meditation. Kaza herself reflects on the kind of ethos, moral ethos included, that silence and darkness create: "In the softened transitions at dawn and dusk the senses grow attentive to subtlety....The dark night presents the primordial emptiness, the cosmos before light and sound. The night environment becomes the Green Buddha teaching. We listen for the great horned owls calling across the fields, we walk carefully in our small bodies on the larger body of the earth. Outside the small center the hills remain dark and wild, an invitation to taste their original state of mind. Many of the traditional forms of Zen practice emphasize restraint, moderation, and simplicity. Taking care of silence and darkness helps to develop the capacity to realize the interconnected nature of all beings and activities. Simply by not talking, by not obscuring the darkness, one increases the chances of engaging directly with other existences. Out of this grows a natural love and sense of personal relationship with land and water. The practices of observing silence and darkness bring one spiritually and literally closer to others, helping break through the fundamental delusion of the isolated self" (Stephanie Kaza, "Cultivating the Empty Field," *Fellowship* 61, nos. 3–4 [(March/April 1995]: 21).

6. Renaissance writers and painters regularly pictured the color of earth as black, as a reminder to humankind of that from which we come, are made, and to which we will return. John Harvey's *Men in Black* (Chicago: University of Chicago Press, 1995) offers a fascinating cultural history of black and blackness.

> Then rest at cool evening
> Beneath a tall tree
> While night comes on gently,
>   Dark like me — ...
> Rest at pale evening...
> A tall, slim tree...
> Night coming tenderly
>   Black like me.[7]

The humble point? Living means lighting a light *in* the darkness, not *against* it.[8] Souls see best in the dark.

SABBATHTIME

Darkness Saving the Day

Because we get the symbols wrong, we get reality wrong. Christians are supposed to know the meaning of Sabbath. But it takes Jewish friends to teach us anew.

We live in the most anti-Sabbath society in the history of the world. Stores boast, rather than repent, that they are open twenty-four hours, open to the "max" any day allows. At "convenience stores" and one-stop shops in many a town I can buy sodas and beer and beef jerky twenty-four hours a day, three hundred and sixty-five days a year — as though beer and jerky at my disposal were the sure sign of the good life. In New York I can in fact eat, drink, entertain or be entertained, make money or lose it twenty-four hours a day.

Should any ever desire to live true countercultural lives, I know the secret. It's stop, simply stop. Stop, that's all. Just quit. Quit everything — for one day. No travel, no food preparation, no buying or selling, no planning, no marrying, giving in marriage, or quitting marriage, no reading, writing, taking out loans or repaying them. Rest. Just rest, that's all. Observe Sabbath. If you don't go crazy on your own, someone will, I assure you, soon certify you as such. You cannot both violate the economy and be sane.

But Christians, unlike Jesus, get the basic symbol wrong. We don't know when Sabbath begins. We think it begins by dawn's early light on the first day of the week, the Big Work Week. Jesus knew better. Of course Sabbath begins at dusk, when shadows rise and candles and quiet and night and God come. And when stillness subtly, in the very last hours, saves the day and the week from sure

---

7. Langston Hughes as cited in Michael Perlman, *The Power of Trees: The Reforesting of the Soul* (Dallas: Spring, 1994), 164.

8. This chapter, as meditation, does not undertake historical, cultural, and social analysis of the immensely destructive use of darkness and "black" as markers of caste and race. For a survey, see Emilie M. Townes, *In a Blaze of Glory: Womanist Spirituality as Social Witness* (Nashville: Abingdon, 1995), chap. 5, "Another Kind of Poetry: Identity and Colorism in Black Life," 89ff.

death by efficiency. Of course Sabbath comes as a gift of darkness, and not light and day. For it is by night and candlelight that we cease our cravings and are taught to open ourselves, allow ourselves to be empty and receptive. Our egos, busy all day and all week, somehow miraculously recede in darkness, darkness with prayers centuries old and chants without composers' names, prayers and chants known only by the smooth tones of a communion of countless saints gathered in quiet worship over umpteen centuries. Of course Sabbath comes with night and candlelight, when spite and quarrels somehow seem badly out of place and hatred a very wrong emotion. Soul knits to soul in the dark, against all the daytime impulses to unravel them. Of course Sabbath comes at eventide, when the soft touch of hands held in prayer, or subdued conversation over a good meal, says "welcome," "shalom," and "be well" to every tired cell in the body. Sabbath may in fact be the one gift that could save this society, because it is the gift of rest and inefficiency, life at ease well before life is morning and new light. Jesus rose by dawn's early light, the Christians say. Yes, but only after, and maybe because, he faithfully lived by Sabbath, with its final trust in God and its confidence in Jubilee. Or *did* Jesus rise at dawn? He rose on the third *day*. And in those days "day" began at dusk, as true living does. Maybe, just maybe, then, Jesus rose in the beauty of blackness. By dawn he was long gone.

La Tristeza de la Vida

Not all darkness is rest, well-being, or serenity. There is a certain quiet sadness of life that is also revealed in the darkness — *la tristeza de la vida*. It's always been there, a low, somber hum in the soul we cover as best we can with the noise of the day and cascading, jarring colors, most of them expensive ads in windows, magazines, newspapers, on billboards and television, in the subway and on the side of the bus. But in the stillness the sadness is heard. It is heard most of all when night takes possession of the day. It is the sadness of life that took Jesus into the wilderness, even for long stretches, and eventually to Gethsemane, just as it can take us beside still waters to listen, with God, to the sorrow that has no name, the sorrow of the world that is our own soul's sorrow. People not only suffer from it but die from it if it is not listened to and if we try to bury it in endless pursuits. Worst of all, people kill people and other creatures out of life's persevering sadness, sorrow, and pain. We and others can die from *la tristeza de la vida* if we do not listen in the darkness, listen deeply, and offer it to God in the large or small Gethsemanes, wilderness wanderings, griefs, exiles, and estrangements of our lives. For it is in the darkness that the soul can float *in* its sadness into the holiness of prayer and solitude and acknowledge the low hum there, without being overcome by depression and terror. True, one speaks little or not at all in such times. One nonetheless shares everything, in the darkness, by candlelight, in the sanctuary, wherever the sanctuary may be; and with that, *la tristeza de la vida* subsides for a merciful moment. Then, out of this very darkness, life itself may emerge as it always has, from a womb where no light is, but where all life begins.

Notice this carefully, then, for the sake of earth: new life always begins in darkness, in dark wombs or dark soil, even dark tombs. New life is the gift of darkness.

CONTINUATION

"An argument leads to a conclusion," Erazim Kohák writes, "an evening does not."[9] Not only an evening, however, but any number of other rhythms and forms of nature. These, too, need our receptive attention for an ethic of kinship with all things. A mountain doesn't have "a conclusion," either, at least not like an argument's or a story's. Nor does water, a bird's song, a high mesa, a festive meal, or our bodies. One of the sure tasks of earth ethics is creatively to symbolize anew the wonder of life around us and listen to the guidance of the "little wonders" for a way worthy of earth as the great wonder.

What is at stake? What modernity has learned to its peril: when we lose nature to the senses, we lose not only God but ourselves as well.

> If the mountains die,
> where will our imaginations wander?
> If the far mesas are leveled,
> what will sustain us in our quest to be larger than life?
> If the high valley is made mundane by self-seekers and
> careless users, where will we find another landscape
> so eager to nourish our love?
> And if the long-time people of this wonderful country are
> carelessly squandered by Progress, who will guide us
> to a better world?[10]

---

9. Kohák, *Embers and the Stars*, 129.
10. From John Nichols and William Davis, *If the Mountains Die: A New Mexico Memoir* (New York: Knopf, 1987); no page number given.

Adam, Where Are You?

Unsustainability is a cosmological and moral disorder tied into a cultural, social, and environmental one. It is only initially about recycling programs and cleaning up a fouled nest. Foregoing chapters have turned anew to nature for guiding ecumenical symbols, symbols old (the tree of life) and new (the gifts of darkness). Sustainable communities require different orientations and habits, not only different technologies. The next chapters stay with symbols, and with nature, but take a different tack. They revisit particular, long-established religious traditions — chiefly Christianity and Judaism — to ask whether they might yet become genuine earth faiths. Do they, or might they, connect cosmology to ethics in ways that orient us well to earth as earth? Are their own strong symbols also strong symbols for earth's care?

PLACE AND CALLING

When the habits of everyday, now largely unsustainable life are peeled back several layers, basic questions appear, questions of religious and moral scope: Who are we in the scheme of things? How have we understood ourselves? And how do we muster living that is something other than a hypnotic mask for a dance of death? The Seventh Assembly of the World Council of Churches in Canberra, Australia, 1991, is a convenient screen for projecting the question of humanity's place and calling, at least as Christians mull that over. Granted, some thought that assembly hopped the track, but Canberra did give earth issues high standing; and it addressed these as a diverse religious community facing the global degradations many of its delegates knew firsthand. The assembly theme itself was a prayer and a plea — "Come, Holy Spirit, Renew the Whole Creation!" Section I of the assembly, with its subtheme, "Giver of Life — Sustain Your Creation!" put a large number of delegates to work on specific earth issues.

Taking the scalpel to WCC reports point-by-point is not the best use of delegates' contributions. More promising is an account of their general orientations for our question: Facing earth's distress, how do we now image ourselves?

Nonetheless, the reports set the stage and aptly framed what followed. Both a preassembly consultation in Kuala Lumpur, Malaysia, and Section I of the Canberra report called for a recast theology of creation and a justice ethic of sustainability. These located the agenda as well as any sonar could have. The Kuala Lumpur report, written as a letter and study document for delegates to Canberra, said this about human identity and location (recall the assembly theme: "Come, Holy Spirit, Renew the Whole Creation!"):

227

If created life shares in the uncreated life of God through the all-pervasive presence of the Spirit, then we humans, bonded to one another and the rest of nature, must respect the mystery of life and acknowledge the dignity of all creatures. In our co-existence with the rest of nature, we may understand ourselves in various ways — as the present trustees of the tiny speck of creation called Earth, as servants of the Spirit and the earth, as the priests of creation, as its tillers and keepers, as co-creators, or as that portion of nature come to consciousness of itself in creation's own ongoing life. But in all cases, we ought to let our understanding be informed by the knowledge that we belong to the community of all created life; let our whole orientation be governed by deep respect for this community of life; and let our actions be moved by compassion for all life and accountability to the Spirit, one another, and the rest of the biotic community. We wield an awesome collective power to affect all life in fundamental and unprecedented ways and we pray for a humble and contrite spirit in its use. May the Giver of Life give us life and the attitude of prayer, that we might be worthy participants in sustaining the only creation we know.[1]

With this as orientation and mood, what were the perspectives of Canberra delegates? More importantly, what is the spectrum of working possibilities today, and where did delegates land on this spectrum? What are and might be the ways we image ourselves as distinctive earth creatures about to cross over into another millennium and in search of the fourth great human revolution? What symbols and models do or might we live by as we answer who we are now and what is required of us?

DOMINION

Beginning with the model of dominion as mastery may seem out of season, since it is on the skids in the public debate and was aired in Canberra only for the sake of naming what should happily be left behind. Yet the theology of dominion remains the reigning one where it counts most, *in practice*. For that reason, comment is required.

Social arrangements, especially, and the busy structures of economic life, in particular, still assume the tenets of mastery and live by them: earth exists for us; reality is a collection of objects we give shape and purpose as we form a world of our own making; and earth is stage, resources, waste container, and information for doing so. Both industrial and informational "ecologies" hold these as the practical doctrine of their working cosmologies and sell their promises on the

---

1. "Giver of Life — Sustain Your Creation!" (Report of the Pre-assembly Consultation on Sub-theme I, Kuala Lumpur, Malaysia, May, 1990), 6–7. The assembly reports themselves are available in *Signs of the Spirit*, ed. Michael Kinnamon (Geneva: WCC Publications; and Grand Rapids, Mich.: Eerdmans, 1991).

basis of them. It is the marriage of science and business especially, Daniel Callahan says, and its joining of "a mechanistic view...with an unrelenting profit motive" that have created "that most treacherous of American progeny: commerce masquerading as human liberation."[2] Commerce masquerading as human liberation is not the only surviving form of dominion theology, but it is the most prominent one. This mastery model still surfaces in religious rhetoric and practice as well. At the hyped twenty-year-anniversary gathering for Earth Day in Central Park, New York City, in April 1990, Cardinal O'Connor (clearly no relation to the bird of the same name and color) delivered his punch line, following a condescending remark aimed at all who worry about saving whales and snail darters: "The earth was made for man, not man for the earth."

The cardinal does not lack support. Official papal teaching, even as recently as 1995 and *The Gospel of Life*, continues to propagate dominion theology. *The Gospel of Life* even answers its own question, "Why is life a good?" (referring to human life only), by arguing that human life is "quite different from the life of all other creatures." While human life, too, is formed from the dust of the ground, it "has been given a *sublime dignity*" created by the special bond of humankind to God as God's image. The Book of Genesis affirms this, the pope continues, "when...it places man at the summit of God's creative activity, as its crown, at the culmination of a process which leads from indistinct chaos to the most perfect of creation. *Everything in creation is ordered to man and everything is made subject to him*." After citing biblical chapter and verse, John Paul II draws his conclusion: "We see here a clear affirmation of the primacy of man over things; these are made subject to him and entrusted to his responsible care, whereas for no reason can he be made subject to other men and almost reduced to the level of a thing." Unwittingly no doubt, the pope has repeated and restated Bacon's sharp distinction: all things are properly subjected to human power if they are of nature, with the grand exception of fellow humans, who are of a wholly different order. The implication is also that nonhuman life resides somewhere very near "the level of a thing." Later the same encyclical in fact says that "animals and plants," while also creations of the divine life, "receive only the faintest glimmer of life."[3]

Despite papal reaffirmation of dominion theology, careful argument for mastery has largely left the scene. Lynn White Jr. perched it on the downward slope in 1967,[4] and feminists have pushed it well along since, not least Rosemary Radford Ruether and her straightforward questions: "People of the biblical traditions have to ask more seriously, who gave humans 'dominion'? Is the 'God' of human dominion the God of earth? Has this God not been created on behalf of the

---

2. Daniel Callahan, "They Dream of Genes," *New York Times Book Review,* September 12, 1993, 26.

3. Pope John Paul II, *The Gospel of Life* (New York: Random House, 1995), 60–61, 149.

4. Lynn White Jr., "The Historical Roots of Our Ecologic Crisis," *Science* 155 (March 10, 1967): 1203–7. White actually broached the thesis of this essay twenty years earlier, in "Natural Science and Naturalistic Art in the Middle Ages," *American Historical Review* 52 (April 1947): 421–35.

dominion, not of all humans, but a male elite?"⁵ So with the possible exception
of Rome, hardly anyone reads Francis Bacon anymore without wincing at his
"enlarging of the bounds of human empire, to the effecting of all things possible" by using nature as the "anvil" on which humans "hammer out" a world.⁶
(This is not quite Thomas Huxley's principle from his lecture on ethics and evolution in 1893, but it is close. Huxley wrote: "Human civilization consists in
resistance, and the negation of nature at every step.")⁷ Mastery and control as
the model for our place in the world, and the kind of moral totalitarianism it
implies, found only an occasional echo in Canberra and no resonance at all in
the official reports and recommendations.⁸

STEWARD

Many suckled and raised on mastery have stepped over the threshold to the custodial or steward model. The language itself may still echo of "dominion," but

5. My thanks to James B. Martin-Schramm, "Population, Consumption, and Ecojustice: Challenges for Christian Conceptions of Environmental Ethics" (Ph.D. diss., Union Theological
Seminary, New York), 20, for pointing me to this passage from Ruether. The source is Rosemary
Radford Ruether, "Symbolic and Social Connections of the Oppression of Women and the Domination of Nature," in *Ecofeminism and the Sacred,* ed. Carol Adams (New York: Continuum, 1993),
22–23. Any number of other sources offer the feminist critique as well. See, for example: Irene Diamond and Gloria Feman Orenstein, eds., *Reweaving the World: The Emergence of Ecofeminism* (San
Francisco: Sierra Club Books, 1990); Anne Primavesi, *From Apocalypse to Genesis: Ecology, Feminism,
and Christianity* (Minneapolis: Fortress, 1991); Rosemary Radford Ruether, *Gaia and God: An Ecofeminist Theology of Earth Healing* (San Francisco: HarperSanFrancisco, 1992); Sallie McFague, *The
Body of God: An Ecological Theology* (Minneapolis: Fortress, 1993).

6. Francis Bacon, *Works,* ed. James Spedding, Robert Leslie Ellis, Douglas Devon Heath, 14
vols. (London: Longman Green, 1879), 3:156. Fairness to Bacon requires noting that he also wrote
the following, in *The Great Instauration:* "Lastly I would address one general admonition to all, that
they consider what are the true ends of knowledge, and that they seek it neither for pleasure of
mind, or for contention, or for superiority to others, or for profit, or fame, or power, or any of these
inferior things; but for the benefit and use of life, and that they perfect and govern it in charity.
For it was from lust of power that the angels fell, from lust of knowledge that men fell, but of
charity there can be no excess, neither did angel or man ever come in danger of it" (cited by Charles
Birch and John B. Cobb Jr., *The Liberation of Life* [Denton, Tex.: Environmental Ethics Books,
1990], 231). It should be noted, however, that this moving qualification of the ends of power and
knowledge does not extend "charity" to (nonhuman) nature.

7. Cited from Thomas Berry, CP, with Thomas Clark, S.J., *Befriending the Earth: A Theology of
Reconciliation between Humans and the Earth* (Mystic, Conn.: Twenty-Third, 1991), 49; no source is
given for the Huxley quotation.

8. Not that the WCC had no history of dominion theology as domination. The third assembly
at New Delhi in 1961 included, without significant objection, statements like this one: "The Christian should welcome scientific discoveries as new steps in man's dominion over nature." The place of
nature itself was voiced in a preparatory document for the world Conference on Church and Society
in 1966: "The biblical story...secularizes nature. It places creation — the physical world — in the
context of covenant relation and does not try to understand it apart from that relation. The history
of God with his people has a setting, and this setting is called nature. But the movement of history,
not the structure of the setting, is central to reality. Physical creation even participates in this history; its timeless and cyclical character, as far as it exists, is unimportant. *The physical world, in other
words, does not have its meaning in itself*" (citations from Wesley Granberg-Michaelson, "Creation in
Ecumenical Theology," in *Ecotheology: Voices from South and North,* ed. David Hallman [Maryknoll,
N.Y.: Orbis Books; and Geneva: WCC Publications, 1994], 97, 96).

it is not dominion as mastery and control. Rather, human beings are pictured as *oikonomoi*, trustees of the *oikos* and the tillers and keepers of earth as our patch of creation. The particular accent for this revived image takes account of the modern world and its altered relationship of human power to nature, including human nature. Humans, who often feel so powerless, are recognized cumulatively as wielders of power to affect all of life in unprecedented ways. We possess knowledge to save and knowledge to destroy. This knowledge is outstripped only by our ignorance and its dangers. The moral quest is thus clear: a just, sustaining use of unprecedented knowledge and power, an urgent effort to match wisdom and responsibility. Douglas John Hall, an active participant in WCC circles, has given this biblical notion renewed theological currency,[9] not least by defining "dominion" Christologically. If Jesus is *dominus* (Lord), then the human exercise of power should be patterned on his kind of lordship — a servant stance in which the last are made first, the weak are made strong, and even the sparrow is cherished, so that all might be gathered into covenantal intimacy on equal terms. Similarly, and with an eye to modernity, Jewish writers have revived the biblical legacy and reaffirmed the long trail of Jewish exegesis of Gen. 1:28 ("Be fruitful and multiply, and fill the earth and subdue it; and have dominion..."). The exegetical kernel is that while all living things on earth have some human reference and use, the proper human attitude is one of restraint, humility, and even noninterference, except for matters of necessity (such as daily bread). In fact, the modern capacity to image human beings apart from the rest of nature is largely lost to Jewish exegesis because Rabbinic Hebrew, like Biblical Hebrew, has no word for nature as a realm separate from human being or for creation as a finished state. As with Hall's work, there *is* a human prominence in this exegesis, to be sure. Precisely for this reason, stewardship is a prominent theme. But it is a prominence *within* a single Community of Life, before, under, in, and with *God* (the biblical angle is not homocentric or biocentric, but theocentric). "Dominion" is never, therefore, exegeted by the rabbis as license for exploitative subjugation. It is humble participation with God in ongoing creation as a totally interrelated reality, accompanied with a high sense of moral responsibility for consequences. We are *shomrei 'ădāmâ* — guardians of earth.

In a nice twist, humans will retain dominion *only* so long as humankind is relatively "righteous," or just, according to the rabbis. When humankind is not, the rest of nature reacts, and the ground itself cries out for justice, challenging these stewards to return in humility to the tilling and keeping of the earth proper to their calling. This is the true custodial task (Gen. 2:15).[10]

---

9. See especially Douglas John Hall, *The Steward: A Biblical Symbol Come of Age*, rev. ed. (Grand Rapids, Mich.: Eerdmans, 1990; New York: Friendship Press, 1990), but also his *Lighten Our Darkness: Toward an Indigenous Theology of the Cross* (Philadelphia: Westminster, 1976) and *Professing the Faith: Christian Theology in a North American Context* (Minneapolis: Fortress, 1993).

10. In reporting the Jewish tradition, Michael Lerner says the dominion notion of stewardship emphasizes "the need for human beings to shepherd and to care for, not to transform and remake. Insofar as human beings are meant to interfere, it is only to make the world more deeply based on an understanding of the ultimate harmony and moral imperative within the universe" (*Surplus Powerlessness: The Psychodynamics of Everyday Life and the Psychology of Individual and Social Transformation*

Hebrew conveys this stewardly calling better than English. The charge "to till and keep" of Gen. 2:15 (NRSV) is literally "to serve and preserve" (*l'ovdah ul'shomrah*). To serve means, literally, to cultivate. The connection to stewards (*shomrei*) as guardians, custodians, and preservers of earth is plain.[11]

The steward's intimacy with the soil is also underscored. "Adam" is derived from *'ădāmâ* — a Hebrew noun of feminine gender meaning earth, topsoil, or ground. "Adam" thus encapsulates something of human origin and destiny itself. Adam derives life and livelihood from the "dust" from which Adam comes and to which Adam returns. Likewise, Adam's companion is Hava — Eve in English. *Hava* literally means "living." In the Bible's words: "The man named his wife Eve, because she was the mother of all living" (Gen. 3:20). Together, then, Adam and Eve signify "Soil and Life."[12] This is the identity from which emerges the human vocation of *shomrei 'ădāmâ*.

Michael Lerner, commenting on this exegetical legacy, emphasizes that it underscores "a real humility in which we see ourselves as part of the totality of Being, understand that nature itself is permeated with the spirit of God, and recognize that the chosenness of the human species, our ability to develop a certain level of self-consciousness, is at the same time an obligation toward compassion, caring, and stewardship."[13]

Yet finally in the rabbinical tradition, it is Sabbath and not dominion that symbolizes the proper relationship of humans to the rest of nature and of all creation together to the creator. Indeed, Sabbath, and not the creation of humans, is the crown and climax of the creation story itself, papal exegesis notwithstanding. Thus while stewardship has real tenure over two millennia of Jewish teaching, and dominion denotes real human prominence in all creation accounts, this has lacked the odor of chauvinism and unrestrained utility that it came to hold in Christian practice after the commercial globalization by the West and the industrial revolution.[14]

[Atlantic Highlands, N.J.: Humanities Press International, 1986], 297). Lerner then cites the rabbis: "Thus, Noah is said by the Rabbinical tradition to have spent much time on the ark trying to quiet down fighting among the animals — but never to have been involved in throwing any species off the boat for not fitting his ideal."

11. Daniel Hillel, *Out of the Earth: Civilization and the Life of the Soil* (Berkeley: University of California Press, 1991), 12, 286n. See also Theodore Hiebert, "Rethinking Traditional Approaches to Nature in the Bible," in *Theology for Earth Community: A Field Guide,* ed. Dieter Hessel (Maryknoll, N.Y.: Orbis Books, 1996,) 23–30.

12. Ibid., 14.

13. Michael Lerner, *Jewish Renewal: A Path to Healing and Transformation* (New York: Putnam, 1995), 416.

14. Helpful overviews are provided by David Ehrenfeld and Philip J. Bentley in "Judaism and the Practice of Stewardship," *Judaism* 34 (summer 1985): 301–11, and Ismar Schorsch in "Learning to Live with Less — A Jewish Perspective," available from the Jewish Theological Seminary, 122nd and Broadway, New York, NY 10027.

A portion of the Section I report of the Canberra assembly is worthy of citation here: "For Jews and Christians together the institutions of the Sabbath, the sabbatical year and the jubilee year provide a clear vision on economic and ecological reconciliation, social restoration and personal renewal. Sabbath reminds us that time, the realm of being, is not just a commodity, but has a quality of holiness, which resists our impulse to control, command and oppress. In the concepts of the sabbatical and jubilee year, economic effectiveness in the use of scarce resources is joined to

The ambiance of stewardship and dominion-talk at Canberra was fascinating and instructive. With support from Asian, African, Pacific Islander, and Latin American Protestants, Eastern Orthodox delegates were emphatic that the language of Section I's report should encompass the place of humans as stewards. By contrast, most Europeans and North Americans stood sober before "dominion" and were fearful that "steward," too, might easily slide into the imperial mastery they knew well. Indeed, Christians from the Two-Thirds World were quick to join interpretations that rejected Western domination as an expression of Christian "civilization" and dominion. At the same time they clamored for the distinction and high office of the human as "steward" or householder and "tiller and keeper." Power relationships and their history are no doubt as crucial as exegetical ones in this instance. But the exegesis probably sided with the Orthodox and their supporters as well. In the terms "the image of God" and "dominion" both Jews and early Christians heard the message of affirmation and dignity married to moral agency and responsibility for history. They heard a task, a calling, a vocation, a commission to be the "responsible representatives" of none less than God.[15] From the perspective of the globally less powerful, to be named by God the custodians of creation and God's appointed is an empowering word.

In Canberra this message swung over to something like this: "We've been beaten up by others for centuries, in Europe, Asia, Africa, the Americas, and the Pacific. We're tired of it now; we've been tired of it for five hundred years. And we are here to say that this is not the reality we live from! The reality we live from is that we are created in the image of God, and we are the subjects, not the objects, of history." Or, to recall Chung Hyun Kyung's comment in floor debate (to applause): "We are new wine. You will not put us in the old wineskins." In such a context as this, backed by centuries of subjugation to colonization and neocolonization and resistance to them, to be dubbed the custodians and guardians of creation by the divine is a gospel word, just as being crowned "the prince of creation" (Orthodox language) is heady status.

Due sensitivity to historical context and power did not settle certain sticking points of the steward symbol, however. Its affinity for a human-centered cos-

---

environmental stewardship, law to mercy, economic order to social justice. It is not production and consumption that sustain our earth but rather the ecological systems that have to support human life" (from "Report of Section I: 'Giver of Life — Sustain Your Creation!'" in *Signs of the Spirit*, ed. Michael Kinnamon [Geneva: WCC Publications, 1991], 59–60).

15. James Nash, *Loving Nature: Ecological Integrity and Christian Responsibility* (Nashville: Abingdon, 1991), 105. Nash earlier notes that "image of God" and "dominion" have exercised disproportionate influence, despite their virtual absence in the Bible. Both *imago dei* and "dominion" are associated exclusively with the "royal theology" (P, or "priestly," theology) segments of Genesis and are never mentioned again in the Hebrew Bible, with one exception for "dominion" (in Ps. 8:5–8). In the New Testament the concept of divine image is attached to Jesus Christ only, but without an "ecological" reference, and dominion in the sense of Genesis 1 is not mentioned at all (Nash, *Loving Nature*, 102). However, these images have a lively and important lineage in postbiblical religious ethics where "image of God" especially has been a grounding for movements of social justice. The cry for basic human rights by racial-ethnic minorities, women, and religious minorities has often taken its cue from *imago dei*. Energy for the same movements has also come from the sense of moral agency and responsibility for the world that "dominion" carries. Exactly this lineage showed itself in the debate in Canberra, reported above.

mology and ethic ("anthropocentrism" or "homocentrism") is one of them. We met it earlier in John Paul II's dominion theology. Yet his papal message of January 1, 1990, for the celebration of the World Day of Peace, carries a different tone. Interestingly enough for a message on peace, one continual refrain is "due respect for nature." "The ecological crisis as a moral crisis" runs throughout as well. And the "ultimate guiding moral norm" "for any sound economic, industrial or scientific progress" is nothing less than "respect for life," the pope writes, anticipating *The Gospel of Life* itself some five years later. Then comes the key qualification: respect for life, but "above all for the dignity of the human person."[16] Pope John Paul II, even in broadening the moral scope to "respect for life," aligns critical moral discussion with *the* norm of official Catholic social teaching, "the dignity of the human person." His message is thus most accurately struck in the sentence, "The human race is called to explore this order [the 'cosmos'], to examine it with due care and to make use of it while safeguarding its integrity."[17] Because official Roman Catholic social teaching continues to set a rather abstracted human person at the center of theological and ethical reflection, papal documents and bishops' pastoral letters have come very late to the ecosocial crisis. The larger point, however, is that there is an anthropocentrism about the steward model that is genuinely here, and that anthropocentrism sticks in the craw of many, including many Catholic theologians (Thomas Berry, Rosemary Ruether, Elizabeth Johnson, Leonardo Boff, and Sean McDonagh, for example).

The discussion of stewardship as a fitting model for the needed cosmology will no doubt go on a long while before any resolution on anthropocentrism surfaces, in either secular or religious circles.

An episode at Canberra confirms this. The most contentious paragraphs of the penultimate draft of Section I were these:

16. What is our place as human beings in the natural order? The earth itself, this little watery speck in space, is about 4.5 billion years old. Life began about 3.4 billion years ago. We ourselves came on the scene some 80,000 years ago, just yesterday in the twinkling of the Creator's eye. It is shocking and frightening for us that the human species has been able to threaten the very foundations of life on our planet in only about 200 years since modern industrialization began. So where do we belong in the Creator's purpose?

17. Some say we are one species co-existent with many others. We hold an awesome power — life and death — in the tiny portion of the universe we will inhabit for but a short period in creation's history. As one species among others in a planetary world that is itself one, from an ecological point of view, we are totally dependent upon the rest of creation.

16. "Message of His Holiness Pope John Paul II for the Celebration of the World Day of Peace" (January 1, 1990), 7.
17. Ibid., 8.

We are precious creatures of God living for a season among other precious creatures of God in an awesome universe, and we are creatures with extraordinary power and responsibility.

18. But what of our traditional Christian theology? Many Christians contend that human beings, as created in the image of God and the crown of creation, occupy a special place among creatures, that there is no intrinsic value or morality in nature apart from human beings, that it is for us to introduce morality into nature, and that the redemption of creation will come about through God's redemption of humanity.

These paragraphs were deleted after speaker upon speaker said they offended the biblical dignity of human beings and the divine calling to be earth's stewards. While many readers will not find them offensive, as some delegates did not, there is no question they threaten a homocentrism that lives deep in Christian theology and much popular culture. Neither is not about to lie down and die, or even say "uncle."[18]

While stewardship as masking a continuing homocentrism may be the sharpest objection, Canberra touched two others. One is reflected in the substance of the deleted paragraphs and their recollections of "the slow womb" of earth and humanity's late arrival. Delegates attentive to evolutionary sciences were incredulous over the embrace of the biblical theme of humankind as creation's charter creature and a blessing to it. Hearing this prominence in the symbol of steward, they doubted that the symbolization of stewardship had much basis in earth's story.

Indigenous peoples, prominent at the assembly, were another standing criticism of "steward." Little explicit voice was given, but little needed to be. These people knew stewardship as the flipside of its message of empowerment for subjugated peoples. They knew stewardship as the ideology of the "civilizers" who had promulgated neo-European ways laced with imperialistic and racist notions. This was stewardship as spoken of by those "subduing the earth," but as now remembered by the subdued.[19]

---

18. A line traced earlier as the history of sustainable development, with its roots in industrial economics as development, bears mentioning here. It was precisely stewardship ethics that were enunciated by seventeenth-century scientists and theologians in keeping with early commercial and industrial advances, and on the basis of mechanistic science and homocentric philosophy and theology. In the religious versions, God had assigned human caretaking and supervisory tasks. Nature was to be managed responsibly for the benefit of human welfare (see Carolyn Merchant, *Radical Ecology: The Search for a Livable World* [New York: Routledge, 1992], 72). Stewardship ethics were also the explicit theme of development as the scientific management of resources for the benefit of society in the philosophy and practice of Gifford Pinchot and others early in this century. And in fact much sustainable-development discussion at the UN and among NGOs today moves in this orbit. It focuses on good planetary management, with or without explicit reference to a divine mandate entrusted to a distinctive species. Furthermore, the homocentricity of most past stewardship ethics is retained. Sentence one of Principle 1 in the Rio Declaration of the 1992 Earth Summit is as follows: "Human beings are at the centre of concerns for sustainable development" (from "Rio Declaration on Environment and Development," appendix 5, p. 86, of Granberg-Michaelson, *Redeeming the Creation*).

19. I am grateful to James Martin-Schramm for drawing my attention in chapter 2 of his doctoral

Amidst the many-sided debate was one matter of shared consensus. Biblically, "the earth is the Lord's," and humans truly "own" nothing. The steward is the one entrusted with things precisely *not* his or her own. He or she is the employee accountable to the owner for proper handling of domestic concerns, keeping the books, collecting rents, and generally managing the affairs of the enterprise or estate. Stewardship, then, is accountability not for what humans *do* own but for what they do not.

PARTNER

There is a variation on the steward model that attracts those who reject homocentrism and would give humans a humbler place. It is the symbol of partner. St. Francis usually leads the nominees for patron saint. Creation, at least earth, is playfully imaged as a kind of "holy democracy" of all creatures great and small. There is an interconnectedness and interdependence among all things that humans must respect in exercising their considerable power. Other creatures are co-siblings of creation in the drama of a shared life. They must be listened to in order to know what earth is saying and requires. Humankind is partner to otherkind as, in Stephen Jay Gould's phrase, "the stewards of life's continuity on earth."[20]

Humans are decentered in this cosmology. They are not the only subjects of the moral universe, nor does "quality of life" pertain to their lives alone. The value of otherkind goes beyond what W. Godfrey-Smith calls "silo" value (a stock of resources), "laboratory" value (the object of learning), "gymnasium" value (value for human leisure and recreation), and "cathedral" value (aesthetic pleasure and religious emotion).[21] All these still regard humans as "the measure of all things" in a way the partner model, with its moral universe of inherent value for all creation's "agents," rejects. Humans do have a distinctive moral calling and unique moral responsibility. But they are not the only creatures due moral consideration.[22] They are decentered both as the goal of creation and as its moral measure.

Yet in another sense humans are recentered in "partner" cosmology and ethics, and not least in the stewardly tradition of planetary caretaker. With the human species, the argument goes, life itself has become self-conscious and capable of transformations no other species can effect. To fail to recognize this one species's power in the midst of all things is to court illusion. Humans *do* exercise

---

dissertation ("Population, Consumption, and Ecojustice") to these criticisms. See also the previous footnote.

20. From Stephen Jay Gould, *The Flamingo's Smile: Reflections in Natural History* (New York: Norton, 1985), 431.

21. Cited from Birch and Cobb, *Liberation of Life*, 150.

22. See the discussion of Jay B. McDaniel, *Of God and Pelicans: A Theology of Reverence for Life* (Louisville: Westminster/John Knox, 1989), esp. chap. 2, "A Life-Centered Ethic," 51ff.

dominion. Yes, the rest of nature may well hold the final trump card. But short of an endgame, the fact of human power in so much of nature means we *are* responsible. The question, then, is: *How* do we exercise this power? And the answer, reflecting the sober realism carried over from the steward model, is to recenter the powerful partner.

The recentered human partner recognizes that, whatever power we wield as a species, we do not legislate the laws of an encompassing nature. Indeed, we violate them to our own and otherkind's peril, even demise. In WCC language, "the integrity of creation" is utterly basic, with "justice" and "peace" pursued as means to creation's flourishing and fulfillment. Justice and peace aren't human goals and states only, but all creation's; and they can be attained only on creation's terms. Creation's integrity sets the terms and requires "moral considerability" of otherkind in human decisions.[23] The preservation of ecosystem communities is of necessity a first value.

The partnership way could, if Francis does get the nomination, offer an element that stewardship often fails to teach: namely, the revival of a creation-loving asceticism. An asceticism that loves earth fiercely in a simple way of life is desperately needed, above all among the wealthy of the world and others habituated to unsustainable consumption as a lifestyle. There are long-standing traditions here (the Franciscans and other religious orders live on) and some seeds sown more recently. (The movement to simplify lifestyles in the 1960s still has its communities. Some of these may yet be the "anticipatory communities" of the future.) In the ecumenical movement it may be Asian Christians who give leadership, however. They have both an unbroken culture of relating voluntary poverty to involuntary and a view of nature as comprehensive of all things, including us. Not by coincidence it was a Korean, Chung Hyun Kyung, whose electric plenary address in Canberra advocated the shift "from anthropocentrism to life-centrism" within an all-inclusive notion of nature in which humans are wondrous microcosms of the macrocosmos.[24] In any case, this newly emphasized asceticism would be an earth-sensuous asceticism, undertaken not in the

---

23. One example of moral consideration is being worked out in discussion of animal rights. Birch and Cobb list tenets of the Humane Society of the United States with which they, in discussing the intrinsic value of nonhuman creatures, agree: "It is wrong to kill animals needlessly or for entertainment or to cause animals pain or torment. It is wrong to fail to provide adequate food, shelter and care for animals for which man has accepted the responsibility. It is wrong to use animals for medical, educational, or commercial experimentation or research, unless absolute necessity can be demonstrated and unless such is done without causing the animals pain or torment. It is wrong to maintain animals that are to be used for food in a manner that causes them discomfort or denies them an opportunity to develop and live in conditions that are reasonably natural for them. It is wrong for those who eat animals to kill them in any manner that does not result in instantaneous unconsciousness. Methods employed should cause no more than minimum apprehension. It is wrong to confine animals for display, impoundment, or as pets in conditions that are not comfortable and appropriate. It is wrong to permit domestic animals to propagate to an extent that leads to overpopulation and misery" (Birch and Cobb, *Liberation of Life*, 155–56).
24. Chung Hyun Kyung, "Come, Holy Spirit — Renew the Whole Creation," in *Signs of the Spirit*, ed. Michael Kinnamon (Geneva: WCC Publications, 1991), 43.

pursuit of self-denial as such or obsessed with sex, but in pursuit of a joyous participation in earth community in nondestructive ways.[25]

SACRAMENT/PRIEST

The heart of the Orthodox contribution, to return to earlier discussion, is not a stewardship cosmology and ethic, however, or even its variant of partnership. It is a sacramentalist one.[26] Sacramentalism as a model has gained new popularity in recent years. It is rooted in diverse and archaic cosmologies from pre-Augustinian Christianity around the Mediterranean to Celtic Christianity in more northerly climes. Orthodox communions from early centuries onward have consistently understood the sacraments as dramatizations of nature's transfiguration. Humans' high calling is as "priests of creation," referring the creation back to the creator in acts of liturgical doxology. In such praise humans act as representatives for the whole creation, setting loose, in Jürgen Moltmann's words, "the dumb tongue of nature" through human thanksgiving. "So when in the 'creation' psalms thanks are offered *for* the sun and the light, *for* the heavens and the fertility of the earth, the human being is thanking God, not merely on his own behalf, but also in the name of heaven and earth and all created beings in them." This is not meant homocentrically, Moltmann explains, because "everything that has breath" praises God, and "the heavens declare the glory of God" in their own way, even without human beings and apart from them.[27] Nature's tongue isn't wholly dumb. Yet human beings are the singers of the cosmic song and the tellers of its tale in a special way; we can represent creation and give voice to it in a cosmic liturgy of praise and transfiguration.[28] We are mediators, then, but not the center. We intercede for the Community of Life and speak on its behalf before God. We are the *imago mundi* (image of the world) who, as

---

25. There is an abbreviated but suggestive discussion of this in Catherine Keller, "Chosen Persons and the Green Ecumenacy," in *Ecotheology*, 306–11.

26. At Canberra use was made of the lovely publication "Orthodoxy and the Ecological Crisis," published by the Ecumenical Patriarchate assisted by the Worldwide Fund for Nature International, 1990. For a highly instructive and programmatic account of the use of sacramental theology in the ecosocial crisis, an account that incorporates Orthodox insights but is written by one who is not Orthodox, see Kosuke Koyama, "The Eucharist: Ecumenical and Ecological," *Ecumenical Review* 44, no. 1 (January 1992): 80–90.

27. Jürgen Moltmann, *God in Creation: A New Theology of Creation and the Spirit of God* (San Francisco: Harper and Row, 1985), 71.

28. There is a delightful musical instance of this in Joseph Haydn's oratorio, *The Creation*. At the completion of all creation except humans, Raphael sings, "Now heav'n in fullest glory shone," and rather whimsically recounts its fullness. Earth "smile'd in all her rich attire" at the air filled with birds, the waters swelled with fish, and "heavy beasts" trodding the ground. Yet Raphael notes that one being is missing. He cannot yet know who or what that being might be. But he knows what it is to do amidst "Earth in her rich attire." "There wanted yet that wond'rous being," Raphael sings, "that grateful should God's pow'r admire," and "with heart and voice [God's] goodness praise." Raphael repeats this several times, as if to emphasize the "wond'rous being's" task. The aria ends here, without yet the creation of humankind but with its proper role already awaiting it (see no. 23, "Now Heav'n in Fullest Glory Shone," of *The Creation: An Oratorio*, by Joseph Haydn [New York: Oxford University Press, Music Department, 1991]).

also *imago dei* (image of God), voice God to creation and creation to God. Such is the Orthodox notion (and the Hasidic Jewish notion as well).

Sometimes called "panentheism," sacramentalism recognizes and celebrates the divine in, with, and under all nature, ourselves included. The creaturely is not identified *as* God, however. (This is pantheism, not pan*en*theism.) Nature and the world are thus not of themselves divine and are not worshiped. Rather, the infinite is a dimension of the finite; the transcendent is immanent; the sacred is the ordinary in another, numinous light — without any one of these terms exhausting the other. Sacraments themselves are symbols and signs that participate in the very Reality to which they point, but they are not themselves worshiped. To identify something earthly as holy and sacred is not to say it *is* God. Rather, it is *of* God; God is present in its presence.

The natural response to the sacramental in our midst is wonder, awe, amazement, fascination, astonishment, curiosity, and surprise. It is also a sense of being very small amidst a grand Reality. At times it is a sense of being unworthy in the face of holy wonder and the awareness that the pulse of God's energy flows through me as it does through all life. In any event, the moral posture is certainly not mastery, control, and abstracted distance. Rather, it is presence, relationship, and the care and respect due the sacred. The sacramental "emphasizes the tender elements of the world," biologist Charles Birch says, and "the spiritual unity that gives the physical its meaning." This in turn nurtures "a humbling sense that all creatures are fellow creatures and that human responsibility extends infinitely to the whole of creation."[29] Species humility and responsibility are the proper foci of earth ethics.

At Canberra, sacramentalism was endorsed not only by the Orthodox but even more powerfully by indigenous peoples. They burst forth upon the assembly's stage, often literally dancing, and demonstrated once again in their words and actions that the primal vision of peoples of the land is invariably sacramentalist. The entire cosmos is the sacred community, and life should be lived with the respect and treatment due the sacred — it's that simple and profound.

In both the Kuala Lumpur document and the Section I report of Canberra the roll call of teachers of "a deepened understanding of creation" included indigenous peoples. Not least among the lessons learned is the easy flow of the everyday into the sacred and the refusal to desacralize any arena of life. For native peoples generally everything in nature represents transcendent power and order, and all the activities of culture — farming, hunting, cooking, eating, householding — are sacramental. They are visible signs of divine power and presence amidst daily practices.[30]

This same lesson, now applied to environmental policy, is given voice by Wes Jackson of the Land Institute in Salina, Kansas. Jackson objects to setting aside

---

29. This is from Birch's address to the Nairobi assembly of the WCC in 1975, published as "Creation, Technology and Human Survival," *Ecumenical Review* 28 (1976): 77, 79.

30. Recall from the chapter "Message from Chiapas" that somewhere near the center of the rebels' concerns rests a spiritual connection to nature, place, community, and culture. This reflects, broadly speaking, a "sacramental" view of creation.

pristine wilderness while at the same time treating the rest of the land profanely (as a commodity and utilitarian vehicle of profit only):

> I do not object to either saints or wilderness, but to keep the holy iso-
> lated from the rest, to treat our wilderness as a saint and to treat Kansas
> or East Saint Louis otherwise, is a form of schizophrenia. Either all the
> earth is holy, or it is not. Either every square foot deserves our respect,
> or none of it does.... The wilderness of the Sahara will disappear unless
> little pieces of non-wilderness become intensely loved by lots of people.
> In other words, Harlem and East Saint Louis and Iowa and Kansas and
> the rest of the world where wilderness has been destroyed will have to be
> loved by enough of us, or wilderness is doomed.[31]

As with the steward model, sacramentalism is not without its distortions and moral corruption. These are rooted in the picture of society present in many sacramentalist cosmologies. The age-old sacramentalist assumption is that a har-mony of social and natural interests exists somewhere just below the surface and that a soft, nurturing process will bring this precious flower to bloom. Sacra-mentalism's metaphors for society as well as church are thus typically "organic." (Feudalism lives on!) Or, in the Orthodox version, "symphonic." But metaphors of organisms and symphonies don't expose the unequal and corrupted power re-lations of life among human beings, nor between humans and other creatures. They mask the fact that struggle and conflict so often *are* the status quo. "Heal-ing," rather than fundamentally reordered ecosocial relationships, thus becomes the "cure" for "sin" in much sacramentalist theology, whether of established re-ligious traditions or quasi-religious movements (New Age rituals, for example). It is as though earth's basic problem were illness or bad tuning, not injustice or unequal and corrupted power.

John Haught says it differently. Nature, ourselves included, is not yet what it could be, nor fully revelatory of God. Its beauty is only partial, and its ful-fillment remains in the form of promise. Any orientation that is *only* a mystical affirmation of what is, no matter how deep the experience, falls short; and with that it fails to recognize the long and painful distance between the present state of affairs and a better earth. The sacramentalist stance, lost in wonder at what is, can easily incline to such mystical affirmation and its resident shortcomings.[32] It can glory in what is, to the neglect of what ought to be. When it does, it sacrifices its inherent moral and ethical power.

Here again indigenous peoples brought an important contribution to Can-berra. They, together with others of the poor and women from all social ranks, know that sacramental spirituality is empty as an ethic if there is no commit-ment to the political agenda, and that political commitment entails organized

---

31. Wes Jackson, *Ecology, Economics, Ethics,* as cited by Daniel J. Kevles, "Some Like It Hot," *New York Review of Books* 39, no. 6 (March 26, 1992): 34. Kevles does not give the page number from *Ecology, Economics, Ethics.*

32. John Haught, *The Promise of Nature* (New York: Paulist, 1993), 110–12, 113–14.

action along paths of hard resistance as well as soft ones of gentle persuasion. Healing *is* needed, and is itself a powerful metaphor, but so is organized force on the way to reconciliation. To make marginated peoples truly angry, steal their spirituality without joining their political struggle.

Ecofeminism has not yet found its full voice in the WCC. But it may eventually make the greatest contribution and impact of all because its potential constituency is huge and because earth consciousness and women's consciousness often go together.[33] Furthermore, ecofeminism discloses a profound understanding of the specific social and sociopsychological causes of earth's suffering. In its Christian incarnation, ecofeminism represents the twin streams of sacramentalism and liberationism flowing together. In Canberra this development was eloquently presented in the plenary session on the council's Decade of the Churches in Solidarity with Women. The pattern of presentation was women from around the world — South Africa, Russia, Palestine, Sweden, Egypt, Nicaragua — telling their stories of women's struggles in their communities and of God's presence with them in adversity. As each finished her story, she took a branch of green and placed it in a large wooden structure resting at the rear of the stage. The structure's dead wood quietly greened as the stories of suffering and hope, death and life, punctuated the session. When all the tales were told, the women together hoisted the heavy beams upright. It was a cross become the tree of life.

The greening of the cross, it turns out, is an ancient Armenian Orthodox tradition, recalling early Christian symbolism. Here the ancient sacramentalist vision of earth nicely served many women's experiences, but only as merged with their struggles for justice, peace, and creation's integrity. Ironically, the general Orthodox intransigence on women and ministry has concealed the extraordinary parallels between Orthodox spirituality and feminist spirituality, just as it has made it so difficult to recognize common ground in creation theology. Yet there is no doubt that women in the ecumenical movement will continue to voice the sacramentalist possibilities in cosmology and ethics, even when one of their sharpest criticisms of the Orthodox, Roman Catholic, and Anglican communions is that the sacramentalism of these traditions has served to reinforce powerful linkages of patriarchy, social domination, and environmental degradation.[34]

Both indigenous peoples and feminists, then, marry sacramental and liberation postures. They insist that our striving is amidst and about ordinary, daily

---

33. For one set of testimonies, see Elizabeth Dodson Gray, ed., *Sacred Dimensions of Women's Experience* (Wellesley, Mass.: Roundtable, 1988).

34. There are "secular" parallels for all these examples. In this case, a happenstance episode in Riverside Park, New York City, comes to mind. Near Ninety-sixth Street is one of the only large flowerbeds in the park. As I jogged by on my daily constitutional, I saw an elderly, regal woman. She was in a black and copper silk nightgown and was dancing amidst the flowers, letting her face feel the morning sun, and quite oblivious to the steady humming of Westside Highway traffic. When I returned some time later, she was not there. She was on a nearby park bench, conversing with one of the homeless. They were talking about what we New Yorkers have to be doing together for the city these days.

activities and that the preciousness of life is found here or it is found nowhere. Either the good and the holy are at the center of what we do in our lives, or else we must *change the ways of living* that segregate the good and holy from the everyday. Part of this involves our very way of thinking and imagining, since, as noted, what people define and imagine as real is real in its consequences. The conclusion is that the symbols of religious traditions must be scrutinized to see whether in fact they are lodged firmly on the side of life and its fulfillment, work against such liberation and fulfillment, or are blandly oblivious to daily affairs. Process theology, to continue the survey of sacramentalist possibilities, has only recently been lofted into ecumenical orbit. But it will be influential for years to come. Not least, its earth-embracing and dynamic cosmology is highly compatible with the newly influential currents of science that in turn find affinities with sacramentalist cosmology. (More on this shortly.) Perhaps the best pre-Canberra discussion of earth ethics emerged from the process theology camp — *Liberating Life: Contemporary Approaches to Ecological Theology*. It included a report written for the Canberra assembly.[35] The drawback of process work is that it has been slow to link its metaphysics to sophisticated social analysis.[36] It thus misses what many in the WCC have come to believe is a vital part of theological method itself. The aforementioned volume, however, was undertaken explicitly to meld insights from liberation theologies and process sacramentalism in order to bring both to bear on "justice and sustainability."[37] Insofar as it succeeds, it does so in the persuasive way of process theology. Which is to say that God so gathers in all things that their condition affects God. When humans and other creatures are violated and their lives diminished, God's glory is dimmed and dishonored. Conversely, when life is fuller and richer, God's glory is enhanced. Justice, sustainability, and the well-being of creation are therefore intrinsically, not extrinsically, related to God and the divine life. This is the core teaching of process theology and ethics.

PROPHET/COVENANT

Finding a fitting label for the prophet-teacher model and prophet/covenant symbol is difficult. Its devotees are not much interested in the metaphysics that good cosmologies require, nor are they very patient with the careful analysis and argument of good ethics. Perhaps ecoprophetism is less a cosmology than an urgent moral call compelled by threatening planetary conditions as they strike religious consciousness. And perhaps the prophetic symbol should be left with that as a crucial human task for this moment. Whatever the case, it is a stance prominent

---

35. Charles Birch, William Eakin, and Jay B. McDaniel, eds., *Liberating Life: Contemporary Approaches to Ecological Theology* (Maryknoll, N.Y.: Orbis Books, 1990).

36. There are important exceptions. One of the important works in ecological economics is one cited earlier, Herman E. Daly and John B. Cobb Jr., *For the Common Good* (Boston: Beacon, 1990). John Cobb, perhaps the leading figure in process theology, has over the years contributed very substantially to ecumenical theology and ethics. Herman Daly has as well.

37. See the introduction by the editors, pp. 1ff., of *Liberating Life*.

in WCC circles. "The stark sign of our time is a planet in peril at our hands," is the blunt language of Canberra, Section I.[38] The entire "Justice, Peace, and Integrity of Creation" phenomenon culminating in the Seoul convocation of 1990 was a conciliar process of covenanting that issued in ten basic faith affirmations and four covenants on a just economic order, security, a culture in accord with creation's integrity, and resisting racism and discrimination. The final document itself was titled "Entering into Covenant Solidarity for Justice, Peace, and the Integrity of Creation" and has all the marks of prophetic cry. ("Humanity seems to have entered a period of its history which is qualitatively new. It has acquired the capacity to destroy itself. The quality of life is being diminished; even life itself is in peril.")[39] The prophetic voice here seeks to grab people's attention before it is too late and calls unabashedly for conversion — real turning to God and away from destructive ways of life. The prophet also becomes teacher, pointing to root causes of planetary threats and mapping another way.

There is no one school or religious tradition here. This cuts across the ecumenical world. But evangelicals who have gotten the message that there is only one world and earth is our home bring a special fervor to the ecumenical forum. And when they do, the emphasis is usually, like the prophets' own, on *covenantal responsibility*. (Recall the Isaiah passage, "The vine languishes;...they have broken the everlasting covenant.")

This is indeed a rich lode. Some argue that covenant is, at least for "peoples of the Book" (Jews, Muslims, Christians), the master image for the relation of God to humanity and the whole creation. Genesis 9 and the Noachic covenant, at once ecumenical and ecological, states it as such: "God said, 'This is the sign of the covenant that I make between me and you and every living creature that is with you, for all future generations: I have set my bow in the clouds, and it shall be a sign of the covenant between me and the earth'" (Gen. 9:12–13). Furthermore, covenant in these traditions is ethically charged. Consequences for all creation hang on human adherence to covenanted terms. Its particular emphasis is justice for the weak and protection for the vulnerable, on the one hand, and the sure negative consequences of injustice (unrighteousness) for the land itself and its peoples when the covenant is violated, on the other. To violate the covenant is in fact to violate the laws of life itself and come to a sorry end. To keep the covenant is to live long and well upon the land and enjoy its abundance.[40] Thus saith the prophet.

It is worth noting in passing that the covenant and prophet models correct the tendencies of sacramentalism toward ethically deficient ritual and uncritical affirmation of what is. In the prophets' testimony two reasons for injustice

---

38. "Giver of Life — Sustain Your Creation!" in *Signs of the Spirit*, 55.

39. See "Final Document: Entering into Covenant Solidarity for Justice, Peace, and the Integrity of Creation," in *Between the Flood and the Rainbow*, ed. D. Preman Niles (Geneva: WCC Publications, 1990), 164–90.

40. Part of the power of Rosemary Radford Ruether's *Gaia and God: An Ecofeminist Theology of Earth Healing* (San Francisco: HarperSanFrancisco, 1992) is Ruether's constructive use of two long Christian traditions in the interests of "an ecofeminist theology of earth healing." She fastens upon the strands of the covenantal and the sacramental, weaving them together in a liberationist stance.

in Israel are the fascination with false gods and the reliance upon cultic observance to the neglect of the moral. The false gods for Israel were the Baalim,
the gods of the land who guarded the fertility of both the land and its people.
("Men blow kisses to calves" to curry favor with Baal, Hosea says in derision
[Hos. 12:3].) While the fertility rites regularly included the offenses of prosti-
tution and drunkenness, the chief offense was diversion from the demands of
just social living, of "righteousness." Caught up in the recurring cycles of nature, people could forget the hard task of fashioning a just society on covenantal
terms. The gods of nature could be adored in glorious ceremony while widows
were forgotten and orphans abandoned. The smoke and song of the Temple
liturgies could praise Yahweh while oppression reigned in the streets and the
poor were sold for a pair of shoes.[41] Covenant was violated while rites were
meticulously observed. Thus was the prophet moved to rail as God's voice:

> I hate, I despise your festivals,
> and I take no delight in your solemn assemblies.
> Even though you offer me your burnt offerings and grain offerings,
> I will not accept them;
> and the offerings of well-being of your fatted animals
> I will not look upon.
> Take away from me the noise of your songs;
> I will not listen to the melody of your harps.
> But let justice roll down like waters,
> and righteousness like an everflowing stream.[42]

(Amos 5:21–24)

---

41. John Shea, *Stories of God: An Unauthorized Biography* (Chicago: Thomas More, 1978), 104–6.

42. Differently said, and as Amos makes vivid, the battle is over moralities. Baal stands for the
morality of nature as that which is, in recurring patterns that govern life. Integrating human life
with these and conforming to the sacred in and as nature is the human task. This has often been
expressed in the rites and rituals of many traditions, not only Baalism. The prophet and the distinctive covenants with the one God, however, repudiate this in the name of a morality born in what
ought to be, often set against what presently is. Humans, in covenant with God, need to impose
a moral order on largely amoral nature, including their own. Humans here are coparticipants in an
ongoing transformation of creation so that the future might be yet better, from the standpoint of
justice, than the past.

Unfortunately, this important struggle over moralities came to be read as one of "history" in
contrast to "nature" in Judaism, albeit, as with Christians, largely in the West after the industrial revolution. Jewish scholars are now rethinking this. The following is from Ismar Schorsch, chancellor
of the Jewish Theological Seminary, New York: "We must dare to reexamine our long-standing preference for history over nature. The celebration of 'historical monotheism' is a legacy of nineteenth
century Christian-Jewish polemics, a fierce attempt by Jewish thinkers to distance Judaism from the
world of paganism. But the disclaimer has its downside by casting Judaism into an adversarial relationship with the natural world. Nature is faulted for the primitiveness and decadence of pagan
religion, and the modern Jew is saddled with a reading of his tradition that is one-dimensional. Judaism has been made to dull our sensitivity to the awe inspiring power of nature. Preoccupied with
the ghosts of paganism, it appears indifferent and unresponsive to the supreme challenge of our
age: man's degradation of the environment. Our planet is under siege and we Jews are transfixed in
silence" (Ismar Schorsch, "Tending to Our Cosmic Oasis," *Melton Journal* [spring 1991], as cited by
Eilon Schwartz in his paper, "Judaism and Nature: Theological and Moral Issues to Consider While
Renegotiating a Jewish Relationship to the Natural World").

SCIENCE'S CONTRIBUTION

Canberra as a spectrum for cosmology and ethics and a sampling of the shape of human responsibility for earth via various symbols for the human would be incomplete without recognition of the contribution made by scientists. Scientists' influence was important in the Kuala Lumpur meeting. Its preparatory document for Canberra nicely summarizes this:

> From the scientists among us, we have learned dimensions of the crisis of the global ecology. We have learned of the intricacies of nature and of nature's fragility in the face of human onslaught. Our sense of the mystery of life and nature, and our awe and wonder at the Creator's handiwork, has been deepened. In the face of our destructive ways, this has only made more poignant the uniqueness and preciousness of life itself on this planet. From the scientists among us, we have also learned spiritual humility anew. They have shown us the delicately woven interdependencies of the web of life and the human threat to them. They have shown us the antiquity of the creation and our own late, brief moment in time. They have given us a glimpse of a cosmos which dwarfs our minuscule existence and quietly ridicules all our claims to grandeur. They have taught us far more, but the upshot is this: we urge you at Canberra to welcome such truths which science teaches us in this dangerous moment of our ignorance and our arrogance.[43]

From the scientists themselves, and in their own voice, we might recall the "Open Letter to the Religious Community" (cited in an earlier chapter) signed by thirty-four renowned scientists led by Carl Sagan and Hans Bethe:

> As scientists, many of us have had profound experiences of awe and reverence before the universe. We understand that what is regarded as sacred is more likely to be treated with care and respect. Our planetary home should be so treated. Efforts to safeguard and cherish the environment need to be infused with a vision of the sacred.[44]

But what is the significance of science's contribution for a viable cosmology and ethic and for our understanding of the human task now? As might be expected, the major contribution is empirical precision about what is happening and why, in "micro" and "macro" terms. For theological cosmology this means that the sacramentalist model, for example, can be joined to an *evolutionary* sense of nature's life. The sacramental model is thereby corrected of its tendencies to a static and repetitive view and practice. (This would hold for any covenant model

---

43. "Giver of Life — Sustain Your Creation!" in *Signs of the Spirit*, 3.
44. Carl Sagan and Hans Bethe et al., "An Open Letter to the Religious Community," 3. The letter is available from the National Religious Partnership for the Environment, 1047 Amsterdam Ave., New York, NY 10025.

as well.) Here the ecological dynamism of all things is given exquisite detail. Not a single present life-form, or any other form in the cosmos, goes back to the very beginning in the form it now possesses, present science tells us. All have a story, a history, and all are participants in both. "Story" in fact may be a better notion than "law" for the logic of nature's doings. In any case, life itself is a drama of changing forms in this great adventure of some fifteen billion years to date, and we ourselves a small but significant part of the saga. All that we have been, are, or ever can be, body and soul and culture, is here in stunning and changing detail.[45] Science's descriptions become, then, a source for religious awe, respect, and humility. This is substance, including moral substance, for any contemplated symbol of humanity's relationship to the rest of earth.

For ethics the detailed attention to environmental degradation and the scientific knowledge of nature's requirements for its ongoing life also pose conflicts that cannot be shoved aside, at least not without risking self-inflicted blindness. Science thereby injects a note of realism for human responsibility and keeps decisions close to concrete realities. The conflicts themselves run between claims for human well-being and the wider well-being of the planet. This means hard ethical choices within the human community and between humans and the larger network of life of which we are part and upon which we are totally dependent. The clear conclusion is that the expectations and habits of humans — some humans far more than others — will have to be radically adjusted in order to avoid even worse consequences. Not least, these conflicts and ethical choices are sharpened when scientists, on the basis of their knowledge *and its limits,* are unable to define sustainability with precision! Precisely because the line that development must cross to be sustainable *is not* clear to scientists either, no one knows quite when to wave the checkered flag and say the race is over. In such circumstances scientific description of what is happening to the planet helps define the issues as finally they must be defined — as essentially moral and ethical choices when all the choices entail tradeoffs and risks. The question becomes: Given *a, b,* and *c,* and the possible outcomes of *d, e,* or *f,* just what ought we to do, in keeping with what possibilities and in accord with what vision of life? Such a question is not one science, as science, answers, in the end. It is an ethical one, even a religio-ethical one. But it is also one for which science provides invaluable knowledge.

None of the choices is made easier by science's portrayal of the strange unity of life and death that comprises creation's integrity. Earth is the source of all life, yet the abode of death — and both indivisibly. Were a natural harmony the normal result, we might, on our better days, find this not only profound but acceptable. But normally life (and not human life alone) takes advantage of life, exploits the environment, treats the neighbors unfairly, and then leaves it to them to handle the waste in recurring cycles![46] This means intervention and

45. Langdon Gilkey, *Nature, Reality, and the Sacred: The Nexus of Science and Religion* (Minneapolis: Fortress, 1993), 92–102.
46. Ibid., 101.

interference are continuous and inevitable in an order where the creative and destructive are intimate to each other and finally inseparable. "There is no such thing as a free lunch" is the bumper sticker version. But the more encompassing fact is that reciprocity and sacrifice are required at the same time that moral ambiguity surrounds the choices that must inevitably be made. This is, yet again, the ascendancy of ethics for a species whose vocation it is to make just such choices.

So while science does not offer an alternative religious cosmology and ethic of its own, Canberra, and a long stream of WCC work well before Canberra, was right to insist that the scientist's voice is an indispensable and exacting one for any viable, down-to-earth cosmology and ethic.[47]

There is no conclusion to this chapter, only an end. I have made my own choices of emphasis among the symbols and models, as perhaps the reader has. In my judgment, an evolutionary sacramentalist cosmology offers the richest conceptual resources for addressing earth's distress, if infused with a profound earth asceticism and married to prophetic efforts aimed at "the liberation of life: from the cell to the community."[48] But personal choices and commitment aside, the far more important point of the chapter is that the question in the Garden is still being asked, again and anew. The question is God's: "Adam, where are you?" (Gen. 3:9). The answer is ours.

---

47. Readers may still wish to see a somewhat dated report, *Faith and Science in an Unjust World: Report of the World Council of Churches' Conference on Faith, Science and the Future*, ed. Roger Shinn (Geneva: WCC Publications, 1980), vols. 1 and 2. The conference was held at the Massachusetts Institute of Technology in 1979.

48. The phrase is the title of the book by biologist Charles Birch and theologian John B. Cobb Jr., *The Liberation of Life* (Denton, Tex.: Environmental Ethics Books, 1990).

ReBeginnings

Adam, where are you? is not the only, or even first, question. Where have we come from? and Where are we going? impact the question of where we now locate ourselves in the scheme of things. Cosmology and ethics insist upon stories of origin and destiny as guides. To take our bearings we look back and ahead, not just around.

Always saturated with symbols, stories of origin and destiny influence human behavior immensely. They "work" as stories of identity and moral orientation. Are we defiant aliens who must create meanings of our own in a strange and pointless universe that is basically hostile toward our welfare? Is nature an order "guided" by dumb chance and blind force, an aimless reality upon which we must impose whatever meaning and morality there is to live by? If God is dead (Nietzsche) and nature is blind, is not the Promethean defiance of humans who steal the fire of nature for their own benefit the only sensible way to live? Or are we deep at home in the cosmos, rather than contingently "thrown" into it (Heidegger)? Are we integrally continuous with the world around us and tuned in our genes and bones to its rhythms, ourselves a flowering of nature itself? Or are we a being set oddly apart? Do the clues for living reside in nature's essence, an essence that is the very heart of reality (*theologia naturalis*)? Or do we create our own world on terms our own?[1]

Cultures carry the answers to these questions in their formative stories. Only now and then are such stories raised to the level of studied articulation and made the subject of scrutiny. More often, they are submerged in the habits and institutions of a society and reflected in the manner in which children are raised, education is pursued, and history is recounted. They are the substance of art, rituals, music, both popular and "high" religion, and social mores — in short, all that goes into a culture's socialization of who we are and what we are about in a given time and place.

No special insight is needed to realize that such cultural freight shapes much of daily life. Cosmology and ethics in the form of a culture's narratives and way of life fashion identity and morality in the deep places.

This chapter is more modest than this beginning may imply. It has in view only two stories of origin read side by side in order to spark thought. One is the Genesis account of creation in Jewish and Christian scriptures, together with its own "genesis" in the exodus experience of the Hebrews. The other is "the common creation story" offered by current science. We will eye both these stories

---

1. These paragraphs are indebted to the opening pages of Erazim Kohák's *The Embers and the Stars: A Philosophical Inquiry into the Moral Sense of Nature* (Chicago: University of Chicago Press, 1984).

for their contributions to a viable cosmology and ethic as we ponder earth's own destiny in the decades allotted us. More specifically, the question is: What moral world comes into view when these stories are read alongside one another as potential shapers of an earth-friendly culture? What help is offered the question of how we ought to lean into the world now, what we ought to be and do?

Such explicitly moral questions are not coincidental to stories of origin. What grounds a morality is a fundamental vision of the nature of reality and where it is headed. It may even be the case that morally mature adults act the way they do, not so much because incentives and disincentives of all kinds push and pull, but because they do not want to violate the nature and bearing of things as they understand these from a moral point of view. Such understanding is usually seeded in stories of origin and reseeded in their telling.[2] One way Moses showed himself a superior legislator, Philo claimed in the first century c.e., is that he set the laws within a narrative that began with the beginnings of the people *as* a people and even the world's beginning itself.[3] These origins known, the people knew how to address the world and one another. Origins hold a special place in people's picture of the world and its universe, its moral universe included.

THE FORMATIVE BIBLICAL STRANDS

The moral lines of the Hebrew Bible take their cues from the character and presence of the Hebrews' God. This God, unlike many others, was not recognized as simply a power or force in the universe that transcended human powers and suffused all nature with its energy. Rather, this sacred power was a *moral* force that rejected the inevitability of oppression and injustice and commanded and made possible transformation of the world on the terms of community. Community and social justice were the focus of biblical faith at the very outset. Moses was not the first person to discern profound spiritual reality. Rather, he was one of the first to hear in God's voice the imperative to throw off injustice and create a different order.[4]

Hebrew conviction about this godly power included a strong ethical component: human beings are morally responsible before God for the condition of the world. In the ancient world, where forces outside and beyond the control of human beings bore down from all sides and were the province of gods and goddesses, this was an extraordinary affirmation of human agency and freedom. It was also a burden, since it identified and measured a vanguard people by the quality of religiously charged civic righteousness. At times such a stringent sense of human responsibility must have been puzzling. It was common in the ancient Near East to understand nature as the interplay of divine forces and personages

---

2. John F. Haught, *The Promise of Nature: Ecology and Cosmic Purpose* (New York: Paulist, 1993), 20–21.

3. Philo is cited by Robin Lane Fox in *Pagans and Christians* (New York: Knopf, 1987), 265.

4. Michael Lerner, *Jewish Renewal: A Path to Healing and Transformation* (New York: Putnam, 1995), 64–65.

in a polytheistic universe. What was not so common was the monotheistic turn that held human beings accountable for sustaining the inherent powers of earth as created by the One God. Israel could not, like its polytheistic neighbors, invoke fertility, for example. Fertility was assumed from the third day of creation onward, as the work of the One God. But Israel could pollute the land, contaminate it, and render it less fertile. Initially, three cardinal misdeeds polluted the land: murder,[5] wrongful sexual activity, and idolatry.[6] It is striking that none of these is "ecological" in any obvious way. They are not abuse of the beloved land or transgressions against nature. Yet somehow all human immorality and faithlessness affect the vitality of the land itself, in Hebrew cosmology. The condition of society and of the land are locked together in a moral universe that turns on human actions and the state of the heart. If people are faithful to God's moral universe and thereby create a just social order, the land will yield its produce in abundance. But if they re-create the oppression they have known and stray from God's commands, ecological catastrophe will follow.[7]

This "promise" is made again and again, and precisely around the central event of the giving of the law, Israel's "constitution" as a new and vanguard people. Texts abound, but one will suffice:

> I am the Lord. If you follow my statutes and observe them faithfully, I will give you your rains in their seasons, and the land shall yield its produce, and the trees of the field shall yield their fruit. . . .
> But if you will not obey me, and do not observe my commandments, . . . and you break my covenant, . . . I will break your proud glory, and I will make your sky like iron and your earth like copper. Your strength shall be spent to no purpose; your land will not yield its produce, and the trees of the field shall not yield their fruit. (Lev. 26:2b–4, 14–16, 19–20)[8]

In a world of relatively puny human powers and life at the mercy of the powers of nature, this linkage could have been puzzling to the normal notions of a small pastoral people. Yet it made sense as an implication of Hebrew ethical monotheism. In this monotheism, all of nature is unified and provided for by the One God. So if things go awry, cause must reside with that one creature given special powers and province by this God: humans. Likewise, if things go well, the reason must be human faithfulness and justice, or righteousness. Humanity,

---

5. The entire cycle of spiraling violence that ends with the decision of God to create virtually anew, and the Flood of destruction to clear the path for this, begins with the killing of Abel by Cain. In this episode, the spilling of Abel's blood by his brother "curses" the ground, and Cain is told: "When you till the ground, it will no longer yield to you its strength; you will be a fugitive and a wanderer on the earth" (Gen. 4:12). Cain replies that to "be driven away from the soil" is a greater punishment than he can bear (Gen. 4:13–14).

6. Tikva Frymer-Kensky, *In the Wake of the Goddesses: Women, Culture, and the Biblical Transformation of Pagan Myth* (New York: Free Press, 1992), 94.

7. Lerner, *Jewish Renewal*, 71.

8. For other examples, see Leviticus 18 and Deuteronomy 8.

and especially Israel as a vanguard people and a witness to the nations, mediates between (the rest of) nature and God.[9] Thus Hosea cries out in warning:

> Hear the word of the Lord,
> O people of Israel;
> for the Lord has an indictment
> against the inhabitants of the land.
> There is no faithfulness or loyalty,
> and no knowledge of God in the land.
> Swearing, lying, and murder,
> and stealing and adultery break out;
> bloodshed after bloodshed.
> Therefore the land mourns,
> and all who live in it languish;
> together with the wild animals
> and the birds of the air,
> even the fish of the sea are perishing.
>
> (Hos. 4:1–3)

Here, near the beginnings of the recorded human adventure itself, the claim is set down: coparticipation with God in creation *is* the human vocation, and, absent other divine personages and powers, the condition of earth depends upon God and humankind! Few claims could have contrasted more sharply with cultures and traditions that pictured life, largely based in experiences of nature, as subject to great, largely uncontrollable powers that needed to be regularly appeased, prayed to, submitted to, feared.[10] That the God of the Hebrews meant power for transformation toward a moral order in which humans played a vital role was grounded in the people's own transformation. It did not arise from contemplation of nature or even the religious tussles with attractive and fascinating neighbors. It arose from the experience of the people's origins as a people. Exodus *precedes* Genesis. The God who "knew"[11] the suffering of slaves and heard their cries, this God of divine pathos, was the God whose power was experienced as the power for peoplehood and freedom. With this God a ragtag band who were no people became a people. With this God a way was hewn and a new society forged where there was none. The issue wasn't how many gods there were but rather what this (compassionate, empathetic) God was like.

---

9. Frymer-Kensky, *In the Wake of the Goddesses*, 105.

10. From another angle, it is helpful to note that for the vast majority at this time (the neolithic revolution), life *was* essentially fated. But it must also be said that this was the phase in human evolution when grain-based societies around the Mediterranean found ways to modify local ecosystems significantly. Over a long period of time — several thousand years — they thus experienced new human powers, and not only fate. The emergence of ethical monotheism in the ancient Near East probably reflects this evolutionary change in human societies as that is played out religiously.

11. The Hebrew "to know" in the Exodus account means to know in the sense of experiencing deeply, to know from the inside as one who shares the experience, to know in the manner of the "one flesh" experience of a good marriage.

This same God, these ex-slaves came to believe, held them as a people responsible for the shape and condition of the community and world that were theirs. Their vocation as a redeemed and saved people was to give concrete social form to the ways of the God who rescued and redeemed them. They were to do so in ways that made this God's ways of justice and mercy their own.

In short, in the Hebrew stories of origin the governing power of the universe is not experienced or understood in the first instance as the principle of fixed cosmic order and necessity — that might sum up fate and Pharaoh quite well! Rather, the governing power of the universe is understood as the principle of free, responsible moral agency and the healing and transformation of the world. Israel did not need to learn that reality was sacred or that high religion validated the grandeur of the universe. They had lived as slaves in a society that ordered their oppression on the basis of just such knowledge and religion. What they needed to learn was the presence of a power in the universe that did not abide social oppression while beckoning song and incense in praise of the "natural" ordering of things. The ethical as transformative possibility must never be ruptured from the spiritual — this was Israel's learning. Ratifying what *is*, when that is oppressive, is idolatry, a forsaking of the newfound God.

What began with Moses and the first recorded slave rebellion soon becomes, in the biblical account, a paradigm of liberation and transformation as a way of life. "A way of life" is in fact the heart of the Hebrew and Jewish (and early Christian as well) endeavor. This "way" is always an ethic inferred from the character and presence of the righteous and compassionate One who showed them mercy. Yet the specific point at the moment is the conviction that with this God, ordinary people — indeed the apparently powerless — can subvert deeply entrenched powers and help effect a new world. In Michael Lerner's words, Judaism was "not just a religion about how wonderful the physical world is but a religion that insisted there is nothing inevitable about the hierarchies of the social world."[12]

Jesus of Nazareth and his movement maintained the Jewish conviction that with this God new creation can happen at the waiting hands of a small number of very common, even hesitant, but emboldened and Spirit-filled people. Elaine Pagels has shown that until the time of Augustine and Constantine the Christian gospel itself was understood in terms of extraordinary freedom and agency for creating new community amidst a dying culture and epoch. Or, in Pauline coin, a new humanity and new world are created in the very midst of "this passing age."[13]

At the same time, Judaism *was* "a religion about how wonderful the physical world is."[14] What dawned upon the freed slaves, in the gradual development of

12. Michael Lerner, "Jewish Liberation Theology," in *Religion and Economic Justice*, ed. Michael Zweig (Philadelphia: Temple University Press, 1991), 131.

13. Elaine Pagels, *Adam, Eve, and the Serpent* (New York: Random House, 1988). Pagels's chief sources are the early Christian readings of the Genesis creation stories. They exhibit the general lines we highlighted from the Hebrew scriptures.

14. Lerner, "Jewish Liberation Theology," 131.

Israelite monotheism, was that their liberating God was the sovereign creator of the universe itself! Thus the Hebrew Bible again and again celebrates the immense grandeur of creation, gives thanks for the breath of life that animates all creation — it is nothing less than the ongoing presence of the same Spirit that brooded over the waters of first creation (Gen. 1:2) — and marvels at the detail of a good creation, knowing that Wisdom herself has been present from the beginning. A universal moral order is even to be discerned in creation, it is claimed. Humans, while occupying a special place of power and responsibility within the orderings of life, ought properly to stand in awe of this universe, respect and learn from it. This includes learning from the animals, birds, plants, and fish (Job 12). They, too, are part and parcel of creation. Like the earth creature Adam, they too are from 'ădāmâ (earth, soil, ground). And together with human earth creatures, they are destined to return there.[15] This is cause for human recognition both that we are dust and that we belong, with otherkind, to a moral and natural order that surpasses us.

The salient matter, however, is that these two biblical themes weave a single strand in Judaism, a kind of double-helical genetic ribbon of the faith: the power that created the universe and sustains it from day-to-day is the same power that "champions the powerless and creates the demand for a moral universe infused with justice and compassion."[16] Psalm 77 is a striking example of this double theme become a single one. The Psalmist moves with ease between the ways of God as creator and redeemer/liberator.

The Psalmist does not begin in praise, however, but in despondency, racked by earth's distress and hardly able to speak. Most of all, the Psalmist is fearful that the Ancient of Days has exchanged compassion for searing anger. So he or she reaches deep down to remember the character of the God of the exodus and confess the following in awe, fear, and gratitude, despite feeling all the weight of earth's woes:

> Remembering Yahweh's achievements,
> remembering your marvels in the past,
> I reflect on all that you did,
> I ponder on all your achievements.
>
> God, your ways are holy!
> What god so great as God?
> You are the God who did marvelous things

---

15. Christopher Stone, going through his father's papers (his father was I. F. Stone), came upon a nice passage from the Talmud and used it as inspiration for his own book on environmental ethics: "The world was made for man, though he was the latecomer among its creatures. This was design. He was to find all things ready for him. God was the host who prepared dainty dishes, set the table, and then led his guest to his seat. At the same time man's late appearance on earth is to convey an admonition of humility. Let him beware of being proud, lest he invite the retort that the gnat is older than he" (cited in the dedication to Christopher D. Stone, *The Gnat Is Older Than Man* [Princeton, N.J.: Princeton University Press, 1993]).

16. Lerner, "Jewish Liberation Theology," 131.

and forced nations to acknowledge your power,
and your own arm redeeming your people,
the sons of Jacob and Joseph.
When the waters saw that it was you, God,
when the waters saw it was you, they recoiled,
shuddering to their depths.
The clouds poured down water,
the sky thundered,
your arrows darted out.

Your thunder crashed as it rolled,
your lightning lit up the world,
the earth shuddered and quaked.
You strode across the sea,
you marched across the ocean,
but your steps could not be seen.

You guided your people like a flock
by the hands of Moses and Aaron.

(Ps. 77:11–20, JB)

Lerner says it well: "Celebration of the grandeur of creation goes hand in hand with transformation of the social world."[17] He goes on to cite Sabbath as a lovely summary symbol of this. Sabbath is two remembrances together: *zeycher le'ma'aseh b'ereyshit* (in remembrance of the events of creation) and *zeycher le'tziyat mitzrayim* (in remembrance of going out from Egypt).[18] Sabbath is thus a day to give thanks for life and the blessing of creation. All *melachah* (interference in the natural order) is disallowed on Sabbath because the grandeur of the universe is to be appreciated on its own terms, apart from any use whatsoever. "Just to be is a blessing; just to live is holy."[19] At the same time, Sabbath is a day to remember liberation from slavery and the human vocation of healing and transforming the world in common tasks. Sabbath is gracious rest from these and thanksgiving for the daily bread won the week long.[20]

The name for God in the Shema, the central prayer of Judaism, accentuates the intertwining of these themes as finally one strong strand. The opening clause

---

17. Ibid.

18. Lerner, *Jewish Renewal*, 66. Lerner notes that while Sabbath remembers emancipation every week, the holy days of Judaism do not commemorate other events in the life of Israel that might have been considered emancipatory, such as the conquest of Canaan, the dedication of the Temple, the creation of the Davidic monarchy, and the various military victories of the Israelite kingdom.

19. Abraham Heschel, *I Asked for Wonder: A Spiritual Anthology*, ed. Samuel H. Dresner (New York: Crossroad, 1983), 65.

20. Lerner, "Jewish Liberation Theology," 131. Jürgen Moltmann notes that the Sabbath is "the first thing that God sanctified: holy is God and holy is his Name, sanctified is the sabbath, sanctified is the people, and sanctified is the land of Israel. The sanctification is in that order. The sabbath comes before the people and the land" (see Moltmann, *God in Creation: A New Theology of Creation and the Spirit of God* [San Francisco: Harper and Row, 1985], 284). My appreciation to David Wellman for drawing this passage to my attention.

is usually translated, "Hear, O God, the Lord our God, the Lord is One." But this masks the double naming of God in the term "Adonai Eloheynu," at least as that is explained in popular rabbinic theology. "Adonai" (YHWH) points to the side of God that promotes freedom and transcendence and the dynamism of transformation. "Eloheynu" designates God as the God of nature, the creator and sustainer of the universe. The creed of the Shema, Lerner says, thereby "calls upon us to witness that the God of nature is the God of freedom, that the Creator is the principle of transcendence toward all that the world and human beings can and ought to be."[21] One conclusion Lerner draws is that any division of spirituality and ethics from politics makes no sense, since "the Force that we've come to worship in the realm of the spirit is actually the Force that makes possible and necessary the moral/political transformation of the world."[22]

The claim of Sabbath and the Shema that God is praised as both creator and redeemer invites further reflection from within the Hebrew stories of origin. Again, important clues are found in the scriptures, this time in their arrangement as well as their silences.

Israel is redeemed from Egypt. Egypt is portrayed as the historical embodiment of the forces of chaos: that is, the powers of death, the anticreational, antilife forces of the cosmos. And Israel is redeemed for a vocation — to embody redeemed creation as community. Israel is ultimately redeemed for life, even abundant life, by way of just community.

But there is no return to Eden or any other normative golden age. Even the Promised Land, though rich with creation's bounty, is not pictured as the lost garden of the tree of life. This is significant as implied commentary on redemption's relation to creation, just as the canonical choice to place Genesis before Israel's own beginnings in Exodus is significant. God's creative purpose for the whole universe (Genesis 1–11) is set out *first* in sacred scripture; and even the "fulcrum text" of the election of Abraham in Gen. 12:1–3 ties this election immediately back to "all the families of the earth."

Why no Eden, no golden age, no utopia, no portrayal of a fixed, normative created order? And why do Adam and Eve absent themselves from the rest of Hebrew scripture? Is it perhaps because creation is not only not static but incomplete? The divine commands themselves in Genesis — to be fruitful and multiply (given to fish, birds, and human earth creatures) (Gen. 1:22, 28), to till and keep (to humans) (Gen. 2:15), to join in ongoing creation and joyful Sabbath (extended to all creatures) (Exod. 20:8) — would seem to say that even "good" does not here entail completed development or perfection. Divine creating is living, dynamic, and continuing. It is unfinished. Terence Fretheim says that "[f]or the creation to stay just exactly as God originally created it would be a failure of the divine design."[23] God's creating activity is not, Fretheim adds,

---

21. Lerner, *Jewish Renewal,* 66–67.
22. Ibid., 67.
23. Terence Fretheim, "The Reclamation of Creation," *Interpretation* 45 (1991): 358. I draw from Fretheim for this discussion of the relation of creation and redemption.

"exhausted in the first week of the world!"[24] And was never meant to be. All creation participates dynamically in movement from one generation to the next. The universe evolves. Earth is a pilgrim on a life journey.

The 1995 translation of the Pentateuch catches this beautifully in the very first lines:

> At the beginning of God's creating
> of the heavens and the earth,
> when the earth was wild and waste,
> darkness fell over the face of Ocean,
> rushing-spirit of God hovering over
> the face of the waters.
>
> (Gen. 1:1–2,
> *The Five Books of Moses*)

Redemption serves this dynamic creation. Redemption means reclaiming broken or despoiled or unfinished creation *for life*. It is not extracreational or extrahuman, much less extraterrestrial. Redemption consists in Spirited actions, often very ordinary everyday ones, against the anticreational forces that violate creation's integrity and degrade and destroy it. The ultimate goal is "a new heaven and a new earth" (Rev. 21:1). Such is the notion of destiny in this story. Even the most apocalyptic writings understand this, like John of Patmos himself, as a radical transformation *of* the created order and not its utter obliteration in favor of realms literally out of this world.

Redemption means freeing Israel (and all creation) from whatever oppresses or victimizes. Its reach is from inner spirit to sociopolitical and economic spheres to cosmic realms. It means realizing the life potential of all things.

Israel's own vocation is to become, again in Fretheim's words, "a created co-reclaimer of God's intentions for the creation."[25] The form of that vocation is as a "witness to the nations." As such, Israel's story signals for *all* peoples their vocation as coparticipants and "created co-reclaimers" of God's creating for the purposes of sustained abundant life.

But why does God redeem? For one reason only: because life and blessing are not yet what they might be and are not available to all, or are themselves endangered. The object of redemption is to free people and the rest of nature to become what all was created to be. Redemption's ethic, then, is not an earth-denying asceticism or any other exiting from creation but "immersion within the very sphere that [is to be] reclaimed by God's redemptive work."[26] It is earth ethics for creation's well-being.

To summarize and conclude. Jewish and Christian stories of origin root here in the strong biblical sense of moral responsibility before the God who is the

---

24. Ibid.
25. Ibid., 365.
26. Ibid., 362.

power in and of creation and the transcending power who beckons creation's redeeming transformation in the steady direction of compassion and justice. From this God, we receive the gift of life. Before and with this God, we are responsible for it. Such is the core of these traditions.

But of course this theology of life does not end here. Indeed, neither does the Bible. The Bible's own moral trajectory remains incomplete, a matter of which the biblical communities were painfully aware and of which the prophets and Jesus constantly reminded them. The "good news" for some was bad news for others or no news at all. Good news for Abraham and Sarah was bad news for Hagar and Ishmael. Good news for freed slaves became bad news for peoples they took as slaves. The social patriarchy that saturates scripture, even against its own theme of liberation and shared power, continues as bad news for women. Sexual ethics and Greek renditions of gender, nature, body, and mind in Hellenistic Judaism itself — and thus in formative Christian beginnings — continues as bad news for many to this day. Jewish biblical scholar Tikva Frymer-Kensky's conclusion is thus not only correct but profoundly "biblical" when she says that "it is now our task to weave the rest of the Bible's religious faith, ... in particular those [areas] that deal with the incorporation of all aspects of physicality, into our religious view of the universe."[27] That is our task and more for earth ethics now. The purpose of this first story of origin is only to say that while the "moral project" of the Bible remains incomplete, an ecumenical and ecological ethic of life should pursue it within the space marked by these two interwoven themes: the healing, mending, and transforming of the world (*tikkun olam*) and the celebration of the gift of life itself as the gracious creation of an infinitely compassionate God in an awesome universe. The Bible's own theology of life turns on these themes.

For contemporary cosmology and ethics four important terms may serve to illustrate the usefulness of the Bible's theology of life. The four terms are "think withs," the needed "instinctive" ways for thinking about things. The terms are creation, justice, neighbor, and mercy.

Creation. "Creation" isn't a synonym for unperturbed nature as gazed upon by awestruck humans. It is a theological word for the totality of all things, together, before God and in whatever blessed, despoiled, or hapless condition may prevail at the moment. The noun form is largely our doing, since "creation" is invariably found in verb form in the Hebrew scriptures, thereby underscoring the dynamic sense of a yet-unfinished world and cosmos. As noted, the recent translation of the Hebrew Bible renders the beginning of book 1, page 1, verse 1 as "At the beginning of God's creating..."

The energy, dynamism, and processive character of creation arise from the confession that creation is one and is God's. The rest of us are tenant farmers who, like all the generations before us and after, will live a little while and then be recycled to the humus from which all humans humbly come. That all creation is one and its creatures finite and transient is not to say creation is har-

---

27. Frymer-Kensky, *In the Wake of the Goddesses*, 220.

monious, however. That would belie the presence and power of evil and the gross imperfection of things yet awaiting redemption. New Age and other romantic fantasies about harmonious creation often indulge this understandable hope and heresy, oftentimes in Christian form. To say creation is one is rather to assert the "bundled" nature of life together, the irrepressible unity of all things in heaven and on earth, and to insist in the face of evil that all will be redeemed together, or nothing at all. Such is the prophet's and priest's conviction as well as the Psalmist's and Paul's.

The 1972 UN Conference on the Environment, that meeting of shoulders in Stockholm upon which Rio and the Earth Summit came to stand, included the following testimony from René Dubos and Barbara Ward.

> The astonishing thing about our deepened [scientific] understanding of reality over the last four or five decades is the degree to which it confirms and reinforces so many of the older insights of humanity. The philosophers told us we were one, part of a greater unity that transcends our local drives and needs. They told us that all living things are held together in a most intricate web of inter-dependence. They told us that aggression and violence, blindly breaking down the delicate relationships of existence, could lead to destruction and death. These were, if you like, intuitions drawn in the main from the study of human societies and behavior. What we now learn is that they are factual descriptions of the way in which our universe actually works.[28]

The title of their volume, *Only One World,* nicely summarizes much of the Jewish and Christian notions of creation, provided "world" is stretched to "cosmos" and all of it together is seen in and before the Ancient of Days in whom we live and move and have our own, and only, being.

Justice. "Justice" is a puzzling and contested notion for us. We may be adding to that with notions of justice in the distant biblical cosmologies. On the other hand, service is sometimes rendered by mining perspectives markedly different from our own. Help arrives via worldviews strange enough and strong enough to lift us from well-worn ruts.

Justice in those ancient cosmologies was a conviction about God and the dynamics of creation itself. The creating of the world and the vast canopy enveloping it was conceived as a mighty movement from chaos toward more harmonious order. The Bible thus opens with: "In the beginning when God created the heavens and the earth, the earth was a formless void and darkness covered the face of the deep" (Gen. 1:2a, NRSV). Yet harmonious order, at least in most accounts, was never wholly accomplished. Chaos survived, lives on, and continually intrudes. The Psalmist's imagination and devilish sense of humor can now and again make light of it, with chaos blithely playing as a Leviathan in the deep while human beings toil for their meager fare (Psalm 104). Life is

---

28. René Dubos and Barbara Ward, *Only One World* (London: Penguin, 1972), 85.

like that. But the more serious treatment points to an incessant struggle between ongoing creation and living chaos that never ceases in the world as we know it. Its forms are many — famine and disease, hunger and plagues, family discord and community estrangement, mental illness, the wasteful ways of war and corruption, lies and deceit, plain craziness — and in benighted, pre-Enlightenment days these were often named "evil spirits" and "demons." Whatever the names and forms, a pervasive disorder exists and is deeply embedded within creation itself, violating it from within. While Christian theologians sometimes label it "original sin," and Jewish thinkers at times try to pin it down by referring to it as "the evil impulse," most Christians and Jews end up vacillating in their explanations about its source. We have never decided whether it is the work of some inexplicable, half-external evil force that virtually possesses us or the clear consequences of human creation, choice, and responsibility. At the moment the latter would seem to have more explanatory power, at least as viewed from the vantage point of earth. We threaten to supplement original sin with terminal sin. The mystery and history of iniquity remain.

Thinking about justice stems from this Hebrew understanding that cosmic reality struggles, in God, to be a harmonious order of and for all creatures great and small. When the biblical confession says again and again that Yahweh is "just," it means God again and again fashions beneficent order from looming and dooming chaos, restrains the chaos, and balances things anew when chaos triumphs. In the prophetic oracles, whole empires spiral into decline when their injustice violates creation and provokes a countervailing wrath. All this gives rise to the conviction in Judaism that whatever evidence there is for the historical triumph of injustice and chaos — and it is often massive — the universe is nonetheless a moral one bent by God's own struggle to arch in the direction of redemption and eventual harmony, even perfection. (Precisely this biblical conviction about justice sustained Martin Luther King Jr. and the struggle of many African-Americans over decades.)

Though it is strange to minds like ours, wholly formed on this side of heightened human power to alter natural environments radically, Hebrew minds did not separate history from nature. No word for nonhuman nature as a separate reality exists in Hebrew. Justice as a concept thus does not participate in distinctions of nature and history. Justice may refer one time to intrahuman events (enemies are defeated in battle) and another time to events we assign to nature and what the insurance companies still call "acts of God" (the Flood).

If we were suddenly to set aside these biblical lenses and put on an imaginary and imaginative pair borrowed from some current environmentalists and feminists, we would call "justice" simply "right relatedness" and thereby land somewhere very close to the ancient notion. We would then begin to realize that justice pertains to humans and nonhumans alike, as it does in the Bible's inclusive, theocentric understanding of creation. We would next realize that sabbatical days and years and the Jubilee laws (Leviticus 25) begin to make a sense for which modernity has had only contempt. The planet as a whole would suddenly appear as the basic unit for justice itself, in the manner of *oikos*. Trees would have standing; land reform would go beyond arguments about ownership; water would be a precious, guarded, and safe resource; community well-being and the poverty of millions would be defined as environmental issues — only to begin the list.

Furthermore, justice isn't a synonym of numerical equality here. Inherent value and respect are key, but even inherent value isn't the same as equal value. Justice is closer to mutuality and means essentially that we share one another's fate and are obligated by creation itself to promote one another's well-being. This is a notion of justice far more comprehensive than that which has reigned in Anglo-Saxon jurisprudence, social philosophy, moral traditions, and daily habits. These last-named have focused justice on individual liberty and/or equal-ity. As an ideal, justice as liberty has meant guaranteeing the widest range of individual choice commensurate with such choice for others, while justice as equality has meant guaranteeing a comparative allotment of goods, services, and opportunities at a humane level for the largest number of citizens pos-sible. Legislation, policies, and regulations of all kinds have then turned on which concepts and compromises come to prevail in the uneasy mix of these. Vital as liberty and equality are, they share a common anthropocentrism and thus restrict justice to matters of human welfare alone, directly or indirectly. By contrast, justice as comprehensive right relatedness means the rendering, amid limited resources and the conditions of a still chaotic world, of what-ever is required for the fullest possible flourishing of creation in any given time and place. Justice assumes creation's integrity and, against the workings of destructive systems, establishes the prima facie right to the life and flour-ishing of all creatures. This in turn assumes sustainability as a condition (the capacity of life to survive and thrive) and establishes it as a baseline crite-rion.[29] Such a notion of justice is a worthy moral resource, recovered not from some golden age or decade of untainted insight but from a nonetheless gifted season when cosmologies were mythic but whole. They were often also empir-ically very wrong, however, and unjust in the concrete working out of things. The living remnants of patriarchy, for example, testify to this. So the effort is to learn anew *from* stories of origin and their outworking, not simply repeat them. Every cosmology and corollary way of life is subject to moral assess-ment on the basis of concrete practices. Aboriginal accounts do not provide their own validation. Their authority, like all authority, is subject to moral assessment.

Neighbor. Jesus the Jew, in whom Christians see the fullest manifestation of God possible in human form, radicalizes the notion of neighbor. He insists that the enemy is neighbor and that we are to treat all neighbors, including the enemy, with a regard equal to the regard we accord ourselves in love. We are to use the same framework of positive reference when we consider others

---

29. See the discussion of Martin Robra in *Oekumenische Sozialethik* (Gütersloh: Gütersloher Ver-lagshaus, 1994), especially the subsection, "Der grössere Haushalt Gottes," 135–50, and its discussion of justice, using Konrad Raiser's *Ecumenism in Transition.*

as we use for ourselves. This is by now so much the pedestrian repetition of moral catechism that we hardly expect anything explosive, or even relevant, from it.

But consider what happens when neighbor-love is extended in ways illumined by "creation" and "justice" and implied by an ethic of living sustainably in a contracting and crowded world. Then neighbor embraces, as H. Richard Niebuhr argued, "all that participates in being," organic and inorganic, present, past, and future.[30] Neighbor then means a comprehensive responsibility inherited from the ancestors and turned toward posterity. In this sense, neighborly responsibility is infinite in extent, with no preordained boundaries. Neighbor means being entrusted with the wealth of nature and the treasures of society for the sake of plant, human, and other animal life alike. It means responsibility for what neighbor literally means — the nigh farmer or, more loosely, the nigh one. But it means accountability to the far one as well, both in time and space. And it means welfare for the enemy as well as those for whom we willingly sacrifice. As in the parable of the Good Samaritan, class, social standing, function in society, address, and purity rules of all kinds have nothing to do with the definition of neighbor. Neighbor is as neighbor-love does to whomever or whatever is at hand and in need. In a word, the neighbor in a million guises is the articulated form of creation to whom justice, as the fullest possible flourishing of creation, is due. (The British high court got it right in the 1932 decision on bottling a contaminated drink: our neighbor is anyone or anything we ought reasonably to think may be affected by our actions.)[31]

Mercy. "Compassion" and "forgiveness" might have been chosen as well as "mercy." While not precisely the same, they, too, specify that quiet power needed to face a broken, despoiled world and decisively to begin over again without destroying our neighbors and ourselves in the process. *Tikkun olam,* the healing and mending of the world, cannot proceed without the two indispensables of forgiveness and oxygen. For those honest about our condition, our efforts, if lacking mercy, dissolve into blaming, recrimination, self-loathing, and repeated patterns of abuse. Even honesty and candor, if lacking mercy, can mean remorse and grieving for planetary destruction and guilt toward the present and future generations, all without the renewing energy of hope. The biblical testimony is that none less than God's own strength is tapped when in mercy "we accept and have compassion for our own and everyone else's . . . limitations."[32] Mercy, then, is an unusual kind of power found on the home turf of chaos and death itself. It is a tonic to revive the weary, despondent, and overwhelmed, and a means to change poisoned relationships unexpectedly. It, or compassion and forgiveness, is the shape of grace when all need to find their way together, whatever

---

30. H. Richard Niebuhr et al., *The Purpose of the Church and Its Ministry: Reflections on the Aims of Theological Education* (New York: Harper and Row, 1956), 38.

31. Shridath Ramphal, *Our Country, the Planet: Forging a Partnership for Survival* (Washington, D.C.: Island Press, 1992), 211.

32. Lerner, *Jewish Renewal,* 144.

their refined enmities. If the healing of the world is to be had, it will begin in mercy.[33]

These brief sketches from the moral universe of Jewish and Christian stories of origin are only that — brief sketches meant to show that these ancient, living traditions still have contributions to make to cosmology and ethics today. There is vastly more content, of course, and other forms than exposition and commentary. But with this as a brief exposition, we bring this section to a close and turn to another account of origins — present science's.

THE CONTRIBUTION OF SCIENCE

The story of origins offered by science requires a jump from ancient sacred texts to recent discoveries. An earth ethic of life at the cusp of a new millennium on a burdened planet does well to draw from these. The purpose is not strictly scientific, however, but moral: that is, to set before the eyes of the concerned the extraordinary unity and diversity we share with one another and all things, living and nonliving, for the sake of taking responsibility.

In lay terms the common creation story available from current science goes something like this. A famous poet, in the line that gave a famous collection its name, wrote better science than he knew: "I believe a leaf of grass is no less than the journey-work of the stars."[34] We, and all else, are variations on exactly the same thing — stardust. The atoms in the grass, and all atoms everywhere, our bodies included, were born in the supernova explosions that birthed stars. Everything is thus radically "kin" from the very beginning. The "createds" are all "relateds." When you peer at the Southern Cross, Orion, or the Big Dipper, the gnat on your arm, the flower near your path, or the food on your plate, you are gazing at a neighbor who shares with you what is most basic of all — common matter as ancient and venerable as time and space themselves.[35] In fact, if you look at more recent complex forms of life — plants and animals — you observe molecules that are very much the same. The most basic functions of cells are the same in all life-forms. DNA and RNA processes, and cell division, for example, are identical across life-forms. (The DNA molecule specifies the characteristics of all living organisms, from bacteria through human beings.) "The wonderful lesson to come out of biology in the last five years," Victoria Foe explains, "is the same genes, the same parts, turn up again and again, from one species to another." "The important lesson to realize is that we're all made of the same fabric, we're part of the same web."[36] The reason is remarkable:

---

33. For a particularly rich discussion of this, see Donald Shriver, *An Ethic for Enemies: Forgiveness in Politics* (New York: Oxford University Press, 1995).

34. Walt Whitman, "Song of Myself," in *Leaves of Grass*, as cited in *EarthLight* 11 (summer 1993): 13.

35. When the universe began the only elements were hydrogen and helium, leading one scientist to quip that gas, given enough time, turns into people.

36. Victoria Foe, "Drawing Big Lessons from Fly Embryology," *New York Times*, August 10, 1993, C12.

all life-forms apparently share the same ancestor. We (all life-forms) probably emerged from an ancient single-celled being. Life is still coded in a way that displays this.[37]

What could be a more radical and visible ecumenical unity than this! All that is (*ta panta*, in New Testament coin) has a common origin and is related in its most basic being. Charles Darwin put this eloquently well before the discovery of DNA and other evidence of life's common ancestry and diverse unity:

> There is grandeur in this view of life with its several powers, having been originally breathed into a few forms or into one; and that, whilst this planet has gone cycling on according to the fixed law of gravity, from so simple a beginning endless forms most beautiful and wonderful have been, and are being, evolved.[38]

Everything is also radically diverse and unique. Regrettably we have poor eyesight, so we see little of what surrounds us. Our powers of imagination suffer as a consequence. Annie Dillard tells of biologists finding in one square foot of topsoil one piddling inch deep "an average of 1,356 living creatures,... including 865 mites, 265 springtails, 22 millipedes, 19 adult beetles, and various numbers of 12 other forms."[39] (This doesn't include two billion bacteria and millions of fungi, protozoans, algae, and innumerable other creatures that make the topsoil the one inch of *'ădāmâ* it is.)[40]

Of course, "the whole is not the total."[41] What counts here is not the population per se, but synergy, how the various interactions build upon one another's results to yield that which makes for life. The whole and its behavior cannot be understood by knowing and totaling the parts.

This, like any story of origins, is not a story we should hear only once. Stories of origin are meant to be retold and appropriated as moral, religious, and aesthetic ethos. So listen again, this time from Professor of Soil Physics Daniel Hillel. "A romantic poet gazing through his window at a green field outside

37. Charles Birch and John Cobb, both active in WCC studies, write as follows in *The Liberation of Life: From the Cell to the Community* (Denton, Tex.: Environmental Ethics Books, 1990), 45: "The evolution of a living cell from organic molecules may have happened more than once on the earth. But probably only one original cell gave rise to all the rest of life on earth. This seems to be the only possible explanation of the basic similarity of the cells of all living organisms. All use the same DNA code and similar amino acids. The doctrine of evolution holds that from one beginning all the diversity of life on earth, its two billions of species (of which two million known species are alive today) and the many varieties within these species, have arisen. Life is like a great branching tree with one central stem."

38. Charles Darwin, *On the Origin of Species* (Cambridge, Mass.: Harvard University Press, 1964 [1859]), 490.

39. Annie Dillard, *Pilgrim at Tinker Creek: A Mystical Excursion into the Natural World* (New York: Bantam, 1974), 96, as cited by Sallie McFague, *The Body of God: An Ecological Theology* (Minneapolis: Fortress, 1993), 38.

40. McFague, *Body of God*, 38. *'ădāmâ* is the Hebrew for the ground from which God created all creatures.

41. William Ashworth, *The Economy of Nature: Rethinking the Connections between Ecology and Economics* (New York: Houghton Mifflin, 1995), 138–39.

might view it as a place of idyllic serenity," Hillel writes. But a closer look "discerns not rest but unceasing turmoil, a seething foundry in which matter and energy are in constant flux." Heat is exchanged and water percolates through the intricate passages of the soil. Plants suck up some of that water, transmitting it to the leaves, which transpire it back to the atmosphere. The leaves also absorb carbon dioxide and synthesize it with soil-derived water to form the primary compounds of life. Oxygen in turn is emitted by those leaves to make the air breathable for animals, which consume and in turn fertilize the plants. Organisms in the soil recycle the residues of both plants and animals. With that nutrients for life's renewal are released.[42]

The "crucible" of this foundry is the soil (mineral particles, organic matter, gases, and nutrients) that, when infused with water, becomes the biological factory that initiates and sustains life, stores and distributes water, and acts as earth's primary cleansing and recycling medium for the health of the environment.[43]

Soil's population is impressive, most of it bypassing our notice. One acre of topsoil from a temperate zone carries about 125 million invertebrates. Just 30 grams can carry 1 million bacteria of a single type, as well as 100,000 yeast cells and 50,000 fungus mycelium.[44]

Moreover, all of this detail — the organic and the inorganic — is essential and active together. It is not as if earth's soils and other physical forms were first put in place through geological evolution and then somehow life emerged within this context. Rather, the living forms that emerged early on in earth history were themselves powerful agents in creating both soils and the atmosphere, just as the evolutionary turbulence of inorganic matter created conditions conducive to life that then changed nonliving matter. Where geology ends and biology begins is a line that cannot be drawn.[45] The geosphere, the biosphere, the hydrosphere, and the atmosphere are our handy distinctions, not earth's. Earth knows only a complex, dynamic, and awesome unity.

Or, looking in another direction — up rather than down, out rather than in or under — we soon lose the capacity to take in what we can see or imagine. We can take in a mountain or a portion of a forest or the horizon of a desert. We can take in a horse, a hut, corn, and the weather. With help from satellites, we can even understand that earth, teeming with variety, is but one tiny cell in the cosmos. But beyond the solar system we can hardly even imagine our way across our own galaxy. This, our home port, is one hundred thousand light years in diameter and has somewhere between ten billion and one hundred billion stars (our census is crude). Were this not enough, there are likely more stars in the

---

42. Daniel Hillel, *Out of the Earth: Civilization and the Life of the Soil* (Berkeley: University of California Press, 1991), 23.

43. Ibid., 23–24.

44. Clive Ponting, *A Green History of the World: The Environment and the Collapse of Great Civilizations* (Baltimore: Penguin, 1991), 15.

45. For detailed treatment of this, see James Lovelock, *The Ages of Gaia: A Biography of Our Living Earth* (New York: Norton, 1988).

universe than all the grains of sand on all of earth's beaches.[46] We cannot quite imagine what fifteen billion years is (the surmised age of the universe thus far) or the full meaning of Carl Sagan's wonderful comment on how utterly unique the course of evolution is over that period. It took fifteen billion years, he said, to give us apple pie.[47]

Crimped imagination notwithstanding, the upshot is that the enduring unity that has been ours (humankind's and all else's) from the very beginning has evolved as highly complex networks characterized *by staggering differentiation and even implausible individuation.* (How many varieties of mushrooms or coniferous trees are there? How many zebras have the same pattern of stripes? Genetics itself teaches the uniqueness of the individual.) Life — no, life and nonlife, organic and inorganic matter together and continuous — is a radical unity marked by interrelated and interdependent fecundity and variety whose detail and wonder escape our shriveled sight. When we do catch a glimpse, we are moved to psalms. But that, alas, is too rare a response. Most of the time we just assume that life *is* and will go on.

The diversity and individuation we have described in different ways have their own expression within the human species. Our lives are concrete, particular, and different. We differ one from the next. Not one of the nearly six billion of us is a clone of another. We are who we are in different places with different cultural, linguistic, religious, political, economic, and sexual traditions. And if we could be pirates of time and peer back a few thousand years, or ahead the same number, we might find it difficult even to recognize our genetic forebears and posterity as persons with whom we could share an afternoon's conversation about most anything. Life, even life within the same "kind," is a many-splendored thing.

This, then, is present science's contribution as a story of origins: the stunning portrayal of a common creation in which we are radically united with all things living and nonliving, here and into endless reaches of space, and at the same time radically diverse and individuated, both by life-forms and within life-forms. And all of it is not only profoundly interrelated and inseparably interdependent but highly fine-tuned so as to evolve together. We are all — the living and not living, organic and inorganic — the outcome of the same primal explosion and same evolutionary history. All internally related from the very beginning, we are the varied forms of stardust in the hands of the creator God. (This latter judgment is obviously a theological, not scientific, one.) This reality is the most basic text and context of life itself, and thus of a life ethic.

---

46. A *Far Side* cartoon has two tiny insects atop a mushroom peering at a starlit sky. One remarks: "Just look at those stars tonight . . . makes you feel sort of small and insignificant." Cited by McFague, *Body of God,* 207.

47. Reported by Thomas Berry in my Ecology and Ethics class at Union Theological Seminary, New York City.

RUMINATIONS

Hopefully the sense of our home as this *oikos* in this cosmos will stimulate the same sense of creation's grandeur and *eucharisto* (thanksgiving) for the gift of life that biblical communities knew. But we should do far better, because this is in fact a description of immensity and intricacy that far surpasses anything known by those communities or their stories of origin. The weight of the nucleus of an atom of life (1,000,000,000,000,000,000,000,000 times less than a gram), the incomprehensible "genius" of DNA, the vastness of a cosmos we hardly have measures for — all this eluded even the most awestruck psalmists and the most imaginative prophets with their bold visions of creation redeemed. Little wonder Thomas Berry calls the universe itself the primal expression of the divine and the primary revelational event. Little wonder he urges us toward a mystique of cosmos and earth and says that "the universe, by definition, is a single gorgeous celebratory event."[48] Neither theology nor ethics has yet truly fathomed what science presents as an infinitely magnificent evolution, an evolution nonetheless gravely threatened by our presence on the only *oikos* we know and the only one fine-tuned for our survival. We should pray aloud for deep humility before life and bow before all else (such as green plants) upon which we depend for every breath we take and every morsel that passes between our teeth. We should echo the aged farmer in Wendell Berry's "The Boundary" who, unable to make his last walk all the way around the farm, stopped to rest and to gaze upon a stream and pass "across into the wild inward presence" of its life. "Wonders," he thinks. "Little wonders of a great wonder."[49] ("God doest only wonders" is St. Augustine's comment.) The old farmer saw around him what the poet Kabir saw in his own body:

> Inside this clay jug there are
> canyons and pine mountains,
> and the maker of canyons and pine mountains.
> All seven oceans are inside, and
> hundreds of millions of stars.[50]

We post a criterion in passing. Any God-talk in the particular cosmologies and ethical systems of different traditions and locales that does not include the entire fifteen-billion-year history of the cosmos and does not relate to *all* its entities, living and nonliving, ancient forms and very recent ones (such as humans), speaks of a God too small.[51] Any cosmology and ethic that does not assert both

---

48. Thomas Berry, *The Dream of the Earth* (San Francisco: Sierra Club Books, 1988), 5.

49. Wendell Berry, "The Farm," in *Late Harvest: Rural American Writing*, ed. David R. Pichaske (New York: Paragon, 1992), 140.

50. Kabir as cited by Robert Bly in "'Seeing' Poems," in *Earth Ethics* (spring 1994): 8. Kabir's poem is titled simply "Poem 37."

51. This is Sallie McFague's effort to sober down what she calls "our natural anthropocentrism" (see *The Body of God*, 104).

radical unity and radical differentiation as reality and necessity together also fails. And any earth ethic that assumes any one species is a species by itself, or that its own members can be treated as collectives only, is faulty from the outset.

There is an important implication for policy. Sustainable development that works from grand schemes that assume universally applicable knowledge, a world economic and political culture of common dynamics, values, and institutions, and global enforcement working from top down is development cutting across the grain of nature. It misses, as Bruce Rich says, "the extraordinary and irreducible heterogeneity of the biosphere at all levels."[52] Local and regional ecosystems evolve historically, and social and cultural diversity accompanies them. Effective policy thus needs to work with and from this intrinsic heterogeneity and its "complex adaptive systems," human and nonhuman.[53] In the language of earlier discussion, sustainable community, with its sense for the local and, we add, their many stories of origin, is far preferable to the globalism of most sustainable development and its own cosmic myth and culture.

Yet policy implications are not the chief matter at the moment. The chief matter is the kind of general orientation the common creation story of science spins out for those of us looking to get our bearings anew. We are creatures of planet earth before all else, this story says, creatures of earth even before we take on the particular identities fashioned by other stories of origin and destiny, religious and cultural ones. The common creation story is not a *substitute* for these other stories, but neither ought they be read apart from it. "We are not tourists here," Mary Midgley writes. "We are at home in this world because we were made for it. We have developed here, on this planet, and we are adapted to life here.... We are not fit to live anywhere else."[54] In the great vastness it is only on earth, with its peculiar and ongoing evolution of life, that we are "at home," citizens of *oikos*.

In the end, it is a wonder that we live at all. But to live and be aware of life can only be the source of both wonder and obligation, the two basic dispositional ingredients of a worthy earth ethic. Earth as a treasure of a cosmos that is infinite in all directions must be preserved, this story of origin implies. Such is an imperative of earth citizenship itself. Sobering for all this is the fact that this good work is in the hands of earth's most dangerous species.

At this juncture we could posit "reverent respect" or even "awesome respect" as a basic moral norm governing all relationships that impinge on others. Or we could return to the "think withs" discussed earlier — creation, justice, neighbor, mercy — and ask how they accord with the common creation story of science, thereby adding to our moral universe. But perhaps we are best served simply to close with a reminder of what stories of origin do, and then enjoy the charm of a four-and-a-half-year-old's commentary.

---

52. Bruce Rich, "The Crisis of Development," *Earth Ethics* (summer 1994): 6.
53. Ibid., 6–7.
54. Mary Midgley, *Beast and Man: The Roots of Human Nature* (Ithaca, N.Y.: Cornell University Press, 1978), 194–95, as cited by McFague, *The Body of God*, 111.

ENDINGS AND BEGINNINGS

Mexican laureate Octavio Paz writes that "we are living through a change of times" in which we reach for a "rebeginning," "the resurgence of buried realities, the reappearance of what was forgotten and repressed," a return to origins.[55] Paz speaks of the special need for "rebeginnings" at the end of the Cold War, now "that the cruel utopias that bloodied our century have vanished" and the time has come for "a radical reform" of liberal capitalist society and the impoverished nations on the periphery.[56] But if rebeginnings are needed after the Cold War as an epoch of the last half-century, how much more are they needed now that it is clear the entire industrialized and globalized way of life building up over the past several centuries is unsustainable? Our era has witnessed what we did not even imagine possible: that human activities could overwhelm our little patch of creation with chaos, could overwhelm life with death. Nor do we yet comprehend that we have in actual fact taken up a new vocation as co-*un*creators on a scale that outstrips all previous generations. This includes that awful thinning of the great crowd of life, extinction itself. It is nothing less than the loss of modes of divine presence, forever.[57] In such a context, stories of origin are vital because they offer the means to search anew, not for first beginnings, but for rebeginnings. There is no return to first beginnings. We all live east of Eden. The garden is gone. It wasn't our destiny, anyway. Nonetheless, nature still holds "incalculable promise."[58] There need be, and can be, rebeginnings.

Posing themes from the biblical account and themes from the common creation story is one way to explore the substance of stories of origin for earth faith and earth ethics. But it is only one instance. It does not, for example, bring the gifts of indigenous peoples to the fore or continue the presentation of the world's universal religions. Earth faith invites far more of the kind of conversation this chapter tries to illustrate.[59]

Danu Baxter, at four-and-a-half, has her own language for the requisite wonder. Here is how she feels at day's end. We could do worse:

> Goodnight God
> I hope that you are having
> a good time being the world.
> I like the world very much.
> I'm glad you made the plants
> and trees survive with the

55. Octavio Paz, "Poetry and the Free Market," *New York Times Book Review*, December 8, 1991, 36.

56. Ibid.

57. This is a point frequently made in the writings of Thomas Berry.

58. Haught, *Promise of Nature*, 125.

59. A splendid example of critical dialogue between Euro-American Christian discussions of creation and Native American ones is found in George Tinker's "Creation as Beloved of God," in *Defending Earth Mother*, ed. Jace Weaver (Maryknoll, N.Y.: Orbis Books, 1996). This volume was prepared as part of the North American contribution to the WCC's Theology of Life program.

rain and summers.
When summer is nearly near
the leaves begin to fall.
I hope you have a good time
being the world.
I like how God feels around
everyone in the world.
Your arms clasp around the world.
I like you and your friends.
Every time I open my eyes
I see the gleaming sun.
I like the animals — the deer,
and us creatures of the world,
the mammals.
I love my dear friends.[60]

60. Danu Baxter, in *Earth Prayers from Around the World*, ed. Elizabeth Roberts and Elias Amidon (San Francisco: HarperSanFrancisco, 1991), 374.

Returning to Our Senses

In the twelfth and thirteenth centuries, parts of Europe faced serious deforestation. A major movement of Germanic peoples eastward into lands occupied by Slavic peoples had taken place from the tenth to the thirteenth centuries. In contrast to the swidden agriculture, stock raising, hunting, and fishing of the Slavs, the Germanic peoples cleared woodlands and used the heavy plow for croplands. By the thirteenth century, many areas were stripped of woodlands through a combination of clearing land and heavy use of wood products. (Trees of life provided food, fuel, and building materials for shelter, furnishings, and tools.) Dependence notwithstanding, the attitude of the new settlers may well have been reflected in the abbot of Fellarich's declaration: "I believe that the forest which adjoins Fellarich covers the land to no purpose, and hold this to be an unbearable harm."[1] In any event, by 1338 the bishop of Bamberg found it necessary to take a pledge, as did his successor. The bishops vowed to place not only the people but the forests under the protection of Christian faith and the church.[2]

The abbot and the bishop both spoke as Christian leaders. But their testimony and vow complicate, rather than answer, the question of whether Christian faith truly offers protection for an endangered biosphere. If churches worldwide were suddenly to pledge the security of both people and the rest of nature and resolve to order all relations in accord with the integrity of creation, would something salvatory happen thereby? More precisely for our concerns, what resources for a viable earth faith and earth ethic would be forthcoming? Is "the ambiguous ecological promise of Christian theology"[3] genuine promise or only genuine ambiguity?

Like the foregoing chapters, this one shares the effort to find religious and moral substance for a sound earth faith and earth ethic. It differs from preceding chapters, however. Here the quest is not the rejuvenation of old ecumenical symbols that have crossed virtually all religions and cultures (the tree of life). Nor is it the effort to cultivate new symbols as we retap our own experience in view of what is happening to earth (the gifts of darkness). Neither is the attempt one of visiting various stories of origin and various symbols of human vocation in the interest of their power for rebeginnings and altered self-understanding (see

---

1. This account and quotation are from Clive Ponting, *A Green History of the World: The Environment and the Collapse of Great Civilizations* (New York: Penguin, 1991), 123.

2. Cited in James A. Nash, "Ecological Integrity and Christian Political Responsibility," *Theology and Public Policy* 1 (fall 1989): 32, from Clarence J. Glacken's *Traces on the Rhodian Shore* (Berkeley: University of California Press, 1967), 339, 322–38.

3. Paul Santmire, *The Travail of Nature: The Ambiguous Ecological Promise of Christian Theology* (Philadelphia: Fortress, 1985).

the chapters above, "Adam, Where Are You?" and "ReBeginnings"). The turn in this chapter and the next is to established Christian confessional traditions and their symbols in order to see, amidst the debris, what treasures for earth faith and earth ethics may yet lie at anchor in the harbor of the churches' memory.

This is still an agenda too grand. The streams of Orthodox communions, global Roman Catholicism, and florid Protestantism are deep, wide, varied, and rich. To offer a chapter or two that would try to capture the whole would be futile.

Nonetheless, something worthwhile can be done. What can be done is a selection for the sake of illustration: in this case a Lutheran recasting of catholic themes. Impulses from Luther and his theology of the cross will trace one sturdy strand of Christianity's potential for earth faith and ethics. The hope is that other religious traditions, Christian and otherwise, will do the same from their own turf. Earth's distress, after all, is the most ecumenical of issues. As such it requires a religious response of corresponding scope. No one religious tradition will suffice. Religions altogether will not save the planet, for that matter. Yet neither will earth be saved without them. So none should be bypassed or overlooked. In an enlargement of the WCC convocation in Seoul, people will need to come from every nook and cranny of earth, and every religious and ethical tradition, to lay gentle, healing hands on the planet in blessing.

The traffic flows two ways. Earth stands to benefit from religious traditions rallying to its cause. At the same time, earth's distress challenges these traditions to their cores. Most of Christianity, for example, has been human-centered and male-centered in conception and leadership. It has been dysfunctionally and destructively dualistic on many fronts, including fateful splitting of spirit from matter, mind from body, and humanity from the rest of nature. It has encouraged earth- and world-denying spiritualities bent on "incanting anemic souls into Heaven."[4] And from its own scriptures onward, it has been deeply ambivalent about the body and sensuality. Much of it has also contributed mightily to modernity, largely supporting an industrial, technological, and urban world deeply at odds with the rest of the planet.[5]

Earth's distress is a challenge to all this "conniv[ing] in the murder of Creation."[6] Christianity's fundamental symbols and their theological construals are put to the test. John Haught is correct that the task is not only the "reexami-

---

4. Wendell Berry, *Sex, Economy, Freedom, and Community* (New York: Pantheon, 1993), 114.

5. Wendell Berry (ibid., 115) argues vigorously that this is the true guilt of modern Christianity: "It has, for the most part, stood silently by while a predatory economy has ravaged the world, destroyed its natural beauty and health, divided and plundered its human communities and households. It has flown the flag and chanted the slogans of empire. It has assumed with the economists that 'economic factors' automatically work for good and has assumed with the industrialists and militarists that technology determines history. It has assumed with almost everybody that 'progress' is good, that it is good to be modern and with the times. It has admired Caesar and comforted him in his depredations and defaults. But in its de facto alliance with Caesar, Christianity connives in the murder of Creation. For in these days, Caesar is no longer a mere destroyer of armies, cities, and nations. He is a contradicter of the fundamental miracle of life."

6. Ibid., 115.

nation of conscience" that religious traditions need to undergo. It is a corollary examination: Are these faiths open to basic reform in the process?[7]

Christianity's own potential gifts for an earth faith must, therefore, be gauged by the same criterion that measures the required reform. Christian faith must show that it has an available power for meeting creation's travail. If Christianity does not demonstrate a power that addresses earth's distress and makes for sustainability, its claims to be redemptive ring hollow.

I am not certain that the Lutheran theology of the cross — the strand of Christian reflection chosen here — can meet this test. But I do know I immediately understood Kosuke Koyama's testimony about his own experience in the firebombing of Tokyo: "The slow assimilation of the traumatic events of 1945, which only gradually yielded their theological implications, has moved me towards the emotive region of the cross of Christ."[8] Creation's conditions today move me instinctively "towards the emotive region of the cross of Christ." A powerful resonance is there. That resonance is the reason for this chapter and the next.

FINITUM CAPAX INFINITI

If we situate ourselves in the "emotive region of the cross," then what might be said in a Lutheran way about present and prospective suffering, otherkind's and humankind's? Theological affirmations will guide us.[9] The one for these pages is this: being with the gracious God means loving earth.

For Europe, World War I meant massive public suffering and shattered cultural confidence. How do we do postliberal theology for a battered and beaten age? became the question for a generation of European theologians. The answer for many was to reassert the majesty, glory, and power of the transcendent God. Fascinating in all this was the theological divergence of two young minds. The young Karl Barth, desiring to proclaim God's majesty, began by placing God at a remote and awesome distance. An infinite qualitative distance separated divine from human ways. Theology's task was deep and unrelenting cultural critique. The even younger Dietrich Bonhoeffer, desiring to proclaim the same majesty, began by bringing God into the closest possible, but also awesome, proximity. Barth was drawing from the Calvinist insistence that the finite cannot hold the infinite (*finitum non capax infiniti*) while Bonhoeffer insisted, with Luther, that the finite bears the infinite and the transcendent is utterly immanent (*finitum capax infiniti*). "God is in the facts themselves," said Bonhoeffer, asserting his conviction that God is amidst the living events of nature and history. His fa-

---

7. John F. Haught, *The Promise of Nature* (New York: Paulist, 1993), 16.

8. Kosuke Koyama, *Mount Fuji and Mount Sinai: A Critique of Idols* (Maryknoll, N.Y.: Orbis Books, 1985), ix. Koyama was a teenager in Tokyo in 1945. The firebombing of the city and other wartime destruction are the events to which he refers.

9. This section draws from an article that elaborates the theology of the cross more extensively (see "The Community of the Cross," *Dialog* [spring 1991]).

vorite quotation from F. C. Oetinger said much the same: "The end of the ways of God is bodiliness."[10]

The meaning of *finitum capax infiniti* is simple enough: God is pegged to earth. So if you would experience God, you must fall in love with earth. The infinite and transcendent are dimensions of what is intensely at hand. Don't look "up" for God, look around. The finite is all there is, because all *that* is, is *there*.

This is earthbound theology. With it Luther is boldly pan-*en*-theistic. A nice passage from Luther asserts this, despite the offense it gave medieval reason:

> For how can reason tolerate it that the Divine majesty is so small that it can be substantially present in a grain, on a grain, over a grain, through a grain, within and without, and that, although it is a single Majesty, it nevertheless is entirely in each grain separately, no matter how immeasurably numerous these grains may be?[11]

At least until the Enlightenment the common catholic conviction was that God is revealed in two books, scripture and the book of nature. Luther's words reflect this. But to recognize only this in the encapsulation of the divine in a single grain is to miss the radical character of Luther's panentheism. His is a massive protest against a Christian world with Greek philosophical genes and against the kind of idolatry this Christianity fostered, an idolatry that took the form of a speculative "theology of glory." It was the serious mistake of Christianity to think, from the second- and third-century Apologists onward, that the way to God was via the contemplative mind and soul ascending the ladder stretched from earth to heaven, progressively abandoning material reality in a preference for pure spirit. It was a disastrous mistake to Platonize and Hellenize Christianity, Luther concludes, and affirm the split and dual realities of a corruptible body and a transcendent, immortal soul, thereby progressively falling out of love with earth in the course of nurturing soul, mind, and reason.

Rather, one should image God and, like God, fall in love with earth itself. In the course of so imaging God we would simultaneously fall in love with ourselves *as creatures*. Luther utterly rejects any flight from the creaturely and finite as the path to communion with God. God is *wholly* in the grain, and the grain is *holy* in God. Earth is "crammed with heaven."[12] It is to be embraced, not spurned. God is present *to* creation *in* creation.

True, Luther is an Augustinian and a mystic. Yet he emphatically turns back Augustine's contention that Christ descended to help us ascend. He counters that Christ descended precisely to keep us from trying! We should have long

---

10. Dietrich Bonhoeffer, "Aufträge der Bruderräte" (December 1939), in *Gesammelte Schriften* (Munich: Kaiser, 1966), 3:388. The best discussion of the relationship of Barth and Bonhoeffer, these two theological emphases and their resolution, is Charles Marsh, *Reclaiming Dietrich Bonhoeffer: The Promise of His Theology* (New York: Oxford University Press, 1994).

11. Santmire, in *Travail of Nature*, cites this from Heinrich Bornkamm's *Luther's World of Thought* (St. Louis: Concordia, 1958), 189. The original is from the *Weimarer Ausgabe* of Luther's works, 32.134.34–136.36.

12. The phrase is not Luther's, but Gerard Manley Hopkins's.

since tossed off the most pervasive medieval metaphor of all, the metaphor of *ascetic ascent,* in Luther's view. He rejects it as spirituality, as metaphysics, and as ethics. The prevalent view that "the intellectual journey to truth and the moral journey to goodness is one with the journey from bodily beings to disembodied Being"[13] simply goes by the boards as Luther disavows "the great chain of being" of Neoplatonic Christianity.

Why does Luther reject a reigning Christian construct with this much seniority?[14] Because it subtly refuses to accept finitude as good and proper and as the place where none less and other than God is! God is with us as utter flesh. At least so far as humans can know, God is never disembodied. Like the devil and life itself, God is always in the details themselves. Forms of nature are, in Luther's image, the "masks" of God (*larvae dei*). "Darkness" and its fecundity constitute one of those masks, as does the tree of life. So, too, does the rest of nature, fellow humans included.

The essence of sin in this perspective is to try to rise above nature. To repent of our sin is always to "return to our senses"[15] and live in accord with our fleshy humanity and celebration of it. To repent of our sin is also to live in accord with the rest of nature and celebration of it, which we know only through the senses.

Differently said, Luther rejects the gradual, graded severance of heaven and earth because it means that in our thirst for God, we abandon our earthly purpose for being. We are created and saved by grace to rejoice in being who we are, utter creatures of *'ădāmâ* (earth, topsoil, ground). The kind of self-denial and the *contemptus mundi* that reject finitude invariably lead to idolatry or to religion (to Luther's mind they are often the same). Luther's panentheism as whole-earth theology is nicely illustrated in Paul Santmire's discussion. For Luther, the nakedness of Adam and Eve was their "greatest adornment before God and all creatures," and their "common table" included the animal kingdom as part of God's kingdom. This echoes the ancient Jewish belief that before the playing out of the evil impulse all creatures were vegetarians (Gen. 1:29) and will be again when the Messiah's work is done. It is more than an aside, then, to note that in the second creation account, after Yahweh concludes that "it is not good

---

13. The words are Rosemary Radford Ruether's description of the Augustinian perspective so influential for Western Christianity (*Gaia and God: An Ecofeminist Theology of Earth Healing* [San Francisco: HarperSanFrancisco, 1992], 135).

14. God as wholly removed from earth lives on long after Luther, both in official teaching of Christian bodies and in much popular religion. Consider one example only, a text from the First Vatican Council (1890): "The Holy, Catholic, Apostolic, Roman Church believes and confesses that there is one true and living God, Creator and Lord of Heaven and earth, almighty, eternal, immense, incomprehensible, infinite in intelligence, in will, and in all perfection, who, as being one, sole, absolutely simple and immutable spiritual substance, is to be declared really and essentially distinct from the world, of supreme beatitude in and from himself, and ineffably exalted above all things beside himself which exist or are conceivable" (cited by Sallie McFague, *The Body of God,* 136).

15. This phrase is, of course, a common figure of speech. But I have taken it from an essay by Mary Evelyn Jegen and used a form of it to title this chapter. The reference in Jegen's work can be found in her chapter "The Church's Role in Healing the Earth," in *Tending the Garden: Essays on the Gospel and the Earth,* ed. Wesley Granberg-Michaelson (Grand Rapids, Mich.: Eerdmans, 1987), 96.

that the earth creature [*Adam*] should be alone, [so] I will make [earth creature] a helpmate" (2:18), the creation of Eve does *not* immediately ensue. Rather, "From the soil ['*ădāmâ*] all the wild beasts and all the birds of heaven" were created (2:19). They are candidates for a fitting helpmate because, like Adam, they are also '*ădāmâ*.[16]

The fit is not wholly sufficient, however. (God makes his/her first mistake here!) So Yahweh creates another from earth creature's very own flesh, and the match "takes." But the point is *the aboriginal companion character of all creatures.*[17] Theirs is an essential kinship precisely as '*ădāmâ*. They are all earth creatures, born of the same vibrant fusion of dust and spirit. (God's mistake is the kind God can be forgiven and we ought to commit!)

The intuitive correctness of Luther's "common table" of beast and human alike is striking. English, and not only Hebrew, retains a whisper of this. "Humans," as all things of earth, are of "humus," that organic residue of roots, bone, carrion, feces, leaves, and other debris mixed with minerals and organized as a community for life.[18] We ought, then, to be humble and look upon ourselves with some "humor." "Humility" means never trying to outgrow our "humanity" and escape or transcend our earthiness. It means accepting ourselves for what we are — spirit-animated nature. To sin is to overstep and overshoot finitude, deny its potentialities and its limits, and reject creatureliness. This kind of pride corrupts and destroys rather than enhances and empowers. Its corrective is humility/humus/humanity/humor. Not by coincidence the "humble" are those who are rightly related to God. In faith they accept the freely given grace by which they are content to be God's creatures, nothing else, nothing more, and nothing less.

One implication is that we do not *have* bodies, we are "our bodies, ourselves," and should treat our bodies "as the earth we carry" (Augustine's lovely image). This mobile humus, erect on its own two legs, ought trustfully to accept its mortal character as the place God is, among us. The end of God's ways is bodiliness.

Differently put, we are *always* rooted in earth as earth creatures. There is no other possibility. We are utterly dependent on earth at every moment of our lives, for everything — air, water, food, materials, and energy. We are made of earth's materials, beings of dust and breath.[19] We are part of earth's history and

---

16. A later statement of this in the Bible is Eccl. 3:19–22. Granted, the context is complaint — all is vanity and does not endure. The passage is nonetheless extraordinary in the identification of humankind and otherkind: "For the fate of humans and the fate of animals is the same; as one dies, so dies the other. They all have the same breath, and humans have no advantage over the animals; for all is vanity. All go to one place; all are from the dust, and all turn to dust again. Who knows whether the human spirit goes upward and the spirit of animals goes downward to the earth? So I saw that there is nothing better than that all should enjoy their work, for that is their lot; who can bring them to see what will be after them?"

17. Though it is fully consistent with his theology, some might argue that Luther goes too far when, in the lectures on Genesis, he lavishes admiration on the mouse as "a divine creature."

18. See the discussion of Donald Worster, *The Wealth of Nature: Environmental History and the Ecological Imagination* (New York: Oxford University Press, 1993), 80–82.

19. I was given a booklet of letters from August Rasmussen, mailed from Montcalm County,

a fashioner of it. So the issue is not whether we can escape earth; we cannot. We live *in* earth *as* earth. The issue is the suffering caused and harm inflicted by not accepting earth's constraints and possibilities.[20] The issue is "conniv[ing] in the murder of Creation."

Luther is rarely cautious, so it is noteworthy that he is sometimes cautious in this panentheism. He is cautious because our propensities tilt us toward idolatry. We try to capture God *in* the finite and creaturely, just as we also try to manage an end run around our finitude (Luther calls this "the theology of glory"). He thus insists that to experience the transcendent immanently (the only way we can) is *not* to circumscribe the divine presence itself, despite all the valiant and not-so-valiant efforts of religion to do so. For Luther religion is the very best thing idolatry and injustice have going for them, even better than politics, which is the next best thing. Indeed, all efforts to either capture God *in* our terms or to "be like God" by *denying* our death and finitude, including political and economic efforts, eventually turn religious or quasi-religious. That is, we seek power, cosmic power included, to escape the insecure and mortal character of our creatureliness. Or, paradoxically, we evade the burden of being a free and responsible self by shrinking from the considerable powers we do have as wondrous creatures of earth.[21] We abuse power in hubris; or, in self-deprecation, even self-hatred, we fail to claim power. Both are failures to be the kind of earth creatures we ought to be "by nature." Either we do not accept the anxiety and insecurity that inhere in human creatureliness and we overreach,[22] or we do not

---

Michigan, to his family in Denmark over the years 1856–1902. One of the charms of these letters are the phrases of endearment for his family members or in remembrance of them: "May God Bless Your Dust!" "Peace Be With Your Dust!" "May Peace Be With Their Dust!" The ease — and pleasure — with which the letter writer speaks of fellow humans as "dust" is instructive. From August Rasmussen, *Pioneer Life in the Big Dane Settlement, Montcalm County, Michigan, North America, 1856–1902* (no publisher given).

20. See the discussion of Jay McDaniel, *With Roots and Wings: Christianity in an Age of Ecology and Dialogue* (Maryknoll, N.Y.: Orbis Books, 1995), 37–38.

21. I am keenly aware that Luther does not truly understand the lot of people whose own opportunities in life are so meager that they can neither overextend themselves in the exercise of power nor evade the responsibility that choices offer, because their range of choice itself is nil. The exposition here, then, is not intended to be comprehensive and does not include a critique of Lutheran shortcomings. Rather, the effort is to glean what we can from Luther that contributes toward a sound earth ethic. Shortcomings here need to be addressed by strengths from other sources even as all sources are challenged by earth's distress.

22. Among modern theologians the classic analysis of sin as overweening pride is Reinhold Niebuhr's largely Augustinian and Reformation exposition in *The Nature and Destiny of Man*, vol. 1 (New York: Scribner's, 1964 [1941]). The following is particularly appropriate to the discussion here. Niebuhr is speaking of "the pride of power" as "prompted by the sense of insecurity." He says: "Sometimes this lust for power expresses itself in terms of man's conquest of nature, in which the legitimate freedom and mastery of man in the world of nature is corrupted into a mere exploitation of nature. Man's sense of dependence upon nature and his reverent gratitude toward the miracle of nature's perennial abundance is destroyed by his arrogant sense of independence and his greedy effort to overcome the insecurity of nature's rhythms and seasons by garnering her store with excessive zeal and beyond natural requirements. Greed is in short the expression of man's inordinate ambition to hide his insecurity in nature.... Greed as a form of the will-to-power has been a particularly flagrant sin in the modern era because modern technology has tempted contemporary man to overestimate the possibility and the value of eliminating his insecurity in nature. Greed has thus become the besetting sin of a bourgeois culture" (190–91).

accept the potentials of life that come with contingency and insecurity attached and we resign ourselves to something less than what could be. Nonacceptance of our finitude and its real though unsteady powers manifests itself in the fear of death, in volatile sexual identity, in the need to belong somewhere to something and somebody at whatever cost, and in the effort to build worlds that beggar the neighbors. Above all, we falsely suppose that through such arrangements we stave off mortality and insecurity. "Ninety percent of human suffering is caused by trying to avoid the 10 percent that is necessary," is Tolstoy's commentary.[23] Religion ratifies this mad effort to banish insecurity and mortality — the 10 percent — as tenacious traits of finitude itself.[24]

Take death and its denial, for example. Death as mortality belongs to life. All that lives, dies — and must for the sake of ongoing life. Decay and physical death belong to the way of nature and life, as any organic gardener knows. But something in us, argues Luther, takes the reality of death to another level. Death becomes a force understood as opposed to life, which death then casts as its utter opposite. The denial of death then becomes a culture, a culture that misappropriates other lives, excludes others, inflates egos, pushes for fame, fortune, aggrandizement, and makes endless bargains with God for immortality and an exemption from the normal human fate of returning to the dust from which we are fashioned. Far from a biological phenomenon that belongs to creaturehood as such, death — and its denial — becomes a force shaping society. And amidst this, religion frequently represents the farther reaches of a common refusal to accept our finitude and the cessation of life that natural death brings. For Luther, this religion is the opposite of faith. Faith, by God's grace, is the capacity to affirm life in the midst and face of death, be reconciled to its limits (including the tragic limits of the human condition), and accept the whole without despair.[25] In the end what separates us from God and others is not our finitude, but killing. Our problem is not that we are mortal but that we are unjust.

But why can we not circumscribe God, in Luther's judgment? Because we cannot, as creatures, know the fullness of God. Nor can we know the fullness of reality. We can only know how God is *with us,* on terms appropriate to our particular brand of humus. We cannot know how rocks experience God, or how sparrows do. We cannot know how baobab trees know God's power or graciousness, or how tulips and lilies do. We might well surmise that "fields and floods, rocks, hills, and plains, repeat the sounding joy" all in their own way. And we might understand, reflecting Luther's common table of all 'ădāmâ's kin, that all things are indeed creationally connected and share a collective doxology that

23. Cited by Richard M. Clugston, in "Ecological Wisdom," *Earth Ethics* 5, no. 1 (fall 1993): 10; no source given for Tolstoy.

24. What Daniel Maguire says of militarism often holds, he notes, for religion as well. It is "self-destructive, superstitious behavior in the cause of security" (*The Moral Core of Judaism and Christianity: Reclaiming the Revolution* [Minneapolis: Fortress, 1993], 9).

25. While this is indeed included in what faith is for Luther, the phrasing of it in this way is suggested by the discussion of Judith Lewis Herman, *Trauma and Recovery* (New York: Basic Books, 1992), 319. Herman, it must be noted, is not discussing Luther but reflecting upon the dynamics of successfully recovering from trauma.

divinity itself experiences.[26] We might well nurture Shug's sensibility, noted earlier, that "I knew that if I cut the tree, my arm would bleed" or wholeheartedly
affirm Celie's conjecture that God must certainly get "pissed off" if you walk
by the color purple in a field and don't even notice.[27] But we cannot, as the
finite human creatures we are, know the full majesty of the Ineffable One or
the full majesty of nature. We see and know only dimly, in a smoky mirror. All
our theories, including those that "name" God, cannot exhaust the objects of
our knowing. Our sensory capacities themselves are limited. So is our judgment.
In a dynamic, interconnected whole, all our categories leak, all our propositions
have unstable presuppositions, and all our orderings have "uncertain edges."[28]
We should beware, then, of master narratives, comprehensive theories and ideologies, grand strategies, language with no "give," and what Ivone Gebara calls
"overly precise" images and symbols.[29] We should be equally beware of under-
representation — voices not present and heard from, perspectives overlooked or
excluded. In the end we must be content to let God be God and let ourselves be
a glorious, partial, refracted, imprecise, limited image of divinity.

As if to anticipate all this, Luther's next lines in the God-in-a-grain passage
are these:

> And that the same Majesty is so large that neither this world nor a thou-
> sand worlds can encompass it and say: "Behold, there it is!...." [God's]
> own divine essence can be in all creatures collectively and in each one indi-
> vidually more profoundly, more intimately, more present than the creature
> is in itself; yet it can be encompassed nowhere and by no one. It encom-
> passes all things and dwells in all, but not one thing encompasses it and
> dwells in it.[30]

Both motifs of grain theology are succinctly put in Luther's commentary to the
sacrament of the eucharist:

> God in his essence is present everywhere in and through the whole cre-
> ation in all its parts and in all places, and so the world is full of God and

---

26. This protest against human pretension to know what nonhuman experience of God might
be like is not meant to discourage identification with nonhuman life. On the contrary, the effort
is to *curb* human pretensions to know, in order *better* to listen to the rest of nature. As we face a
planetary environment in decline, we need far more materials like Ray McKeever's song, "Even the
Stones Will Cry Out."

27. Alice Walker, *The Color Purple* (New York: Pocket Books, 1982), 178.

28. This is a theme in the work of a Lutheran ecological theologian of the twentieth century,
Joseph Sittler. See his *Essays on Nature and Grace* (Philadelphia: Fortress, 1972). The above citation
is from p. 3.

29. Ivone Gebara, "The Face of Transcendence as a Challenge to the Reading of the Bible in
Latin America," in *Searching the Scriptures: A Feminist Introduction,* ed. Elisabeth Schüssler Fiorenza
(New York: Crossroad, 1993), 177. Charles Birch says something of the same, with respect to Chris-
tianity in particular. Christian doctrines make "good sign posts" but "bad hitching posts" (cited
from Anne Primavesi, *From Apocalypse to Genesis: Ecology, Feminism, and Christianity* [Minneapolis:
Fortress, 1991], 75).

30. Santmire, *The Travail of Nature,* 130, citing from Bornkamm, *Luther's World,* 189.

God fills it all, yet God is not limited to or circumscribed by it, but is at the same time beyond and above the whole creation.[31]

GOD'S MASKS

Much of this can be said differently and helpfully with the use of Luther's image of the "masks of God" (*larvae dei*) or God's "wrapping" (*involucrum*). Here Luther follows the Jewish conviction that we cannot wholly know the Ineffable One, look upon the Divine Presence directly, or utter the unutterable Name above all names. Our only knowing comes via the finite ways God is concealed, or "wrapped." We only know the creator, redeemer, and sustainer in finite disguises, as the Ineffable One is hidden behind one mask or another.

Nature is God's disguise. Nature is not God, but the source of the signs, the metaphors, the symbols of the *deus absconditus* (the hidden God). Nature is how and where the hidden God is revealed (*deus revelatus*). Here, in "the majesty of matter,"[32] we know God in the only way we, as finite creatures, can. Rain or a fruit tree or a child at the breast is a disguise of God's. None is God, but all are an epiphanic presence of God's presence. As with the grain, God is wholly present in them, and they holy in God, but they do not exhaust the Inexhaustible One.[33]

Little wonder Luther speaks now and again of "reverence in the face of life" or "awe before life" ("*Ehrfurcht vor dem Leben*").

Luther here means awe before life in all dimensions. One of his convictions — discussed in detail in the next chapter — is that God is sometimes revealed "under the opposite" (*sub contrario*). So where we expect light to bear life, and darkness death, we find darkness fecund and light tyrannical. Virtues are turned on their heads and serve vices while seeming dilemmas lead to breakthroughs. Even the consummate image of death-dealing — torture and crucifixion — has a strange truth-bearing power: just where the end is expected and apparent, deeper resources for life marshal their forces against insuperable odds.

Yet whether in "opposite" form or not, nature is the medium of God's presence and the only medium by which we, as creatures of nature, can know the Unknowable One. God "fills it all," Luther says. And even though God is not limited to it, we are, even when we are not less than "the image of God."

---

31. From Martin Luther, "That These Words of Christ — This is my Body, etc., Still Stand Firm against the Fanatics," *Luther's Works* (Philadelphia: Muhlenberg, 1961), 37:59.

32. Luther, *Weimarer Ausgabe*, 39.2.4,32.

33. After concluding this chapter I received a rich discussion of Lutheran reflections on the doctrine of creation: *Tro and Tanke: Svenska Kyrkans Forskyningrad* 5 (1995); this is a special issue, edited by Viggo Mortensen, entitled *Concern for Creation: Voices on the Theology of Creation.*

IMAGO DEI

This acknowledged — that God fills the earth but is not circumscribed by it — there remains the matter of saying what *imago dei* would and would not mean. Humans "in the image of God" *would not* mean to possess some substantive faculty or quality that at one and the same time makes us like God and "above" other creatures — reason or will or freedom, to cite three contenders from the history of Christian thought. Nor would creation in God's image mean, on a limited planet, trying to escape the realm of necessity that nature imposes, trying to repeal the laws and limits of finitude. Freedom and the good life must be fashioned *within* the realm of necessity in accord with creation's integrity. Image of God thus means that precisely *as* the creatures we are, situated in threefold relatedness to God, other human creatures, and otherkind, we would be turned toward God. In this stance we would mirror God's way in our own way.[34] This might be thought of in the manner of the priestly vocation discussed earlier. That is, "to stand before the Creator on behalf of all creation (intercession), and, in turn, to interpret the good intention of the Creator to and for all."[35] (George Herbert dubs us "secretaries of God's praise.")

Imago dei is a moral matter as well. Imaging God is acting in a godly way toward one another and other creatures. Imaging God is loving earth fiercely, as God does. Understood Christologically, as Luther and the classic creeds do, this means exercising dominion in the manner of Jesus, who is *dominus* (Lord). The manner of Jesus' lordship is noted in the next chapter; the point here is that *imago dei* is understood relationally and dynamically, as human imaging of God's way and as humans turning toward God on behalf of creation as priest, trustee, servant, partner. These stances all carry moral consequences.

To summarize: *finitum capax infiniti* — the finite bears the infinite — is grassroots earth theology. It is earthbound and limited. That is God's way, among us. The body, nature, is the end of God's path. God is not a separate item, even a very large one, on an inventory of the universe, but the universe itself is God's "body." (Such an image necessarily risks symbol and metaphor, as all religious language must.) God is not totally encompassed by the creaturely, but the creaturely is the one and only place we know the divine fullness in the manner appropriate to our own fullness. Experiencing the gracious God means, then, falling in love with earth and sticking around, staying home, imaging God in the way we can as the kind of creatures we are. The only viable earth faith is thus a biospiritual one. Earth ethics is a matter of turning and returning to our

---

34. The discussion of *imago dei* is helped along by Douglas John Hall's book, *Imaging God: Dominion as Stewardship* (Grand Rapids, Mich.: Eerdmans, 1986).

35. From the second draft of "Justice, Peace, and the Integrity of Creation," prepared for the world convocation of the WCC in Seoul, Korea, March 6–12, 1990. I should add that in the document the discussion is about the nature of human vocation in creation without speaking directly of *imago dei*. I use the quotation to connect the two matters.

senses. The totality of nature is the theater of grace. The love of God, like any genuine love, is tactile.

Such is the turn needed to resolve the ambiguity suspended between the views of the abbot of Fellarich and the bishop of Bamberg, the turn that declares Christian faith and ethics unambiguously and joyfully earthbound.

The Cross of Reality

> But ask the animals, and they will teach you;
>   the birds of the air, and they will tell you;
> ask the plants of the earth, and they will teach you;
>   and the fish of the sea will declare to you.
> Who among all these does not know
>   that the hand of the Lord has done this?
> In [God's] hand is the life of every living thing
>   and the breath of every human being.
>
> (Job 12:7–10)

Is not such wisdom enough? Is not nature, masking and wrapping God, *the* teacher of all that is essential? If we took this advice to heart and no more, would we not have everything necessary for an adequate earth ethic? Many environmentalists, religious and otherwise, think so. Luther thinks not. Is he right, or they?

Finitum capax infiniti — the finite bears the infinite — expresses itself as a rich panentheism. Yet Luther does not make nature at large *the* focal reality of the hidden God's revelation. He loves earth fiercely and celebrates even the house mouse as a divine creature with lessons to teach us. At the same time, Luther polemicizes against every cosmology that grants fallen and broken nature (the only nature we have) inherent value, and he rejects it as *the* highest source of moral instruction even when it is a "book" revealing the divine. Nature is not yet what it ought to be and is too casual about the kind of suffering that means death rather than life. It will not of itself suffice as the single, inclusive source and authority for ethics.

Rather, for Luther *the* compelling glimpse of God is in a particular historical event in nature. It is found in the humanity of a particular poor Jew from Nazareth in the region of Galilee during that convulsive season when Augustus happened to be the Roman Caesar. What we can most reliably know of God's own way we know in the way of this Jesus, in his birth, ministry, crucifixion, and resurrection. God is like this Jesus. God is not more divine than God is in Jesus' humanity, Luther insists. God is not more powerful than God is in the power seen in this Jesus. God is not more majestic than God is in the "self-emptying" (*kenosis*) of God in this Jesus. God is not greater than God is in this servant.[1]

---

1. This is a paraphrase of Jürgen Moltmann's *The Crucified God: The Cross of Christ as the Foundation and Criticism of Christian Theology* (New York: Harper and Row, 1974), 205. I am grateful to Lisa Stoen Hazelwood for bringing this passage to my attention anew.

As this particular Jesus is, so also is God. Yes, God *is* the ultimate life-source of the entire universe, its creator, sustainer, redeemer; and this God is disclosed in the cosmos as a whole. But, in the manner appropriate to *human* experience and knowing, this life-source is disclosed most compellingly in Jesus.[2] This Jesus is the incandescence of God in human form.

The disclosure of God in Jesus as a source for an earth faith and ethic is the focus of this chapter. More precisely, attention is to the central symbol of Christianity, the cross and resurrection. Does this symbol add substantively to the foregoing chapter's argument that "being with the gracious God means loving earth"? Is anything more, and vital, said by also claiming that "being with the gracious God means loving Jesus"? How is the cosmology and ethic for which we search, in quest of sustainable community, thereby affected?

Being with the gracious God means loving Jesus — *crux sola nostra theologia est.* Though Luther's claim that the source and redeemer of life is wholly present in the life of Jesus of Nazareth is first a Christian New Testament claim, the pattern is the Hebrew one met in our chapter entitled "ReBeginnings." Just as the Hebrews moved from the experience of a redeeming God to the awesome realization that this very One was also the creator of the universe and of life itself, so Luther moves from the redemptive presence of the transcendent God in the human Jesus, as experienced by his followers, to the awesome presence of this God in all things great and small. The universe is itself the primordial revelation of God. But the path to understanding God as creator comes by way of the distinctive, though not exclusive, revelation of the suffering and redeeming God seen in Jesus and his way.[3] That redemption's scope is all creation is not at issue, then. But the route by which it is known is via the formation of a people whose mission is to display redeemed creation as just community. Such is the pattern for both the formation of Israel and the "People of the Way" of Jesus (Acts 4:32–35). This is Luther's argument for Jesus as *the* masked clue to the revelation of the Ineffable One. A humanly experienced historical event opens onto an apprehension of all reality.

This Jesus is wholly of earth. He is not a fleeting docetic visitor, nor a ghostly bearer of gnostic truth, but real, mortal flesh and blood from the countryside. Joseph tickles his bare belly button and covers his bare bottom; Mary puts his hungry mouth to her bare breast. (These are Luther's images.)

Jesus is, to be sure, not the *exclusive* revelation of the ubiquitous immanence of God. *All* creation manifests God. But Jesus is the single most definitive revelation. Thus while "God is [always] in the facts themselves," including all the

---

2. In the course of the Reformation and beyond, this Christological focus ironically worked counter to Luther's attack on earth-denigrating cosmologies. It became part of an antinature train of thought in Protestantism that subverted his own rich panentheism and contributed to the Protestant derailing of a rich sacramentalist doctrine of creation.

3. While I respect the experience of both Luther and the Hebrews, the fallout has often been, for some streams of Christianity, to make the Second Article of the Creed the First, that is, to obscure the fact that creation is primary and redemption is wholly the redemption *of* creation. God is first "maker of heaven and earth"; God is redeemer in order to save and serve creation's destiny.

facts of nature, the facts are best "read" via God-in-Jesus. In Jesus we see the kind of God God is.

This is so with special power in the cross that ended Jesus' short life. Jesus on the cross is a very strange revelation, however. In one sense, a dying Jesus is akin to all revelation of God in that it is only indirect — itself a rather strange attribute for "revelation"! But that's the way it is with God, Luther insists. Just as Jesus is a human male of a particular time, place, family, and people, so all evidence of God and God's presence is masked. It is hidden in something else, something finite and creaturely, borne of nature itself and limited to time and space as we know them. We have no direct evidence of God at all, no manifestation that is not via nature. We see only what Luther calls the "rearward parts" of God (*posteriori Dei*). The reference is to Moses and the Jewish notion that no one would survive a direct encounter with the majesty of God.

Yet the cross is not only indirect exposure of God; it is, as noted in the foregoing chapter, God's presence "under the opposite" (*sub contrario*). This is not only the rearward parts: it is the indecent exposure and scandal of a God who is crucified as well as hidden (*Deus crucifixus et absconditus*). God is concealed in a vilified and broken human being. Jesus is God made poor and abused. God is revealed not only indirectly in nature but in a broken and scandalous condition.

Reason and all "theologies of glory" (Luther's phrase) expect and insist upon something else of God: namely, God in power, majesty, and light, in triumph, ecstasy, mighty deeds, wild success, and pure unadulterated experience. God is found, however, says Luther, in weakness and wretchedness, in darkness, failure, sorrow, and despair. God is not found *only* there, to be sure, but God *is* there in a special, crystallized, and saving way. God is present in a certain kind of suffering love and as a certain kind of power on the home turf of deadliness, brokenness, and degradation itself. God is present in twistedness and pain, and not in beauty and health alone. God is affected by creation's turmoil and earth's distress, reaches into it and takes it into divine being itself. When nature suffers degradation — any of nature — God suffers.

But what does a crucified and hidden God — a God present in degraded as well as flourishing nature — mean for ethics? Gazing from the foot of the cross to the man of sorrows, how does the heart of the universe manifest itself here for the well-being of all creation? To return to our question, isn't falling in love with earth sufficient to discern the way of God? Why is it necessary to fall in love with Jesus as well?

Jesus, as the way of God among us and as "a model of the godly life" (see the Lutheran Book of Worship), shifts the attention in ethics decisively in a way the opening verses from Job of themselves do not, even if read in a panentheistic way ("In [God's] hand is the life of every living thing, and the breath of every human being"). For that matter, this is true not only of Job but also of most other ethical systems. Standard attention in modern ethics concentrates upon our own resources to effect good, to see what we might do to leave the world less a mess than we found it. The attention is to me and my capacities as a morally autonomous and responsible agent armed with natural powers. The way of Jesus, however, means, as the starting point itself, entering into the predicaments of others who suffer.

There is a major moral and a major theological assumption here, but they merge so as to be indistinguishable in practice. The moral assumption is that the farther one is removed from the suffering present in creation, the farther one is from its central moral reality; and the closer one is to the suffering, the more difficult it is to refuse participation in that afflicted life, humankind's or other-kind's. Compassion (suffering-with) is the passion of life itself in this view, even as joy is. Both are corollaries of the fact that the only way we can be human is to be human together and with otherkind; and this integral connectedness includes the pain of unredeemed relationship. In fact, compassion is not, as we often think, something high and religious. Compassion, as the Dalai Lama says, is

> the common connective tissue of the body of human life. . . . Without it children would not be nurtured and protected, the slightest conflicts would never be resolved, people probably would never even have learned to talk to one another. Nothing pleasant that we enjoy throughout our lives would come to us without the kindness and compassion of others. So it does not seem unrealistic to me: compassion seems to be the greatest power.[4]

"A-pathy" contrasts with compassion. It is the denial of the senses and of our inherent connectedness to all things. It is a rejection of our constitutional sociality and of the pathos of life. The corrective is a return to our senses and, in Richard Rorty's words, to "the imaginative ability to see strange people as fellow sufferers." (Rorty's "strange people" should be extended to include other strange creatures.)[5]

As if to support Rorty with empirical findings, Sharon Daloz Parks's study of persons who possess significant levels of moral imagination and moral courage shows the following. These persons possess "a capacity for empathy" that is it-self "the ground of compassion" and "the driving energy in the formation of the ethical imagination." Empathy is not, she explains, pity or sentimental identifi-cation. It is the ability to "see from multiple perspectives, be affected by them, and take them into account."[6] It leads into compassion and action.

Chesed is the Hebrew Bible's term for much of this. Often translated "loving-kindness," the better choice is some combination of "compassion" and "justice." *Chesed* means to recognize that we are flesh and blood, weak and vulnerable, and often trapped in social realities we cannot transcend. There are limitations and constraints that beg for responses of gentleness and loving-kindness. Entering such reality and knowing it from the inside out as one wounded healer among

---

4. H. H. the Dalai Lama, "Tissue of Compassion," *Cathedral* 5, no. 1 (December 1989): 5.

5. Rorty as cited in Anthony Walton, "Awakening after Boston's Nightmare," *The New York Times*, January 10, 1990, A27.

6. Sharon Daloz Parks, "Is It Too Late? Young Adults and the Formation of Professional Ethics," in *Can Ethics Be Taught? Perspectives, Challenges, and Approaches at the Harvard Business School*, ed. Thomas R. Piper (New York: McGraw-Hill, 1993), 53.

others is part of redeeming action itself. At the same time *chesed* is moral obligation. It is not a conclusion on the order of "I'm OK, you're OK," a conclusion that might well let the world rest where it is. Compassion as *chesed* is fired with the desire to move from where we are to where we ought to be. The goal is creation's repair (*tikkun olam*) and its fullest possible flourishing (justice).[7]

Compassion (suffering-with), then, is the key virtue for a Christian ethic and an earth ethic as well, with empathy and solidarity (standing-with) as the key means. Paul says it most succinctly: "If one member [of the body] suffers, all suffer together; if one member is honored, all rejoice together" (1 Cor. 12:26).

The quest of a religious earth faith in this understanding is precisely for a power that overcomes suffering by entering into it and leading through it to abundant life, with abundant life pictured as the Sabbath condition of redeemed creation. God's goal is newness of life; but God's means is overcoming by undergoing; and God's way — to recall Luther — is seen in Jesus as a living parable of God's compassion in human form.

What is discovered via Jesus is this: only that which has undergone *all* can overcome all. In this sense, cross and resurrection ethics is an utterly practical necessity. Suffering, in its many expressions among its many creatures, will not be redemptively addressed apart from some manner and degree of angry,[8] compassionate entry into its reality, some empowerment from the inside out, some experience of suffering as both a burden and a burden to be thrown off, some deep awareness of it as unhealed but not unhealable. Of course it's frivolous to call this "no pain, no gain" ethics. But it's not frivolous to recognize, with Kozah Kitamori, that until our pain is intensified at the sight of creation's pain, as God's is, there is no movement toward redemption. Until we enter the places of suffering and experience them with those who are entangled there, as God does, our actions will not be co-redemptive.[9]

The simple logic here is that any power that does not go to the places that community and creation are most obviously ruptured and ruined is no power for

---

7. From the discussion of *chesed* in Michael Lerner, *Jewish Renewal: A Path to Healing and Transformation* (New York: Putnam, 1995), 112–15. After drafting this chapter, I came across the compact discussion of these themes by Kathleen Talvacchia. It nicely restates the above. Citing the etymology of "compassion," she notes the Latin *compati*, which combines *com* (together with) and *pati* (to bear or suffer). Talvacchia says: "Thus, compassion combines the act of personal suffering with an act of solidarity. *Webster's Third International Dictionary* provides a fuller description, defining compassion as 'deep feeling for and understanding of misery or suffering and the concomitant desire to promote its alleviation.' Compassion can be thus understood as a willingness to risk suffering in solidarity with another or others, accompanied by the desire and commitment to remove the cause of the suffering. It goes beyond sympathizing with another's pain; alleviation of the pain or suffering is the ultimate goal" ("Learning to Stand with Others through Compassionate Solidarity," *Union Seminary Quarterly Review* 47, 3–4 [1993]: 181).

8. The reference is to the line from Augustine that hope has "two lovely daughters, anger and courage"; anger at the way things are and courage to set them right. For a profound treatment of this in Christian ethics, see Beverly Wildung Harrison, "The Power of Anger in the Work of Love: Christian Ethics for Women and Other Strangers," in *Making the Connections: Essays in Feminist Social Ethics*, ed. Carol S. Robb (Boston: Beacon, 1985), 3ff.

9. The reference is to the work of the Japanese theologian Kozah Kitamori, and especially *The Theology of the Pain of God* (London: SCM Press, 1966).

healing at all. The failure to make this trek is the impotence of what we wrongly name "power" — wealth, fame, legions of soldiers and ships, triumphalist ideologies, and arrogant, wasteful ways of life. Such "power" is strangely impotent; it does not understand and cannot address the disasters of the spirit, the catastrophes of the psyche, and the acidity of rain, soil, mind, or household. Thus it cannot help work wounds back to health *from within battered flesh itself.* Or, if such "power" does recognize the normal pathologies of everyday life — they are, after all, as unavoidable as death itself — it treats them as rabid leprosies and as sectors of life to be quarantined out of sight, beyond notice and beyond feeling. These common "powers" are not the powers of deep bonds. They are abstracted, distant powers that violate them.

The only power that can truly heal and keep the creation[10] is power that intimately knows its degradation. The degradation may be physical, social, psychosocial, or environmental. It "squeezes out life" and traps the soul. It is suffering that yields "no discernible good" and is "a more or less attenuated equivalent of death."[11] Against such degradation, the only effective power is power instinctively drawn to these hellish sites, there to call forth from the desperate and needy themselves extraordinary yet common powers that they did not even know they had. It is the discovery of the Filipino farmers: "We're the people we've been waiting for!" This is the power seen in Jesus, Luther argues. It is strength in weakness, life in the midst of death, joy within suffering, and grace where only wrath and pain and the rearward parts of God are most obvious. The ironic thing, one worthy of reversing standard accounts of wisdom and foolishness, is that this weak kind of power, learned in suffering, expressed as empathy and compassion, and deeply sensed and felt, is what wins space for joy and abundant life. Perhaps this is why, in the emotive region of the cross and in the awful silence before a dying Christ, one hears the seismic whisper of none other than the power of God. It feels like that. Even temple curtains are torn from top to bottom in that moment. The stones themselves cry out (Matt. 27:51).

In short, the kind of earth ethic that reads the presence of the divine from the flourishing of nature on its better days, or that reads the presence of the divine off the beauties of nature as we observe or imagine those (e.g., Job 12), speaks of God in true but limited ways. God *is* present in creation's beauty. The holy most certainly *is* mediated in nature's grandeur. But God is also present in the crosses of pain and twistedness and whatever other ways by which creation is violated.[12] If God were present only in the beautiful and graced and not in the blighted and disgraced, and if *we* were present only in a redeeming way to creation's beauty

---

10. The phrase is a reference to the excellent work of the Presbyterian church, *Keeping and Healing the Creation* (Louisville: Committee on Social Witness Policy, Presbyterian Church [U.S.A.], 1989).

11. These comments on degradation are from Elizabeth A. Johnson, *She Who Is: The Mystery of God in Feminist Theological Discourse* (New York: Crossroad, 1993), 261. For a profound treatment of them in biblical scholarship, see Phyllis Trible, *Texts of Terror* (Philadelphia: Fortress, 1984).

12. See John Shea's discussion in *Stories of God: An Unauthorized Biography* (Chicago: Thomas More, 1978), 148–61.

and not in its plunder and rape, then broken creation would never be healed. The only creation saved would be that infinitely small portion that does not stand in need of it. Moreover, avoiding despoiled creation absents us from the very reality for which we are responsible. And it numbs the very senses that are the only sure signs we are alive. Absenting ourselves from creation's pain, ours included, and not expecting anything of God there when it is our pain, leaves us despoiled as well.

There is another dimension to all this.

Luther's attention to the cross is a certain way of looking reality in the eye. God revealed as *sub contrario* and hidden means we should be suspicious of our own sense-making. What we call "wisdom" may in fact be "foolishness," and vice versa, to recall Paul's words to the Corinthians (see 1 Corinthians). The cross is the symbol that shatters humanly generated coherences in reading the world. It is suspicious of all master narratives and says instead that we live in a world of half-light and half-truths posing as full light and full truth. Our understandings are provisional, tentative, reflective of changing time and place. We do not possess the powers of universal voice and should be especially suspicious of those who speak thus, above all when they do so in the name of God. Faith, by contrast, begins in darkness, where it does not see well. It is the great *un*knowing, avoiding the pretense that lays *our* truth upon reality in the hopes of controlling it. Faith's only claim is less a firm knowledge claim than an ardent trust, a trust that God is there. Luther's confidence is thus not in reason's grip on the world as abstracted from its suffering, but in identifying with the ways of a passionate and compassionate God who knows earth's distress from the inside out. And this Luther sees in Jesus of Nazareth, even in Jesus' death by torture.

SUFFERING

More precision is gained by saying *which* cross is the cross of Jesus as the parable of the suffering God. This is not ancient Rome's cross. It is not the imperial power's cross of the crucifixion of the Jews (Jesus included) and of others branded as rebels and criminals. But neither is it the Christian cross of Constantine, that fusion of Christianity and imperial rule that eventually played out, in one form, as the "civilizing" by the Christian West around the world through the establishment of neo-Europes. (This is the cross noted earlier in the processional murals of the Palacio Tiradentes in Rio de Janeiro.) Nor is it the cross of bearing suffering we can do nothing about ("We all have our crosses to bear"). It is, by contrast, the cross of resistance to oppression and suffering that need not be. Rather than a cross that subtly sanctions suffering and passivity to harm inflicted, it is the price sometimes paid for resisting the culture of death in the name of life, even radically new life where it is not expected.

To trust, in faith, that God is present even in a horrendous death like that of Jesus means that more yet must be said about this cross and creation's suffering. The cross opposes most suffering. The struggle of Jesus and his movement is to end crucifixions. There are to be no more Golgothas, no more Places of the Skull and dead trees, but only the resplendent city with its rivers of crystal waters and trees of life lining their verdant banks. Yet the cross does not oppose *all* suffering, since not all suffering is negative.[13] Some suffering belongs to finitude itself. This suffering may make life difficult, but we cannot *live* without it. It belongs to our creaturely constitution to struggle in the process of becoming, to bump and grind and muddle our way through various stages of life, to know anxiety, temptation, and loneliness as pain, and to face limits of all kinds, including the ultimate limit of death itself. Such suffering is not, of itself, negative. That is, it is not necessarily life-destructive. It may in fact be integrative in that it furthers our development as mature and responsible beings. Even death itself may not be destructive suffering, although far too many deaths are and a great deal of dying is. But death as the cessation of all bodily functions belongs to life itself, and when it comes at the close of a life lived fully and well, it is not evil.

Cross and resurrection theology is bold here and a little unsettling because it holds that death belongs to finitude and finitude per se is not evil. When shalom reigns in fullness, creatures will continue to die. When creation in its integrity realizes its own promise, creatures will continue to experience suffering, too, but only that suffering that is in the service of life itself, including the "good death" that comes as the final stage of a full life. The suffering the cross opposes is the suffering that negates life and destroys the realization of creation. Such suffering is the particular kind of death that is "the wages of sin" (Paul). Its forms are multiple — psychic, physical, political, economic, cultural, familial, sexual, racial, religious, environmental. What they share in common is disintegration of that which is created to be whole within its finite limits. What they need in common is restorative healing. *This* disintegrative suffering is the kind the cross, in *its* suffering, opposes.

In Douglas John Hall's summary: the cross focuses upon suffering and pain, not from a morbid interest in pain, "but because there *is* pain, because disintegrative pain is not part of what *should* be, and because there can be no healing

---

13. I am keenly aware that moral discourse about suffering is fraught with danger. It is one of the chief reasons I question the Lutheran legacy in ethics. Talk about suffering is especially dangerous when the symbol is the cross and is always perilous when the cross is used as a sign of victory by those (like Constantine's progeny) who make history rather than simply "take" it. W. H. Auden rightly says that "only the sated and well-fed enjoy Calvary, as a verbal event." And Kosuke Koyama rightly worries about what happens when "the well-salaried and well-caloried" find this symbol at the center of their culture, rather than a shocking sign from its periphery. Too, there is a perverse legacy of the cross among the poor and among many who are socialized to be self-denigrating. This is the deep strain of oppression and sadomasochism in Christianity as expressed in pieties of crucifixion (in Roman Catholicism) and in cross theology as "I-am-a-worm theology" (in both Lutheranism and Calvinism). In Lutheranism this legacy of legitimized suffering has been reinforced with a patriarchal orders-of-creation theology that survived Luther's antidualist and antihierarchical polemics against the medieval cosmology of "the great chain of being." The dynamics of Luther's radical teaching on justification and his cross theology did not materialize in such a way as to recast his social theory. In social theory and ethics Luther remained largely Constantinian, medieval, patriarchal, and anti-Semitic. This legitimized oppression rather than resisted it, often in the name of the cross itself.

that does not begin with the sober recognition of the reality of that which needs to be healed."[14] This is the disintegrated world we experience as contradictory to its own well-being. The way of the cross as the ethic of Jesus fastens, then, on that which negates and threatens creation's life. What is held foremost is the integrity and promise of creation and the moral mandate to help realize, rather than obstruct, that promise. The suffering entailed in pursuing justice, peace, and creation's integrity serves this. This suffering is God's struggle, through incarnational solidarity with the world's suffering, to turn history and nature toward life rather than death. It is the suffering that issues in resurrection.

Theologians Dorothee Sölle, Dietrich Bonhoeffer, Martin Luther King Jr., Kozah Kitamori, Kosuke Koyama, and Jon Sobrino have all explicated in profound terms the way of the cross and its relationship to suffering, pain, justice, peace, and creation's healing.[15] While we cannot detail their contributions here, we can note their common insistence. Their common insistence is that the cross is the cross of reality. It neither justifies suffering nor denies it.[16] But where it finds suffering caused by the culture of death, it willingly enters into it for the sake of life. Such suffering is evoked by the vision of creation restored and renewed. "I will rejoice in Jerusalem and be glad in my people!" (Isa. 65:19), Sölle recalls from the prophet, in order to add: "We understand God's annunciation only when we hear the tears in God's voice."[17]

It is not enough then, these voices contend, to say simply, "Being with the gracious God means loving earth." We must also say, "Being with the gracious God means loving Jesus." This means Jesus on the cross and the way of the cross and resurrection as God's ethic and ours. Love earth, yes, emphatically. But redeeming the planet means embracing its distress as the movement of that love. It means going to the places of suffering to find God and God's power there. God in Jesus strangely, offensively, makes the margin the center and hefts the rejected stone to set the corner itself. The cross is erected "outside the gate" (Heb. 13:12), beyond the zone where salvation is commonly expected, in the place of the damned. There redemption, renewal, and restoration begin, the unexpected eruption of new life "after its evaporation has left us desolate."[18] Such

---

14. Douglas John Hall, *Thinking the Faith: Christian Theology in a North American Context* (Minneapolis: Fortress, 1989), 33.

15. See Dorothee Sölle, *Suffering* (Philadelphia: Fortress, 1975); Dietrich Bonhoeffer, *Letters and Papers from Prison* (New York: Macmillan, 1972); Martin Luther King Jr., *A Testament of Hope: The Essential Writings of Martin Luther King, Jr.*, ed. James M. Washington (San Francisco: Harper and Row, 1984); Kozah Kitamori, *Theology of the Pain of God*; Kosuke Koyama, *Mount Fuji and Mount Sinai: A Critique of Idols* (Maryknoll, N.Y.: Orbis Books, 1985); Jon Sobrino, S.J., *Christology at the Crossroads* (Maryknoll, N.Y.: Orbis Books, 1978); idem, *The Principle of Mercy* (Maryknoll, N.Y.: Orbis Books, 1994). For a discussion of Bonhoeffer's themes, see my volume with Renate Bethge, *Dietrich Bonhoeffer: His Significance for North Americans* (Minneapolis: Fortress, 1989), chaps. 7 and 8, "Divine Presence and Human Power" and "An Ethic of the Cross."

16. This theme is also explored by Sally B. Purvis in *The Power of the Cross: Foundations for a Christian Feminist Ethic of Community* (Nashville: Abingdon Press, 1993).

17. Dorothee Sölle, "Suffering," in *Dictionary of the Ecumenical Movement*, ed. Nicholas Lossky et al. (Geneva: WCC Publications, 1991), 963.

18. Langdon Gilkey, *Nature, Reality, and the Sacred: The Nexus of Science and Religion* (Minneapolis: Fortress, 1993), 135.

life issues in "subversive gladness,"[19] since its discovery is of a power that fights for life from grounds the principalities and powers cannot control. Even massive historical failure is, then, not the last evidence of what the future holds.[20] This is the meaning of resurrection itself.

CONCLUSION: GOD'S WORK MUST TRULY BE OUR OWN

Lawrence Durrell once quipped that "theology is very old ice cream" and "very tame sausage."[21] Much is. There certainly is, to shift images, a kind of fatigue in many theologies using standard Western categories and classical sources. We have nonetheless tried to show that one old strand — the Lutheran theology of the cross — still has more bite than flat sausage and more life than very old ice cream.

But where, precisely, does this grayed tradition within a wider Christianity land in our efforts for a constructive earth faith and ethic? Considered with a view to what is now happening to the planet, Lutheran cross and resurrection theology falls somewhere near the intersection where liberation theologies and creation theologies themselves cross, though with a very chastened sense of creational limits and human propensities to deny them idolatrously. Lutheran cross and resurrection theology, like both liberation and creation theologies, is curiously optimistic. It has seen the worst and discovered a mighty power for life there, smack in the midst of death, "even death on a cross" (Phil. 2:8). This power is not an alien one. On the contrary, it is the power of God that inheres in all creation. It belongs to us at the very core of our being, even as it belongs to the rest of earth. It is in our bones, and it places sustainability, even redemption, within reach because it says there is nowhere in earth and heaven where it is out of reach. This power is the power of the Holy Spirit, *Spiritus creator* itself. The peculiar twist of cross and resurrection theology is that, like Jesus himself, it moves in the power of the Spirit to the places of negative suffering in order to discover and uncover power for life there. As an ethic of compassion and solidarity it seeks out the places of oppressive suffering in order to overcome suffering's demonic, or disintegrative, manifestations. It goes to victims in order to stand with them in their reality. Its quest is not for victims but for the empowerment needed to negate the negations that generate victims. Its goal is the end of victims, human and otherwise. It insists that environmental justice is also social justice and that all efforts to save the planet begin with hearing the cry of the people and the cry of the earth together. Not least, it is a power that receives our violence without responding in kind so as to multiply it.

---

19. Gustavo Gutiérrez, *The Power of the Poor in History* (Maryknoll, N.Y.: Orbis Books, 1983), 107.

20. See Elizabeth A. Johnson, *She Who Is*, chap. 12, "The Suffering God: Compassion Poured Out."

21. Cited from Shea, *Stories of God*, 163.

Whence the power for this? As a panentheistic ethic, cross and resurrection theology, like much feminist theology and creation theology, says, "In creation itself, where God is." A power is present in the cosmos and in nature as we know it. It is God's own power, and it is sufficient for the redemption of all things (*ta panta*). We are not exactly cocreators — an insufferably arrogant notion — but we are coparticipants, together with all else "in heaven and earth and under the earth."

Power as cosmic energy identified as God's own and as an expression of the Spirit is no new Christian theme. It is an ancient catholic one, perhaps most nurtured over the centuries by millions of Christians in the Orthodox communions. The Lutheran variation here only insists on a steady focus upon the crucified, human Jesus as the place and way this power concentrates for the redemption of creation, as that relates to human knowing and human participation. The focus on Jesus, and not on nature apart from the revelation of a compassionate God, is essential in a very practical way. Without attention to creation crucified, most rich worlders will work to save nonhuman nature but not creation. They will sever environmental from social justice and treat the environment in ways that sustain their interests alone.

Differently said, the issue for an adequate environmental ethic is not, finally, an upgraded view of nature, even a religiously sensitive one. That might be no more than a romantic rendition of *finitum capax infiniti*. (Remember that romanticism's theme, and the source of its illusions, is its love of directly experienced nature.) The issue for earth ethics is the discovery of a power throughout creation that serves justice throughout creation.[22] This is the power manifest in "the erotic excitement of the earth and the intensity of life"[23] that Luther names as God's presence "in, with, and under" all things finite. And it is seen and felt with piercing presence in Jesus' life, death, and resurrection. It is a matter of returning to our senses both to know the good and to do it, with compassion for earth's suffering as the single strongest sense.

To close: Jesus' crucifixion should have been the last one, and he should have been the last victim. But it was not, and he was not, and God's sovereignty and Christ's lordship remain contested to this very day and hour. Thus, Paul says,

---

22. Recent scholarship has underscored the presence of this power as a theme of Jewish and Christian scriptures. Most significant has been the many dimensions of Wisdom as the presence and power of God in creation and the revelation of this. Just as significantly, the God revealed is the Holy and Compassionate One. In Christian New Testament materials, Wisdom is seen in Jesus. As summarized by Denis Edwards: "The Wisdom that shapes a world of more than a hundred billion galaxies reaches out to embrace struggling and suffering and sinful creatures in love which finds its final expression in death for us [that is, the cross]" (Edwards, *Jesus the Wisdom of God: An Ecological Theology* [Maryknoll, N.Y.: Orbis Books, 1995], 76).

23. The phrase is Ivone Gebara's in her chapter, "The Face of Transcendence as a Challenge to the Reading of the Bible in Latin America," in *Searching the Scriptures: A Feminist Introduction*, ed. Elisabeth Schüssler Fiorenza (New York: Crossroad, 1993), 181. I have borrowed the phrase to describe Luther's theology. Gebara is not discussing Luther. She is describing the kind of "theology of life" we need to address "social-eco" realities. Such a theology could "court the erotic excitement of the earth and the intensity of life which arises everywhere in an absolutely unpredictable and mysterious way." In my judgment, this describes selected themes in Luther well.

the sufferings of disciples are now added to Jesus' own as they join in the same cause of earth's redemption. Jesus suffered ahead of us, rather than instead of us, and the way of the cross and resurrection as God's own way remains ethically paradigmatic so long as creation suffers.

Exercising the redemptive power found in creation and seen in Jesus is a profoundly human and divine task. As Luther insisted, human actions are among God's masks in nature. Long before Bonhoeffer's famous call to mature moral responsibility and its exercise for the sake of future generations, a moral responsibility we would learn only by living "as though [the] God [of religion] did not exist," Luther himself proposed the same. In general, he wrote, civil authorities "should proceed as if there were no God and they had to rescue themselves and manage their own affairs."[24] He similarly addressed another group of responsible citizens, "Nor will God perform miracles as long as [people] can solve their problems by means of the other gifts [God] has already granted them."[25] In short, and like both liberation theologies and creation theologies, Lutheran cross and resurrection theology asks us to find God deep in the gifts we naturally possess or those we might develop as the exuberant creatures we are. This means returning to our senses and developing our powers within the boundaries of a nature spirituality that is also a "sensate spirituality," a spirituality that experiences God "not by blocking the senses and transcending them" but in and through the "distractions" of the biophysical world.[26] This echoes a famous line from President Kennedy's inaugural address, a line probably uttered as a mild expression of a theology of glory: "God's work must truly be our own." More importantly, in an era of heightened human impact upon the entire planet, our own work must truly be God's.

For earth ethics formed by cross and resurrection theology, this is anything but Promethean in tone, or a substitution of ourselves for God, even when heavy responsibility lands squarely in human hands. It is, instead, humble power exercised in the sobering shadow of the cross and in full view of the considerable powers of destruction we have and wield. It is keenly aware of sin as refusal to act as responsible representatives (*imago dei*) who value the lives and well-being of other creatures and their habitat. Of sin as the injustice of grasping more than our due as corporate bodies, as nations, as individuals, and as a species, thereby depriving other corporate bodies, nations, individuals, and species their due. Of sin as the arrogance of treating earth as property at our disposal. Of sin as denial of creaturely limitations upon our ingenuity and technology and their uses.[27]

The temper here knows that history and nature together are broken, tragic, and ironic. It recognizes the Achilles' heel of human creativity — that the same

24. Martin Luther, "Exposition of Psalm 127, for the Christians at Riga in Livonia," in *Luther's Works* (Philadelphia: Muhlenberg, 1962), 45:331.

25. Martin Luther, "To the Councilmen of All Cities in Germany That They Should Establish and Maintain Christian Schools," in *Luther's Works*, 45:56. I am grateful to Elizabeth Bettenhausen for drawing this passage and the foregoing one to my attention.

26. James Nash, *Loving Nature* (Nashville: Abingdon, 1991), 115.

27. Ibid., 119.

human powers that shape earth can destroy it.[28] But it also knows the hope and exhilaration that spring from mended places and the affirmation of human participation in that mending. The temper is like that of two quite different passages. The first is Reinhold Niebuhr's in the closing paragraph of *The Children of Light and the Children of Darkness*, written during the very worst days of World War II:

> The world community, toward which all historical forces seem to be driving us, is mankind's final possibility and impossibility. The task of achieving it must be interpreted from the standpoint of a faith which understands the fragmentary and broken character of all historic achievements and yet has confidence in their meaning because it knows their completion to be in the hands of a Divine Power, whose resources are greater than those of men, and whose suffering love can overcome the corruptions of man's achievements, without negating the significance of our striving.[29]

The second is a portion of the "Letter to the Churches" of Pentecost Sunday, 1992, sent by members of the WCC conference at the Earth Summit:

> The Spirit is the giver and sustainer of life. All that fosters life, such as justice, solidarity, and love, and all that defends life, such as the evangelical commitment to stand with the poor, the struggle against racism and casteism, and the pledge to reduce armaments and violence, concretely signifies living according to the Spirit. This is more than a political act for the Christian; it is spiritual practice.... Where we must always begin is with veneration and respect for all creatures, especially for human beings, beginning with those most in need. The Spirit teaches us to go first to those places where community and creation are most obviously languishing, those melancholy places where the cry of the people and the cry of the earth are intermingled. Here we meet Jesus, who goes before, in solidarity and healing.[30]

---

28. Ibid., 157.
29. Reinhold Niebuhr, *The Children of Light and the Children of Darkness* (New York: Scribner's, 1944), 189–90.
30. "Letter to the Churches," appendix 1 of Wesley Granberg-Michaelson, *Redeeming the Creation, the Rio Earth Summit: Challenge to the Churches* (Geneva: WCC Publications, 1992), 72–73.

Song of Songs

In April 1943, Dietrich Bonhoeffer, his sister Christine, and her husband, Hans von Dohnanyi, were arrested by the Gestapo. Bonhoeffer was taken to Berlin's Tegel Prison. He remained there until after the failed plot on Hitler's life on July 20, 1944. When his part in the conspiracy was suspected, he was transferred to the Gestapo prison in the Prinz-Albrecht-Strasse. Throughout the Tegel stay, he carried on a lively and often uncensored correspondence, thanks to the smuggling courtesies of a friendly guard, Corporal Knobloch. That correspondence is published as *Letters and Papers from Prison* and *Love Letters from Cell 92*. The latter, his exchange with his fiancée, Maria von Wedemeyer, includes a letter from August 1943. The prisoner relishes the fact he gets to write in the sunshine of the prison yard and voices his wish to once again walk in the woods, go boating, enjoy a swim, and then "lie in some shady spot" with Maria to listen to her, just "listening...for ages without saying a word [myself]." "As you see," he continues, "my desires are quite earthy and concrete, ... and I'm temporarily giving free rein to an equally natural and lively dislike of my present state [prison confinement]. The sun has always attracted me, and I've often been reminded by it that human beings were taken from the earth and don't just consist of thin air and thoughts."[1]

A letter five months later voices similar sentiments: "When I picture our first reunion I don't see us talking together in a room; I instinctively see us walking in the woods, seeing and experiencing things together, in contact with the earth and reality."[2] What sounds like a passing phrase in a touching love letter — "I instinctively see us...in contact with the earth and reality" — is more than that. Nor is the comment of August 12, that "our marriage must be a 'yes' to God's earth,"[3] only a chance expression. "A 'yes' to God's earth" is Bonhoeffer's theme from beginning to end, as we shall see.

The fuller text is the following. "Earth" is a matter of "faith":

> When Jeremiah said, in his people's hour of direst need, that "houses and fields [and vineyards] shall again be bought in this land" [Jer. 32:15], it was a token of confidence in the future. That requires faith, and may God grant it to us daily. I don't mean the faith that flees the world, but the faith that endures *in* the world and loves and remains true to that world in spite of all the hardships it brings us. Our marriage must be a "yes" to God's

1. Dietrich Bonhoeffer and Maria von Wedemeyer, *Love Letters from Cell 92*, ed. Ruth-Alice von Bismarck and Ulrich Kabitz, trans. John Brownjohn (Nashville: Abingdon Press, 1995), 68–69.
2. Ibid., 162.
3. Ibid., 64.

earth. It must strengthen our resolve to do and accomplish something on earth. I fear that Christians who venture to stand on earth on only one leg will stand in heaven on only one leg too.[4]

This is well-planted, two-legged fidelity to earth, even earth in its distress at the hour of people's "direst need." (Bonhoeffer is writing during the war's worst days.)

But what is fidelity to earth and where does it lead? How do the themes of the foregoing chapters — the finite bearing the infinite as a sacramental understanding of earth, and the ways of a passionate and compassionate God as strong lines for an earth ethic — come together in one life lived amidst the twentieth century's deadliest convulsions? And what other themes come together in this gifted life of one who died much too young at thirty-nine?

This last chapter of part 2, then, is not another symbol discussed. It is episodes from a life, told as a concrete instance of earth faith and ethics already lived. Of course, Bonhoeffer was not privy to the discussions that preoccupy us — a basically altered human relationship to earth, sustainability, the integrity of creation and its requirements. Yet his was an earth faith and ethic held by someone who knew he was living as one age was dying and another was struggling to be born. As such Bonhoeffer is instructive. In any case, just as we need to review and critique traditions for their contributions to a viable earth faith and ethic, so we need also to find those individuals who have already shown the way in word and deed. Bonhoeffer, like Ben-Gurion, is one of these.

EARTH AND ITS DISTRESS

Dostoyevsky's account of Alyosha[5] found him running from old Zossima's casket to throw himself flat upon the earth, to embrace it, to kiss it, to weep over it, sob and drench it with his tears and "vow frenziedly to love it, to love it for ever and ever." " 'Someone visited my soul in that hour!' " he testified later, this "little man of God" who "had fallen upon the earth a weak youth, and rose from it a resolute fighter."[6]

There is an echo of this theme in the youthful Bonhoeffer. He was twenty-two years old and serving as vicar in a German Protestant congregation in Barcelona. The year was 1929. In a precocious address he speaks of the giant Antaeus, who, like Alyosha, also rises from the earth "a resolute fighter." Antaeus's strength is no more than that of "a weak youth," however, if Antaeus is lifted from the ground. If he loses contact with earth, Antaeus loses his strength. Like Bonhoeffer's comments to Maria about Christians, both feet must be firmly planted. One foot on earth and the other in heaven doesn't work.

4. Ibid.

5. Recall the Introduction above, pp. 6–7.

6. Fyodor Dostoyevsky, *The Brothers Karamazov*, translated and introduced by David Magarshack (Baltimore: Penguin, 1963), 426–27.

But let Bonhoeffer tell it. After arguing that "[e]thics is a matter of earth and of blood, but also of the One who created both,"[7] he says:

> A glimpse of eternity is revealed only through the depths of earth.... The profound old saga tells of the giant Antaeus, who was stronger than anyone on earth; none could overcome him until once in a fight someone lifted him from the ground; then he lost all the strength which had flowed into him through his bond with the earth. The person who would abandon the earth, who would leave its present distress, loses the power which still holds him by eternal, mysterious forces. The earth remains our mother, just as God remains our Father, and our mother will only lay in the Father's arms those who remain true to her. Earth and its distress — that is the Christian's Song of Songs.[8]

"Earth and its distress" as the Christian's "Song of Songs" seems an odd association. The Song of Songs is the biblical paean of sensuous love celebrating earth and flesh as uninhibited, redeemed Eden.[9] No distress mars sheer erotic blessing or overshadows it in this poem. Earth is pure celebration of desire and pleasure. It is tender eros unbound.[10]

Yet the Barcelona wording is earth "and its distress." This is earth and its afflictions, its misery and degradations, not only its blessing, pleasures, and passion. Bonhoeffer loved life and is continually thankful for all earth's gracious powers and pleasures of good. The prison letters find him dreaming of the woods and Maria, longing for colors, flowers, and the song of birds, grateful for all that friendship means, relishing good food and a good book, even a good cigar, happily remembering family, and enjoying music. Yet earth in the Barcelona address is more than this and, for that matter, more than Alyosha's vault of heaven, the still night, and precious ground. It is earth comprehensive of its suffering and agony.

Since masochism wasn't a strain of Bonhoeffer's character, something other than *Schadenfreude* explains this embrace of earth. Perhaps his sentiment anticipates Dag Hammerskjold's conviction about citizens and their country: that we

---

7. Dietrich Bonhoeffer, "Grundfragen einer christlichen Ethik," *Gesammelte Schriften* (Munich: Kaiser, 1966), 3:56; trans. mine.

8. Ibid., 57–58.

9. See the stunning exegesis of Phyllis Trible in *God and the Rhetoric of Sexuality* (Philadelphia: Fortress, 1978).

10. A reference fifteen years later approaches this. God desires our love, Bonhoeffer writes from prison to his closest friend, Eberhard Bethge, but "not in such a way as to injure or weaken our earthly love." Rather, love of God is meant "to provide a kind of *cantus firmus* to which the other melodies of life provide the counterpoint.... One of these contrapuntal themes (which have their own complete independence but are yet related to the *cantus firmus*) is earthly affection.... Even in the Bible we have the Song of Songs, and really one can imagine no more ardent, passionate, sensual love than is portrayed there (see 7.6). It's a good thing that the book is in the Bible, in face of all those who believe that the restraint of passion is Christian (where is there such restraint in the Old Testament?)" (Dietrich Bonhoeffer, "Letter of May 20, 1944," in *Letters and Papers from Prison* [hereafter *LPP*] [New York: Macmillan, 1972], 303).

must *love* what we are seeking to change or the whole enterprise is lost from the beginning.[11] Or perhaps Bonhoeffer is anticipating what we have newly discovered: that earth now is earth as all things caught up together in a *single* complex, evolving, conflicted reality. It is nature and culture together, history, society, and nature of a piece, social and environmental degradation and well-being together. It is damnation and redemption as the damnation or redemption of all things or none. Earth's riches, sorrows, and delights cannot easily be disentangled. Life must be met as it comes and taken from there, at least if we are to be responsible and not fall into the illusions of religious otherworldliness. "I don't mean the faith that flees the world, but the faith that endures *in* the world and loves and remains true to the world," he had written to Maria.[12] "The Christian, unlike the devotees of the redemption myths," Bonhoeffer wrote in another letter from prison, "has no last line of escape available from earthly tasks and difficulties into the eternal, but, like Christ himself ('My God, why has thou forsaken me?'), he must drink the earthly cup to the dregs."[13]

In short, both the younger and older Bonhoeffer profess an indefatigable love and embrace of earth, its distress included. "The earth remains our mother, just as God remains our Father, and our mother will only lay in the Father's arms those who remain true to her."[14] The "earthly cup" must be drunk "to the dregs." There is no other way to new life. "It is only when one loves life and the earth so much that without them everything seems to be over that one may believe in the resurrection and a new world."[15]

EARTH AS BLOOD AND SOIL

Bonhoeffer's language in the prison letters of the mid-1940s differs from the youthful vicar's of 1929, however. The reasons are plain enough, and important. The story itself is complex, however, almost as complex as the period itself — Bonhoeffer's adult years exactly match those of the Third Reich. The story merits the telling.

In the early 1930s, a demonic earth faith and ethic arose. It was appealing and destructive in equal, massive measure. It was Nazism, and it articulated its faith as nothing other than "a matter of earth and of blood" (to recall Bonhoeffer's own phrase). *Blut und Boden* — blood and soil — became the rallying cry of a fevered German nationalism.

Bonhoeffer's own blood, too, could be stirred by the nationalism of the late 1920s. He, too, could speak of *das Volk* — the people — with sympathy. Yet neither he nor the rest of his remarkable family was ever tempted by Nazism's celebration of nature as mysterious, inexhaustible energies that coursed through

11. Dag Hammerskjold, *Markings* (New York: Knopf, 1964), 77.
12. See the citation above, p. 295.
13. Bonhoeffer, letter of June 27, 1944, *LPP,* 337.
14. From the Barcelona quotation, cited above.
15. Bonhoeffer, "Advent 2," in *LPP,* 157.

the veins of an Aryan people drawing deeply from its own eternal soils. Fascist vitalism, which combined a romantic presentation of nature and culture with a valorization of a primitive state set against the corruptions of civilization,[16] was never Bonhoeffer's, and he would warn against it in ways that meant trouble for him as early as 1933. Yet he never wavered from his loyalty to earth and immersion in its agonies and ecstasies as the only place God is met, faith is lived, and eternity is glimpsed. He simply gave up the jingoistic language that would have been heard as fascist nationalism and nature romanticism.

What was Bonhoeffer's alternative? And how is it he saw with such clarity? What took the place of an ethic "of earth and of blood" was earth already taken up into God via a cosmic understanding of Christ. Like Paul's, Bonhoeffer's was, from the early 1930s on, a cosmic Christ "in whom all things [*ta panta*] cohere" (Col. 1:17) and in whom "the fullness of God was pleased to dwell" (Col. 1:19). Like Paul as well, Bonhoeffer understood earth and all creation as reality already reconciled in Jesus Christ and striving to be realized in our own lives and throughout creation. He would write in the posthumously published *Ethics:* "In Christ we are offered the possibility of partaking in the reality of God and in the reality of the world, but not in the one without the other."[17]

Yet for Bonhoeffer the whole is not self-revealing. It has a name, Jesus Christ. Jesus Christ is the center of "nature, humanity, and history" for Bonhoeffer.[18] The drama of redemption for him, as for Irenaeus in the second century, is one in which the whole of creation is reconstituted in such a way that enslaved nature and broken humanity are redeemed together. This Christic grandeur has the face of the suffering God, however. God enters into the anguish of creation itself, into earth's distress and degradations, winning space for new life in the places of death itself as well as the places of ongoing, carefree life.

But what does this turn to Jesus Christ mean for Bonhoeffer's repudiation of an equally relentless fascist earth faith? For one thing, nature seen in the suffering God can never be deified for Bonhoeffer as it was for Nazism. Nature is not "blood and soil" rooting and uniting a people in mysterious forces they share with the forests, mountains, sea, and their own mythical past. Moral ideals cannot, for Bonhoeffer, be moral absolutes embodied in the natural order in a way that only awaits the impassioned appropriation by *das Volk* as a superior people who have the right to wield those absolutes. And nature does not offer, as it did for the Nazis, new powers for a nation's rebirth and the suspension of

---

16. See the chapter entitled "Nazi Ecology" in Luc Ferry's *The New Ecological Order*, trans. Carol Volk (Chicago: University of Chicago Press, 1995), 90ff. Among the fascinating arguments of this chapter is that three laws on the protection of nature and animals, enacted in 1933, 1934, and 1935, were favorites of Hitler and among the first pieces of environmental legislation of the modern world. Nazi ecology was, it turned out, not incompatible with genocide.

17. Dietrich Bonhoeffer, *Ethics,* ed. Eberhard Bethge, 6th ed. (New York: Macmillan, 1965), 195.

18. See especially Dietrich Bonhoeffer, *Christ the Center,* trans. John Bowden (New York: Harper and Row, 1978), 61–67. Note the following: "To sum up, we must continue to stress that Christ is indeed the centre of human existence, the centre of history and now, too, the centre of nature; but these three can be distinguished only in the abstract. In fact, human existence is always history, is always nature as well" (67).

critical judgment as an expression of vitalism. Vitalism, Bonhoeffer says, arises from the absolutizing of a genuine insight: namely, that "life is not only a means to an end but an end in itself." Yet fascism construes this truth in such a way that "life which posits itself as an absolute, as an end in itself, [becomes] its own destroyer."[19]

Nor, to strike another chord, dare human beings take on the stature of the outsized *Übermensch* for Bonhoeffer, those larger-than-life, dynamic human beings who create history with the power of their own self-assertion. This larger-than-life notion of the strong and virile, with whom the Nazis identified themselves, denies real human frailty and limits and leaves us striving "to become a god against God." Rather, the presence of God in Christ in, with, and under all reality means that each human being "is at liberty to be the Creator's creature," and no more or less. Truly responsible action, Bonhoeffer goes on to say, is limited "by God and by our neighbor." It is "not its own master, because it is not unlimited and arrogant." By contrast, responsible action is "creaturely and humble."[20]

Blood-and-soil vitalism and larger-than-life figures who drive history end up idolizing death, Bonhoeffer continues. Supposedly reveling in human strength, they show contempt for human beings. In Nazi hands the "life and the earth" Bonhoeffer loves "so much that without them everything seems to be over" become toys of arrogance and condescension. Nazi nature romanticism soon goes hand in hand with genocide.

Bonhoeffer penned a precise description of Nazism and its consequences at a time when Hitler was at the height of his popularity. A portion follows:

> Boastful reliance on earthly eternities [blood and soil] goes side by side with a frivolous playing with life. A convulsive acceptance and seizing hold of life stands cheek by jowl with indifference and contempt for life. There is no clearer indication of the idolization of death than when a period claims to be building for eternity [the thousand-year reich] and yet life has no value in this period, or when big words are spoken of a new man, of a new world and of a new society which is to be ushered in, and yet all that is new is the destruction of life as we know it.[21]

In this earth faith, Nazi idolization of the human (some humans far more than others!) went hand in hand with the idolization of death. Bonhoeffer's antidote to this aggressive rendition of nature as a people's identity, morality, and destiny is also fidelity to earth, but it is fidelity as a matter of faith in the suffering God pegged to earth in the cross and resurrection of Jesus. Echoing Luther, Bonhoeffer says that "one takes of life what it offers, not all or nothing." Neither "cling[ing] convulsively to life nor cast[ing] it frivolously away, . . . one allows to

---

19. Bonhoeffer, *Ethics*, 149.
20. Ibid., 235.
21. Ibid., 78–79.

death the limited rights which it still possesses."[22] Jesus Christ, in whom we graciously have earth *with* God and God *with* earth,[23] but never one without the other, shows another, humbler, more vulnerable and compassionate way.

In Bonhoeffer's own life, that way meant engaging life where it was most obviously violated and degraded, as well as enjoying common pleasures as they were available. When he studied at Union Theological Seminary in New York in 1930–31, it meant confronting white racism as *the* American issue. But at home it was anti-Semitism and the treatment of the Jews. When the synagogues of Berlin went up in flames and the glass of Jewish homes and shops shattered in the streets, Bonhoeffer marked the date, November 9, 1938 (Kristalnacht), in the margin of his Bible, beside Psalm 74: "[T]hey burned all the meeting places of God in the land." He underlined the next verse, adding an exclamation point: "*We do not see our signs; there is no longer any prophet, and there is none among us who knows how long.!*" He took these and other passages (Zech. 2:8; Romans 9–11) to his course at the Finkenwalde Seminary together with a dictum the students would remember across the years ahead: "Only those who cry out for the Jews dare sing Gregorian chant." Later he wrote in *Ethics* that those who drive the Jews from the West expel God in Jesus Christ, "for Jesus Christ was a Jew."[24] Bonhoeffer's own participation in the plot on Hitler was in large part because of German crimes against the Jews, just as, in the end, his own death on the gallows was traced not only to the plot per se but to the (treasonous) action of falsifying papers to help Jews escape Germany. For him, this was discipleship — the form that following Jesus Christ called for if one were a Christian German of that period. And it was penance — the action required to end German war crimes and show the face of a different Germany. Bonhoeffer thought of himself not as a martyr but as a patriot whose death was as a German *for* Germany.[25]

When he himself, a son of privilege, became one of the defiled and experienced earth's distress as a victim, he was astonished at the insight and clarity it brought. He did not wish himself or anyone else to be a victim or think like one. But he did understand how Jesus on the cross and his solidarity with the perishing yielded "new eyes" and an indispensable perspective. In an essay entitled "After Ten Years," a Christmas gift for fellow conspirators that passed along the lessons learned in ten years of resisting Nazism, he included a paragraph that has become a *locus classicus* of liberation theology:

---

22. Ibid., 79.

23. Sometimes Bonhoeffer says "earth" (*Erde*) and sometimes "world" (*Welt*). The latter is usually the realm of human history and our daily life and its requirements, without much direct mention of its embeddedness in the rest of nature. "Earth" by contrast is comprehensive of all life together as a single totality. Bonhoeffer's imprecision in using both terms almost interchangeably is telling. It is a characteristic of his thought as a whole that he moves both within and outside the cosmologies and ethical traditions he has inherited.

24. Bonhoeffer, *Ethics*, 90.

25. See the poems "Night Voices in Tegel" (in *LPP,* 349ff.) and "Stations on the Way to Freedom" (in *LPP,* 370ff.) as well as the letter of July 21, 1944 (in *LPP,* 369–70).

There remains an experience of incomparable value. We have for once learnt to see the great events of world history from below, from the perspective of the outcast, the suspects, the maltreated, the powerless, the oppressed, the reviled — in short, from the perspective of those who suffer. The important thing is that neither bitterness nor envy should have gnawed at the heart during this time, that we should have come to look with new eyes at matters great and small, sorrow and joy, strength and weakness, that our perception of generosity, humanity, justice and mercy should have become clearer, freer, less corruptible. We have to learn that suffering is a more effective key, a more rewarding principle for exploring the world in thought and action than personal good fortune. This perspective from below must not become the partisan possession of those who are eternally dissatisfied; rather, we must do justice to life in all its dimensions from a higher satisfaction, whose foundation is beyond any talk of "from below" or "from above." This is the way in which we may affirm it.[26]

Elsewhere he wrote that people ought to be judged more in light of what they suffer than what they do.[27]

Here Bonhoeffer's theology of the cross intersects his experience. "The view from below" and the cross constitute the entry point for radical critique of the reigning arrangements. They constitute the place to understand the dynamics of modernity that kill. They are where the violations of community and creation are most obviously played out. And they are the loci for knowing God's compassion for the suffering of the world. For Bonhoeffer, the humiliated one on the cross knows those "below" and stands with them.

In sum, Bonhoeffer's language of Christian ethics as earth ethics abandons ethics-talk "as a matter of earth and blood"[28] because Nazism's own earth faith and ethic captured mass, perverse appeal. His own fidelity to earth is not therefore abandoned, however. Rather, it holds to Jesus Christ as the suffering God who humbly drinks "the earthly cup to the dregs."

While Bonhoeffer was answering Nazism and a whole people's spiritual and moral deformation, other enemies of a true earth faith surfaced. They, too, had to be faced down. Three little-noticed documents from 1932 bear careful reading. Perils to a viable earth faith come from many sides at once.

WORLD-WEARINESS AND OTHERWORLDLINESS

Bonhoeffer was twenty-six and a young professor at Humboldt University in Berlin in 1932. In November he delivered an address to the students and faculty of a school in Potsdam-Hermannswerder. At university he was lecturing on

---

26. Bonhoeffer, "After Ten Years," in *LPP,* 17.
27. Ibid., 10.
28. The phrase is from the Barcelona address of 1929.

"creation and fall," but for the Potsdam occasion he chose the first petition of the Lord's Prayer. He titled his remarks "Thy Kingdom Come: The Prayer of the Church for God's Kingdom on Earth."

The opening paragraph stipulates who may rightly pray this petition: "Only wanderers who love both earth and God at the same time can believe in the Kingdom of God."[29] One temptation is otherworldliness:

> We have been otherworldly ever since we hit upon the devious trick of being religious, yes even "Christian," at the expense of the earth.... Whenever life begins to become painful and oppressive, a person just leaps into the air with a bold kick and soars relieved and unencumbered into the air. A person leaps over the present, disdains the earth, is better than it, and has his easy eternal victories next to the temporal defeats.[30]

Such flight from earth and its distress may "make it easy to preach and speak words of comfort" during a time of growing world-weariness, near anarchy, and economic collapse, Bonhoeffer acknowledges.[31] But the price paid is the Christian's failure to recognize the true Christ. The true Christ is incarnate in earth and offers a certain earthbound way as creation's own. In passages foreshadowing more famous ruminations from prison a dozen years later, Bonhoeffer explains that the suffering God becomes weak in the world so that we may become strong and assume full worldly responsibility. Otherworldliness is the temptation of those who cannot bear earth as it is.[32] In their weakness they flee earth's turmoil and travail. Such weak ones should not be denied help, Bonhoeffer goes on, and in fact Christ offers it. But "Christ does not will this weakness." Rather, "he makes people strong." "[Christ] does not lead people in a religious flight from this world but gives them back to the earth as loyal sons. Be not otherworldly, but be strong!"[33] he tells the students and their teachers. (This was far from the theology that students were asked to imbibe in a school catechism of the 1930s: "Jesus built for heaven; Hitler, for the German earth.")[34] From prison he would write that "human religiosity makes people look in their distress to the power of God in the world: God is the *deus ex machina*" while "the Bible directs humans to God's powerlessness and suffering; only the suffering God can help."[35]

The Potsdam address may exhibit a young professor feeling his own powers, with all the attendant enthusiasm, but at least there is no misunderstanding him. "Whoever evades the earth does not find God. They only find another world; their own, better, more beautiful, more peaceful world." "Whoever loves God,"

---

29. Dietrich Bonhoeffer, "Dein Reich komme! Das Gebet der Gemeinde um Gottes Reich auf Erden," *Gesammelte Schriften*, 3:270; trans. mine.
30. Ibid.
31. Ibid.
32. Ibid., 271.
33. Ibid.
34. Cited by Geffrey B. Kelly and E. Burton Nelson, eds., *A Testament to Freedom: The Essential Writings of Dietrich Bonhoeffer* (San Francisco: HarperSanFrancisco, 1990), 117; source not given.
35. Bonhoeffer, *LPP,* 361.

he goes on, "loves God as Lord of the earth as it is; and whoever loves the earth loves it as God's earth. Whoever loves God's kingdom, loves it wholly as God's kingdom, but also wholly as God's kingdom on earth. They do so...because God has blessed the earth, and created us from earth."[36]

That was 1932. The evil was sufficient unto the day: 1932 saw the intensification of fascism and anti-Semitism, the persistence of economic depression, and political turmoil. Bonhoeffer allows that we might well imagine conditions and occasions when the church can endure those who would turn from the world's troubles and seek their salvation apart from earth's distress. But he adds cryptically, "[T]his is not one of them." Rather, "[T]he hour in which the church today prays for the kingdom is one that forces the church, for good or ill, to identify completely with the children of earth and of the world, binding itself with oaths of fealty to the earth, to misery, to hunger, to death.... The hour in which we pray today for God's kingdom is the hour of deepest solidarity with the world, a time of clenched teeth and trembling fist."[37] This is indeed earth's distress as the Christian's Song of Songs!

It is also straightforward earth patriotism, "with oaths of fealty to the earth" and a worldly solidarity of "clenched teeth and trembling fist" as the hour's vocation. Yet something else stirs for Bonhoeffer: "Only where earth is fully affirmed can its curse be seriously broken through and destroyed."[38] Which is to say: only those who love the earth fiercely, _in_ its distress, will effect whatever redemption it might know.

Even this isn't the deepest of reasons for Bonhoeffer, however. As we have seen, it is God who, long before us, fully affirmed earth, identified with it, and, in Jesus, died and rose for creation's new life. Ultimate earth fidelity, then, is God's. "God is always both the one who binds himself to the earth and the one who breaks its curse."[39] If we are called to till and keep and tend and heal, it is because our ways should conform to this God's passion for earth and its flourishing. Fidelity to earth is imitation of God.

The culprit that cuts the cord of responsibility is religion as otherworldliness, a "religious flight from the world" as troubles mount. Bonhoeffer's developed critique of religion is more than we can unravel here, but the following must be said. His is not an argument with "religion" as earlier described, using John Haught as mentor. Bonhoeffer means religion as deliverance and a certain psychosocial dynamic that deflects or deadens human responsibility.

God is the deus ex machina for some people when life turns grim. In the plays of antiquity, whenever the normal course of events could not muster some action essential to the plot because the players had gotten themselves in a fix, a god or goddess intervened and did his or her job. Life could then move on to the next episode. Bonhoeffer says people of religious consciousness turn to

---

36. Bonhoeffer, "Dein Reich komme!" 273.
37. Ibid., 274.
38. Ibid., 278.
39. Ibid.

God and religion when their human resources fail to secure solutions to problems that elude or exhaust them, in order somehow to rescue them from dangers they cannot face or control. For Bonhoeffer, this bleeds out and leaves anemic the ethics of responsibility just when accepting full human responsibility for earth's distress is most crucial. Thus he admires people of what he calls "world-come-of-age" consciousness. These are persons who fully understand that the increase of human powers to affect all life, a phenomenon emerging with full force in this century, means that human destiny and that of other creatures fall into human hands in greater measure than ever, and we are thoroughly accountable. Where there is failure, as there will be, the buck is not deposited with "God," "circumstances," "the economy," "fate," or any other religious or quasi-religious account. For genuine earth ethics people accept responsibility for the consequences of their actions, short- or long-term, seen or unseen. They are "of age" in a "world-come-of-age." There is no returning to an adolescent dependence on a father — even a heavenly Father — on whom final responsibility falls. Humankind's and otherkind's future or lack of it rests in human hands, especially now when, in Bonhoeffer's terms, "organization" has largely replaced "nature" as our immediate environment and the means of channeling human powers.[40] The aim of the West, he wrote in *Ethics*, is to be independent of nature. Enlightenment reason[41] sought this through a "technical science of the modern western world" that did not seek "service" but "mastery," "mastery over nature."[42] This was a "new spirit," "the spirit of the forcible subjugation of nature beneath the rule of the thinking and experimenting man." The outcome, Bonhoeffer continues, is that "technology became an end in itself," "with a soul of its own." Its symbol "is the machine, the embodiment of the violation and exploitation of nature."[43] A couple years later, from prison in 1944, Bonhoeffer would write that Western technical and organizational success has been extraordinary, so much so that human beings "have managed to deal with everything, only not with themselves." "The spiritual force is lacking" just at the historical moment

---

40. Bonhoeffer, "After Ten Years," in *LPP,* 380.

41. In light of what has transpired in the 1980s and 1990s, Bonhoeffer's critique of Enlightenment ethics is noteworthy. Writing in the early 1940s, he said that the Enlightenment notion of the ethical was "as a universally valid rational principle which implied the invalidation of all concrete factors, all limitations of time and place and all relations of subordination and authority, and which proclaims the equality of all men by virtue of their innate universal human reason. One must be quite clear about the fact . . . that the actual goal of this new conception of the ethical, which was to establish a universal union of mankind in the place of a fossilized form of society characterized by the antagonism of privileged and unprivileged members, has not only not been achieved but has turned out to be exactly the opposite of what was intended. The ethical, in this sense of the formal, the universally valid and the rational, contained no element of concretion, and it therefore inevitably ended in the total atomization of human society and of the life of the individual, in unlimited subjectivism and individualism. When the ethical is conceived without reference to any local or temporal relation, without reference to the question of its warrant or authority, without reference to the concrete, then life falls apart into an infinite number of unconnected atoms of time, and human society resolves itself into individual atoms of reason" (*Ethics,* 272–73).

42. Ibid., 98.

43. Ibid.

when everything "turns on human beings."[44] Yet there are world-come-of-age people who recognize this and face up to it, without succumbing to religion and God as the place to deposit responsibility for errant civilization.

God as the deus ex machina contributes to the dulling down of earth responsibility in another way. The God of deliverance is not experienced by people in their achievement and strength but in their weakness, their resignation, and their flights into otherworldliness. Since God here sits on the margins and is turned to only when people are in trouble or their powers peter out, people are most "religious" when they are most exhausted, defeated, perplexed, or self-denigrating. They are most religious when they are most self-preoccupied. But on their own, in their exercise of responsibility and strength with a sense of world-come-of-age maturity, they are largely "nonreligious." God then is experienced by people in their moments of weakness. But God is not experienced in the exercise of their strength in earthly responsibility "in the midst of our life."[45]

Given this critique of religion, it is easy to see why Bonhoeffer would counter with a picture of Jesus Christ as the incarnation of the suffering God "who makes us strong." The presence of Jesus is to aid in the development of human powers for responsible, earth-oriented action. It is also easy to see why Bonhoeffer contrasts faith at the center of human life and strength with religion at its boundaries. Jesus at the center frees those whose life he shares to find their own responsible answers and does so through participation in their struggle. He confronts the sins of weakness in people in order to make them strong. He also confronts the sins that arise from people's strength in order to render power accountable to the welfare of others, including future generations.[46] Jesus affirms people's "health, vigor, [and] happiness" and does not "regard them as evil fruit; else why should he heal the sick and restore strength to the weak?" Bonhoeffer asks. Jesus "claims for himself and the Kingdom of God the whole of human life in all its manifestations," he goes on.[47] Or, in another phrasing, he says that "the religious act" is always something partial while faith is something whole, involving the whole of one's life. "Jesus calls," he concludes, "not to a new religion, but to life."[48]

There is more to Bonhoeffer's indictment of religion.[49] But this suffices to illumine his comments in 1932 that religious otherworldliness is an anti–earth ethic wholly inadequate as a response to the turmoil that the Nazis' own earth ethic capitalized upon. For Bonhoeffer, there is irony here because conscientious, secular humanists were, in their godlessness, closer to the God of the Bible's call to become fully responsible for their world than were religious Christians

---

44. Bonhoeffer, *LPP,* 380.

45. Ibid., 282.

46. In the essay for fellow resisters Bonhoeffer writes, "The ultimate question for a responsible man to ask is not how he is to extricate himself heroically from the affair, but how the coming generation is to live" (*LPP,* 7).

47. Ibid., 341–42.

48. Ibid., 362.

49. See my volume with Renate Bethge, *Dietrich Bonhoeffer: His Significance for North Americans* (Minneapolis: Fortress, 1990), especially the chapter "Worship in a World Come of Age," 60–67.

in their alliance with the God of religion. Bonhoeffer is particularly critical of clergy who use all their "clerical tricks" to retain religion as a special sector of life. They do this by marketing certain questions and problems — death and guilt and distress, for example — "to which [supposedly] only 'God' can give an answer." Bonhoeffer goes on to say that he regards the attack by Christian apologetics on the adulthood of the world (world-come-of-age) "in the first place pointless, in the second place ignoble, and in the third place unchristian." It is pointless because it seems "like an attempt to put a grown-up man back into adolescence." It is ignoble because it seems an attempt to exploit human weakness, and it is unchristian because "it confuses Christ with one particular stage in man's religiousness."[50]

Fidelity to earth, then, calls for an earth faith and ethic that avoid, on the one hand, vitalistic nature romanticism and, on the other, any ethic tinged with an otherworldliness and world-weariness that dumb down a full sense of human responsibility for earth's distress and its cure. By contrast, the kind of participation Bonhoeffer is looking for is already to be found, he argues, in the compassionate, empowering way seen in Jesus.

Just such a view of earth and participation is also in accord with our own deepest nature. That is also Bonhoeffer's subject in 1932, this time in the university lectures on "creation and fall." Sentences tumble over one another in this exegesis of Genesis 1–3, and we probably best let them do so. But only after noting their context.

OUR BODIES, OUR SELVES

Creation as a theme had practically no previous play in Bonhoeffer's (limited) teaching. The rise of fascism, however, brought renewed life to the mythic origins that surface during every epoch of transition and new movements. Reconsidered creation accounts were part of these "rebeginnings." In theological circles themselves, the "orders of creation" became a subject of debate precisely as a way of adjudicating cultural identities and claims. In 1931, the prominent Lutheran theologian Emanuel Hirsch, soon the darling of the nazified "German Christians," published *Creation and Sin,* with its focus on "the natural reality of the human individual." It was but one example, albeit a noticed one, in what became a battle of articles and treatises on human origins and the human condition and what these meant for emergent Germany in the 1930s. Bonhoeffer himself experienced moral restlessness about the treatment of creation and human nature in these debates and, in what became his typical theological move, argued that Jesus Christ stood at "the center of life." Christians ought not understand creation in ways that put someone or something else at the center.[51]

---

50. Bonhoeffer, *LPP,* 326.

51. This information about the setting draws from Eberhard Bethge's magisterial biography, *Dietrich Bonhoeffer: Man of Vision, Man of Courage,* trans. Eric Mosbacher et al. (New York: Harper and Row, 1970), 161–64.

It is in this setting that Bonhoeffer speaks to the university students of human "being" and the earth. It is remarkable that he risks an unqualified earth identity and bond in the face of — and against — nature romanticism, orders of creation theology, and the assertion of peoplehood that was serving fascism within the churches and without. Here are the sentences that tumble over one another: "Darwin and Feuerbach themselves could not speak more strongly" than Genesis in recognizing we are "a piece of earth" and that our "bond with the earth belongs to (our) essential being." It is God's earth from which we are taken and from which we have and are "body." Our body belongs to our "essential being." We *are* our bodies, ourselves.

The significance cannot be exaggerated: "The one who renounces his body renounces his existence before God the Creator." "The essential point of human existence," Bonhoeffer argues, "is its bond with mother earth, its being as body."[52] "Escape from the body" is thus escape from being human and "escape from the spirit as well. Body is the existence-form of spirit, as spirit is the existence-form of body."[53]

Bonhoeffer ends this section by saying that humanity "is the image of God not in spite of but just because of its bodiliness. For in bodiliness humankind is related to the earth and to other bodies, humankind is there for others, dependent upon others. In bodiliness humanity finds its brother and the earth. As such a creature of earth and spirit humanity is in the likeness of God."[54] Elsewhere, Bonhoeffer had echoed Friedrich Christoph Oetinger that the end of God's own ways "is bodiliness."[55]

Writings ten years later in *Ethics* lie close to these of 1932. But the language turns to "rights," not least because Bonhoeffer had, through involvement in the conspiracy, learned of Hitler's grotesque violation of human beings — Jews, "Gypsies," gays and lesbians, communists and other political dissidents, as well as prisoners of war. (Bonhoeffer's brother-in-law, Hans von Dohnanyi, was keeping a chronicle of Nazi atrocities. It was found after the failed assassination attempt of July 20, 1944, and was part of the evidence used against the conspirators.) Bonhoeffer may in fact be the first German Protestant theologian to speak of "rights."[56]

How he does so is as telling, and as crucial, as what he says. Fighting the conventional Protestant devaluation of creation, he speaks of "natural life" as "formed life" and, within this circle of discourse, talks of "the right to bodily life" or bodily integrity as "the foundation of all natural rights without exception."[57] The reason is not difficult to discern: "The living human body is always

52. Dietrich Bonhoeffer, *Creation and Fall: A Theological Interpretation of Genesis 1–3* (New York: Macmillan, 1959), 44–45. Language altered slightly: "one" for "man."

53. Ibid., 45–46.

54. Ibid., 46. Language altered: "humanity" and "humankind" for "man."

55. Bonhoeffer, "Aufträge der Bruderräte," in *Gesammelte Schriften*, 3:388.

56. Incidentally, Bonhoeffer is also one of the few modern theologians to write of the tree of life. See *Creation and Fall*, 49–52, 96.

57. Bonhoeffer, *Ethics*, 156.

the person himself/herself."[58] Bonhoeffer goes on: "Rape, exploitation, torture and arbitrary confinement of the human body are serious violations of the right which is given with the creation of humankind."[59]

Natural rights, then, reside in bodily requirements and bodily integrity. That which is necessary for bodily flourishing — and that certainly includes its protection against violation — merits a right secured by law. These rights are grounded in creation itself and belong to life's requirements for flourishing, since our bodiliness is our unbreakable bond with earth and all its creatures. In different language for the same thing, the rights of natural life are "the reflected splendour of the glory of God's creation."[60]

It is not surprising that Bonhoeffer's abbreviated discussion of bodily rights (he didn't live to complete *Ethics*) leads to a restatement of otherworldliness as an errant response to earth's distress. It goes like this, though the language is ours as much as his.

Using the argument of our bodies as ourselves in *Creation and Fall*, Bonhoeffer considers otherworldliness dysfunctional and destructive because it splits mind and spirit from body and our bodies from the rest of nature. When these dualisms take over, we act as though our primordial intimacy were not with earth itself but with some other environment and world. Usually, to recall Bonhoeffer, this is some world of our own creating that "disdains the earth, is better than it, [where] easy eternal victories [lie] next to the temporal defeats."[61] Then we no longer have feeling rapport with the rest of the natural world or even ourselves. Our bodies are no longer sources of deep knowledge about the world; they are no longer experienced as microcosms of the macrocosm; and we may even come to have no essential relationships with other bodies and the earth. We soon forget that created life is, in Bonhoeffer's words in *Ethics*, "the community of men and creatures" that comprises an "indivisible whole," that "participation in [this] indivisible whole" is the very "sense and purpose of the Christian enquiry concerning good,"[62] and that "the good demands the whole."[63]

Strange things then transpire as a result of this human split-off. We may hardly notice the air is bad, the water polluted, and the soil degraded, even when these assault our health. Or, feeling that assault, we may in our otherworldliness not register its magnitude and significance. Or we may do our best to escape it in lifestyle enclaves as distant as possible from earth's agonies. Our consciousness is no longer resident in body and in earth. Otherworldly reality of our own making numbs us to the most essential relationship of all.

Tucked in his protest is Bonhoeffer's comment in the prison letters that he

---

58. Ibid., 183. I substitute "person," rather than "man," for the German *Mensch*.

59. Ibid. I am indebted to Wolfgang Huber, now bishop of Berlin-Brandenburg, for this discussion of Bonhoeffer on human rights, from his address at Union Theological Seminary, February 14, 1994.

60. Bonhoeffer, *Ethics*, 151.

61. Bonhoeffer, "Dein Reich komme!," 3:270.

62. Bonhoeffer, *Ethics*, 193.

63. Ibid.

finds himself preferring the Christian Old Testament to the New, or at least that Christians should not get to the New Testament too soon. Like *Creation and Fall*, the prison letters also name flight from the bodily as the enemy of fidelity to earth and our own being. The Hebrew understanding of redemption is slighted when Christians rush to the New Testament and get stuck there. Without the First Testament and the healthy Hebrew sense of earthiness, it is less clear that redemption is the redemption of all things, nature and society together, as that happens in history and to earth, and in ordinary, everyday, this-worldly relationships. Precisely these constitute the place God is God and the drama of life is played out. All other notions of redemption are otherworldly; they are docetic opiums passing as Christianity. They forget that "the good demands the whole."[64]

Bonhoeffer's protest runs deep, since otherworldly redemption is already an affliction borne in the formation of early Christianity itself, as it is in the Hellenistic Judaism of the day. Otherworldly Greek cosmologies, from Plato's *Timaeus* forward, were enormously influential for Christianity and Western thought. We ought, Bonhoeffer contends, to take far more of our basic orientation from Jesus' own scriptures, the Hebrew Bible, its Song of Songs included.[65]

Yet the point is less the preference of Testaments than the insistence upon bodiliness and fidelity to earth. A basic irony emerges for all who forget that "body is the existence-form of spirit, as spirit is the existence-form of body."[66] By ignoring full embodiment and "the essential point of human existence" as its "bond with mother earth,"[67] we slowly create circumstances that have the potential not only to assault the senses but to annihilate bodies and spirits themselves. Learning to be truly in our bodies and to respect earth's body as the rest of *us,* or us as a glorious instance of it, is imperative. The right to bodily life as an expression of natural rights, together with repudiation of otherworldliness of all kinds and a return to our "essential bond" with earth, is a combination that furthers this learning.

What must be added, of course, is that Bonhoeffer's final appeal is not to bodiliness per se even when it is our very existence. Bodiliness as the end of *God's* path is the more fundamental ground for Bonhoeffer, as is earth as the place of God's chosen abode.

The prison letter of June 27, 1944, renders all this theological in the manner typical of Bonhoeffer. In writing to Eberhard Bethge, he complains that Christianity has become one of the "religions of redemption" where the emphasis is "on the far side of the boundary drawn by death." "But it seems to me," Bonhoeffer goes on, "that this is just where the mistake and the danger lie."[68] Redemption now means "redemption from cares, distress, fears, and long-

64. Ibid.
65. See n. 10 of this chapter.
66. Bonhoeffer, *Creation and Fall*, 45–46.
67. Ibid., 44–45.
68. Bonhoeffer, *LPP,* 336.

ings, from sin and death, in a better world beyond the grave," Bonhoeffer says, only to go on to ask: "[Is this] really the essential character of the proclamation of Christ in the gospels and by Paul?"[69] His answer is that it is not. He then explains:

> The difference between the Christian hope of resurrection and the mythological hope is that the former sends a man back to his life on earth in a wholly new way which is even more sharply defined than it is in the Old Testament. The Christian, unlike the devotees of the redemption myths, has no last line of escape available from earthly tasks and difficulties into the eternal, but, like Christ himself ("My God, why has thou forsaken me?"), he must drink the earthly cup to the dregs, and only in his doing so is the crucified and risen Lord with him, and he crucified and risen with Christ. This world must not be prematurely written off; in this the Old and New Testaments are one. Redemption myths arise from human boundary-experiences, but Christ takes hold of a man at the centre of his life.[70]

Permit an aside from a wholly different source. We can listen in on another echo of Alyosha and Antaeus and their embrace of earth, and another affirmation, like Bonhoeffer's, of body and flesh. This echo is also about the solidarity of clenched teeth and trembling fist Bonhoeffer asked for in the 1930s, and about fighting. In this case it is about fighting degradation on every side in just about the only way slaves could. While not about Alyosha under the starry heavens, it is about weeping and watering the earth with the tears of gladness and loving those tears; and about ecstasy and trembling souls like Alyosha's in contact with other worlds in the very same moment they embrace and kiss this one. It is about a congregation gathered, the congregation in Toni Morrison's *Beloved,* to hear Baby Suggs testify and preach. This is about slaves meeting secretly in the "Clearing":

> Finally [Baby Suggs] called the women to her. "Cry," she told them. "For the living and the dead. Just cry." And without covering their eyes the women let loose.
> It started that way: laughing children, dancing men, crying women and then it got mixed up. Women stopped crying and danced; men sat down and cried; children danced, women laughed, children cried until, exhausted and riven, all and each lay about the Clearing damp and gasping for breath. In the silence that followed, Baby Suggs, holy, offered up to them her great big heart.
> She did not tell them to clean up their lives or to go and sin no more. She did not tell them they were the blessed of the earth, its inheriting

69. Ibid.
70. Ibid., 336–37.

meek or its glorybound pure. She told them that the only grace they could
have was the grace they could imagine. That if they could not see it, they
could not have it.

"Here," she said, "in this here place, we flesh; flesh that weeps, laughs;
flesh that dances on bare feet in grass. Love it. Love it hard. Yonder they
do not love your flesh. They despise it. They don't love your eyes; they'd
just as soon pick em out. No more do they love the skin on your back.
Yonder they flay it. And O my people they do not love your hands. Those
they only use, tie, bind, chop off and leave empty. Love your hands! Love
them. Raise them up and kiss them. Touch others with them, pat them
together, stroke them on your face 'cause they don't love that either. *You*
got to love it, *you*! And no, they ain't in love with your mouth. Yonder,
out there, they will see it broken and break it again. What you say out of
it they will not heed. What you scream from it they do not hear. What
you put into it to nourish your body they will snatch away and give leavins
instead. No, they don't love your mouth. *You* got to love it. This is flesh
I'm talking about here. Flesh that needs to be loved. Feet that need to rest
and to dance; backs that need support; shoulders that need arms, strong
arms I'm telling you. And oh my people, out yonder, hear me, they do not
love your neck unnoosed and straight. So love your neck; put a hand on it,
grace it, stroke it and hold it up. And all your inside parts that they'd just
as soon slop for hogs, you got to love them. The dark, dark liver — love
it, love it, and the beat and beating heart, love that too. More than eyes
or feet. More than lungs that have yet to draw free air. More than your
life-holding womb and your life-giving private parts, hear me now, love
your heart. For this is the prize." Saying no more, she stood up then and
danced with her twisted hip the rest of what her heart had to say while the
others opened their mouths and gave her the music. Long notes held until
the four-part harmony was perfect enough for their deeply loved flesh.[71]

Commentary would spoil this, as would paraphrase. Suffice to say this is a
protest against the degradation of bodies and spirits together, an affirmation of
redemption as a matter of earth and of flesh and of ethics as a matter of earth
and of blood in a precious, sensuous, utterly nonfascist and nonromantic way.
In fact it exemplifies what Bonhoeffer says ethics itself is to foster and promote:
namely, heightened participation in life,[72] a participation that, like the hope of
resurrection, sends one back into life on earth "in a wholly new way."[73] It is
also congruent with Bonhoeffer's insight that gaining moral strength happens
in the spontaneous and ordinary experiences of joy and earth's goodness as well
as in the solidarity that enters into distress and suffering in conscious efforts to

71. Toni Morrison, *Beloved* (New York: Knopf, 1987), 88–89.
72. See Bonhoeffer, *Ethics*, 269–70.
73. Bonhoeffer, *LPP*, 337.

combat and alleviate them.[74] The goodness of life together and the reciprocity learned in genuine community create moral agency and responsibility.

COMMUNITY AND AGGRESSION

The last document from 1932 can be treated briefly. "The Right to Self-Assertion" was an address given at the Institute of Technology in Berlin when Bonhoeffer was a chaplain there. He asks questions about the right to become a nation or people. Does such a right exist? If so, where is it grounded? And what would such a right mean for issues of force and collective self-assertion? Few questions were more pressing as German nationalism surged and a defeated and rather anarchic Germany strove to gather itself as a people and stand tall once again among the nations.

To address these questions Bonhoeffer sketches two ways of understanding ourselves in the larger scheme of things, trusting that "cosmological" frameworks might steer an approach to fundamental political issues. For the first framework he cites *tat tvam asi* from the Hindu Upanishads, translating this as "All that is you" or "You are all that." That is, you are the cosmos; the cosmos is you. Human beings belong to the universe as microcosms of the macrocosm.

The ethic here is an ethic of cosmic community in which minimal harm is done all creatures and suffering is accepted in deference to the larger whole, its rhythms and well-being. Bonhoeffer then praises Gandhi as one who, with this cosmology, found a way to turn it to the task of building a nation and a people. Gandhi channeled the energy of this cosmology into positive protest and community building. He and emergent India thus offered one answer to the question about peoplehood and collective self-assertion.[75]

The other answer, striking for a young German in 1932, as was his attraction to Gandhi, is what Bonhoeffer calls "Euro-American civilization." If India's is a "history of suffering," the West's is a "history of war" in which "war and industry," or "the machine and war," are the chief means of self-identity, self-assertion, and problem solving.[76] Bonhoeffer asks about the spiritual and intellectual sources of this and concludes that it is rooted in the West's battles "to turn nature to its service." "Human conquest of nature is the foundational theme of Euro-American history."[77] Or, in words we cited earlier from his prison thoughts: the West's aim is to be independent of nature and substitute "organization" as the immediate and controlled environment.

Yet this is not only a battle against nature. It becomes a battle "against other

---

74. Ivone Gebara's work with poor women in Recife, Brazil, substantiates this. Like the experience Morrison captures from the "Clearing," the moments of joy and shared goodness strengthened moral agency and resolve, Gebara reports. Conversation at Union Theological Seminary, spring semester 1994.

75. Bonhoeffer, "Das Recht auf Selbstbehauptung," in *Gesammelte Schriften*, 3:262–63.

76. Ibid., 263.

77. Ibid., 264.

human beings" as well. "[European] life in its essence means 'to kill,'" Bonhoeffer says bluntly.[78] Western civilization, fragmented from nature in its very consciousness, destroys both natural and human communities (to use his distinctions). Needless to say, Bonhoeffer rejects this kind of collective self-assertion and problem solving.[79]

Bonhoeffer romanticizes India here. Gender and caste analysis never enter his appraisal, and his address slides too easily into an unqualified affirmation of sacrifice for the larger community. Thus the concrete meaning of sacrifice — who sacrifices what for whom within what relationships — eludes Bonhoeffer. It is nonetheless intriguing that a young German of privilege saw, in the early 1930s, that Western culture and its aggressive technology and institutions rested in a spiritual worldview destructive of human community and the rest of nature. It is also intriguing that a German theologian of this period became sufficiently interested in Gandhi to secure an invitation to live and study with him, for several reasons. There was, Bonhoeffer concluded in the early 1930s, a kind of exhaustion of Western Christian spirituality and ethics. They were inadequate to the day's troubles and challenges. He thought Gandhi and the East might embody what he called "Christianity in other words and deeds."[80] Too, Bonhoeffer thought the arts of the ashram might prove to be the arts of life together in community. "Community" and "life together"[81] were continuous themes for Bonhoeffer because he understood human beings as essentially social, all reality as essentially relational, and both as matters of earth. And he thought Gandhi had something to teach about community and its ordering. Lastly, Bonhoeffer desired to learn the way of nonviolence.[82] He was always of the conviction that the truth of the gospel cannot be separated from the life of the community, and he hoped peaceful ways of resistance might be of use for the Confessing Church's resistance to the encroachments of Nazism. (Bonhoeffer did receive an invitation from Gandhi in 1935. But it came just at the time he was called to head up the Confessing Church's new seminary at Finkenwalde. The latter was compelling, and the trip was postponed indefinitely.)[83]

Yet the fascination with Gandhi and the East, and the tantalizing foray into a cosmology of cosmic community as politically relevant for the rights of peoples, is two steps to the side of the real significance of Bonhoeffer's address. The crucial insight is that Western modernity, with its master image as the image of mastery itself, first destroys other worlds and then its own. Its very "essence"

---

78. Ibid.

79. Ibid., 263–65.

80. Dietrich Bonhoeffer, "Brief an Helmut Rößler," in *Gesammelte Schriften* (Munich: Kaiser, 1965), 1:61; trans. mine.

81. The title of one of Bonhoeffer's books is *Life Together.*

82. For a discussion of Bonhoeffer's interest in Gandhi, see my article, "Gandhis Einfluß auf Bonhoeffer," *Die Zeichen der Zeit,* 3:69.

83. The Confessing Church was the portion of the Protestant church that broke from the rest in protest of the Aryanizing of the churches themselves. Because it did so, it had to set up its own institutions even though this was illegal. Bonhoeffer was called to direct Finkenwalde Seminary and did so until the Gestapo closed it down.

is "to kill" by way of industry and aggression, even when it thinks itself to be "civilizing" in the process.

Here Bonhoeffer's later insights about Nazism are already implicit and in due course become part of his critique of the deformation of the West. That is, the combination of fascist vitalism and aggressive human beings who reckon themselves larger-than-life fashioners of history and nature is not a Nazi disease alone. It is a variant on the drive that impels the modern West in many guises. (Social Darwinism is another such prominent earth faith.) It belongs to a Western arrogance whose sense of sovereign mastery finds the very notion of finite limits repulsive. Bonhoeffer in turn, we might be reminded, found this way of leaning into the world irresponsible because responsible action is limited "by God and by our neighbour," "is not its own master because it is not unlimited and arrogant" but "creaturely and humble."[84] He even calls this "apostasy" because it rests in an understanding of human nature that rejects our essential creatureliness, the very creatureliness God created and loves. Nazism was thus, for Bonhoeffer, the most recent chapter in the decay of the West; it was not sui generis, nor the origin of the deformation.[85]

FAITH AS PARTICIPATION

A portion of a crucial prison letter finishes the glimpse of Bonhoeffer as one instance of an earth faith and ethic forged during the worst years of this deadliest of centuries. The letter shows that while the language of fidelity to earth as a matter of earth ethics changes from 1929, the substance deepens and expands. Ethics is no longer a matter of "earth and of blood and the One who created both" but rather is about "the profound this-worldliness of faith." Only a few words of introduction are necessary.

On July 20, 1944, the plot against Hitler attempted by Count von Stauffenberg and others failed. Bonhoeffer learned of the disaster that evening in Tegel Prison. Because the conspirators had successfully occupied government offices in Berlin for a few hours, key players and their networks became known immediately. Many were summarily executed. Bonhoeffer now had to reckon with the loss of everything: his engagement to Maria and their plans for a life together, his family and friends, his career and work, his country and his deeply patriotic hopes for postwar Germany. In the confines of his cell he composed a poem, "Stations on the Way to Freedom." The stanzas are, in sequence, "Discipline," "Action," "Suffering," and "Death." The next day, certain his own death would not be long in coming, Bonhoeffer sent the poem and a letter to Eberhard Bethge. The letter is stunning in its composure, its resignation to his likely fate, and its gratitude for what he has learned:

---

84. Bonhoeffer, *Ethics*, 235.
85. For a comparison of Bonhoeffer and post-Holocaust Jewish thinking on this theme, see Rasmussen and Bethge, *Dietrich Bonhoeffer*, the chapter entitled "Divine Presence and Human Power."

I remember a conversation that I had in America thirteen years ago with a young French pastor. We were asking ourselves quite simply what we wanted to do with our lives. He said he would like to become a saint (and I think it's quite likely that he did become one). At the same time I was very impressed, but I disagreed with him, and said, in effect, that I should like to learn to have faith. For a long time I didn't realize the depth of the contrast. . . . I discovered later, and I'm still discovering right up to this moment, that it is only by living completely in this world that one learns to have faith. . . . By this-worldliness I mean living unreservedly in life's duties, problems, successes and failures, experiences and perplexities. In so doing we throw ourselves completely into the arms of God, taking seriously, not our own sufferings, but those of God in the world — watching with Christ in Gethsemane. That, I think, is faith; that is *metanoia;* and that is how one becomes a human being [*Mensch*] and a Christian. How can success make us arrogant, or failure lead us astray, when we share in God's sufferings through a life of this kind?

I think you see what I mean, even though I put it so briefly. I'm glad to have been able to learn this, and I know I've been able to do so only along the road that I've travelled. So I'm grateful for the past and present, and content with them.[86]

To embrace earth and live unreservedly in life as it is given us is to land in the arms of God. It is to embrace God and be embraced by God, the suffering God. Such living — throwing ourselves into life, embracing earth and its distress — is in fact the way of faith itself. It is responsible "participation in the sufferings of God for the life of the world" and a faith that both saves life and savors life. It is earth ethics and our experience of "the power which still holds [us] by eternal, mysterious forces."[87] It is life lived to the fullest. It may, in evil times, include the cross of resistance in the name of life itself.[88] This is earth and its distress as "the Christian's 'Song of Songs.'"

---

86. Bonhoeffer, letter of July 21, 1944, in *LPP,* 369–70. I have substituted "a human being" for "a man" as the translation of *Mensch.*

87. Cited from the Barcelona address.

88. The parallels between Bonhoeffer and Martin Luther King Jr. should be noted. Some are coincidental. Both met a violent end at exactly the same age of thirty-nine years and two months. But beyond such coincidence, both were what E. F. Schumacher later called "homecomers." Responsibility and community were concrete, finite, tied to time and place and a people, with a genuine life with and for others in earthbound loyalty. It was rooted community responsibility, drawing upon the best of their traditions yet critical of and revising them in the course of engaging the grave issues of the day. King and Bonhoeffer became genuine ecumenical figures, influential far beyond American and German borders, precisely because they were so deeply immersed in the communities that shaped them.

PART III

EARTH ACTION

The last letter of Alice Walker's *The Color Purple* begins: "Dear God. Dear stars, dear trees, dear sky, dear people. Dear everything. Dear God."[1]

Just as this book as a whole is a long echo of Maya Angelou's "On the Pulse of Morning," its concluding part is a commentary on Walker's last letter. Around a certain theme: ethics and community, earth ethics and earth community.

"Dear," the affection that accompanies all that Walker addresses, seems casual enough, the way of common speech. Yet it carries a certain attitude. Earth, life, the world are received as gift. "Dear" is thus not a moneyed term here. None of the treasures on this list are commodities or capital. Life, even amidst its distress, is blessed and received with gratitude. Walker is doing what Paul Hawken, in the closing paragraph of *The Ecology of Commerce,* says we all must learn anew — "how to say grace." How to say grace, "knowing that we *do* take and harm as we live" and "that life is always a moral question that lies before us sweetly." Which is only to say that life is "dependent on our gratitude and constant struggle to cause as little suffering as possible to all and everything around us."[2] (Dietrich Bonhoeffer, writing to Maria and looking to a postwar future, was certain everything would need "to be washed in the purifying waters of contrition and gratitude.")[3]

"Earth Faith," the preceding part, was an attempt to lean into the world in a way that receives earth, with its distress, graciously. Those chapters assumed the earlier ones — "Earth Scan." An assumption of both parts is that human subjectivity over the last centuries was created in tandem with changes wrought by the great human revolutions themselves. "Industry" is a culture, as is "information," and cultures have an interiority and morality that roughly match their exteriority. Against a restless autistic subjectivity, part 2 began the long process of instructing our collective psyche differently. It tried to do so with symbols old, new, rediscovered, and reconsidered, as well as the example of a life. A turn to earth as our own skin was the common theme.

But the ecology of self and soul never works one direction at a time. To foster one interior world while living another is to court frustration, alienation, even madness. An earth faith with no corresponding way of being in the world is destined, at best, for private fantasies indulged off-camera in off-time.

This means that a work in earth ethics is finally confronted not with the questions of how we see things anew and feel about them, essential though they

---

1. Alice Walker, *The Color Purple* (New York: Pocket Books, 1982), 249.

2. Paul Hawken, *The Ecology of Commerce: A Declaration of Sustainability* (San Francisco: Harper Business, 1993), 219.

3. Dietrich Bonhoeffer and Maria von Wedemeyer, *Brautbriefe: Zelle 92,* ed. Ruth-Alice von Bismarck and Ulrich Kabitz (Munich: C. H. Beck, 1992), 176; trans. mine.

be. The question finally is: What on earth is to be done? What configuration of nature and society "without" accords with the configuration of nature and society "within" that we strive for? What manner of social cosmos aligns with the heart's cosmos? What systems in the world we inhabit are congruent with the earth faith that inhabits us? What objectivity and exteriority match a corresponding subjectivity and interiority in such a way that each creates the other? How, in our day-to-day practices, do we become the people we need?

If the analysis is correct that unsustainability is the sure tending of the present course and that a fourth human revolution is required, then these are the vital questions. The rest is journalism. Or, to say it less dismissively: the rest, too, is life, but life stopping abruptly at the water's edge. And to stop there would disqualify this work as one in ethics. An ethic that offers analysis (Earth Scan) and orientation (Earth Faith) but never ventures guidance for actions, policies, and systems is an ethic stillborn.

None of this emphasis on constructive action subtracts from what we have tried to do thus far. In fact, we have sought another way of seeing and understanding precisely *because* Thomas's axiom holds: "What people define as real is real in its consequences." Actions and outcomes follow from our cosmologies and express our faith.

But there is an even larger point that is consistently overlooked by all those who live in the mistaken confidence that we are delivered by our concepts and values and saved by education and character. Actions and outcomes are central even when they do *not* express our cosmologies and dreams or line up with the values and understandings we prize. The way of existence as it is daily *lived*, whether its practitioners embrace it as their own or find it alien to their spirit and welfare, is what finally matters to earth and one another. Systems largely control our lives, and they can be blithely oblivious to intentions and worldviews, no matter how firmly held and well argued.

East Asians, for example, traditionally hold to a cosmology of great power and beauty in which nature is all-encompassing and embracing, and humans are a microcosmic expression of the macrocosm. The passage found on the wall of an eleventh-century administrative official, Chang Tsai, says it well:

> Heaven is my father and earth my mother and even such a small creature as I finds an intimate place in its midst. That which extends throughout the universe, I regard as my body and that which directs the universe, I regard as my nature. All people are my brothers and sisters and all things are my companions.[4]

"Life" and "nature" are wholly inclusive notions here. The individual is organically identified with the universe and the infinite in a vast community. All of us together belong, modestly, to a harmony grander than we imagine.

---

4. From *Earth Prayers*, ed. Elizabeth Roberts and Elias Amidon (San Francisco: HarperSan-Francisco, 1991), xxi.

Yet if the Japanese organize life in a way that deforests the tropics in the interests of a lead role in the global economy, and if they foster a fossil fuel, automobile, and high-tech world of expanding markets and consumption; or if the Chinese erode and otherwise despoil essential and precious land for intensified agriculture, industry, infrastructure, and human habitat; or if South Koreans combine some of both of these; or if Malaysians and Indonesians industrialize agriculture, forestry, and export-oriented manufacturing on an expanding scale, then Asian actions may very well work systematically against traditional Asian cosmology, wisdom, culture, and custom. Encompassing, harmonious nature and human existence as one expression of Asian tradition may still hold a treasured place. But it will be relegated to split-off aesthetic and quasi-religious expression — lovely sculpted gardens and impressive landscape paintings, moving poetry, dance, dress, and song, for example, with all of it an expression set on a different plateau from day-to-day routine and enjoyed precisely *because* of its altitude.

So while cosmologies do matter immensely — we know and act through guiding symbols in significant degree — what matters most are our actions themselves and the world of which they are part and parcel, whether or not they express who we are deep down.

There is yet another point — or rather the earlier one restated. As the inner world shapes the outer ("what people define as real is real in its consequences"), so, too, the outer molds the inner. "Behavioral changes *pre*cede attitudinal ones" is the axiom here. We act our way into new ways of thinking and feeling as often as the reverse. Future Asians (or any other people) will be different as changed ways of doing things effect changed outlooks and temperaments.

The question asked, then, is the right one: What on earth is to be *done*? What is to be done about configuring society and nature together as our immediate world, both our internal and our external world? What arrangements and channeling of human power and normal behavior make for sustainability as a state of society as well as a corresponding state of the collective soul?

The task is pretentious and can only be chiseled at in the manner of a sculptor working with a piece of marble early on. The results will be, at best, something like Michelangelo's "slaves," still trapped in the rock from which they were being hewn. This, too, then, is the kind of beginning, but not ending, the rest of this book is — with some direction, some ideas, some proposals, a lot of fussing, a little poetry and prayer, and a few telling contours in the marble.

So with the necessary modesty that accompanies outsize tasks in the hands of flawed creatures, we turn to the contours of sustainable earth community. "Dear God. Dear stars, dear trees, dear sky, dear people. Dear everything. Dear God" is the right disposition. It is only a place to begin, but it is the right place.

Earth Community

What on earth is to be done? What is to be done if community on various levels — from the biophysical world as a whole to the local neighborhood — is the organizing theme of earth ethics and sustainability itself? Where are the guideposts and samples? What stakes out the account of responsibility we seek? What tilts us gram by gram toward the fourth human revolution?

Ask the Danes. In 1990 three school children in Kalundborg, Denmark, received an assignment. They were to produce a model in environmental studies showing how industrial wastes were being exchanged among several local companies.

Kalundborg, Denmark, is a small city on the shore of a deep fjord on the Great Belt, a body of cold salt water connecting the Baltic and North Seas. A medieval cathedral prominently marks the town center, its streets otherwise lined with the one-, two-, and three-story buildings of solid colors and subdued, practical affluence that typify Scandinavia. Kalundborg is perhaps more industrial than most similar Scandinavian cities, despite its rural surroundings and character.

What the students found was this. A coal-fired power plant, an oil refinery, a pharmaceutical company specializing in biotechnology, a sheetrock plant, concrete producers, a producer of sulfuric acid, the municipal heating authority, a fish farm, some greenhouses, local farms, and other enterprises had all discovered mutually beneficial ways to trade waste. The Asnaes power plant started the process in the 1980s by recycling its waste heat in the form of steam. Before that it had condensed the steam and returned it to the fjord that emptied into the Great Belt. Now the steam goes directly to the Statoil refinery and the Novo Nordisk pharmaceutical company. It also provides surplus heat to greenhouses, a fish farm owned by the utility, and town residents. (Thirty-five hundred oil-burning heating systems in town were shut off as a result.)

The Statoil refinery is connected to the Asnaes power plant in another way. After the water the refinery pulls into its system from the Great Belt is used as a coolant, it is sent along to Asnaes, where it serves as a coolant a second time. Statoil itself produces surplus gas as a consequence of the refining process. The gas was not used prior to 1991 because it contained excessive amounts of sulfur. The refinery has since installed a process to remove the sulfur, and the cleaner-burning gas is sold to Gyproc, the sheetrock factory, as well as to the coal-fired utility plant (saving thirty thousand tons of coal). The retrieved sulfur itself is sold to Kemira, a chemical company. At the Asnaes power plant the process that removes the sulfur in the smokestacks also yields calcium sulfate, which in turn is sold to Gyproc as a substitute for mined gypsum. Fly ash from coal

322

generation is used for road construction and concrete production. Waste heat from the refinery is used to warm the waters of a fish farm that now produces 250 tons of turbot and trout annually. (The fish grow more rapidly in warmer water.) Fish sludge goes to local farmers as a natural fertilizer.

Meanwhile, Novo Nordisk, an internationally known creator of insulin and enzymes, has developed a process to use the seven hundred thousand tons of a thin, nitrogen-rich slurry it previously dumped into the fjord and thus into the Great Belt. The slurry is now pumped free to local farmers who, after the addition of chalk-lime and processing at nine degrees centigrade for an hour to kill any remaining microorganisms, use it as a fertilizer. The farmers in turn grow biomass for Novo Nordisk's fermentation vats. Yeast cake from the vats goes back to the farmers as food for their hogs.[1]

The students detailed all this and then drew an analogy wiser than they knew. They likened Kalundborg's industrial ecology to food webs in nature. They were correct. Kalundborg had unwittingly heeded the basic earth principle that "waste equals food."[2] By being a *closed system*, nature recycles everything in such a way as to contribute to further cycles. Robert Frenay, who with Paul Hawken reports the case of Kalundborg, quotes the poet George Meredith at this juncture: "Earth knows no desolation. She smells regeneration in the moist breath of decay."[3] Said differently, nature's own blueprints go beyond linear thinking and cradle-to-grave designs.[4] Nature uses cradle-to-cradle designs instead.[5] The "moist breath of desolation" smells of "regeneration." The students didn't state it that way, but describing nature's food webs is much the same.

Sustainability, if it happens, will issue only from careful listening to nature and mimicking its basic design strategies. Learning from nature to mimic its design strategies is the first answer to the question of what on earth is to be done.

Yet this begs the real question: *Which* nature and how is it *understood*?

Ask the Danes. The industries of Kalundborg did not set out to mimic nature. Their actions simply found it beneficial to the interests of all to think ecologically (sideways, around corners, cyclically, and in spirals) about the community, its economic livelihood, its land, fjord, and sea. The students, in their momentary fit of genius, simply showed the industries and town how close to the rest of nature's webs their own had evolved.

Differently said, Kalundborg half-stumbled onto the basic requisite for sus-

---

1. This example is a composite of information provided by Paul Hawken, *The Ecology of Commerce: A Declaration of Sustainability* (San Francisco: Harper Business, 1993), 62–63, and Robert Frenay, "Biorealism: Reading Nature's Blueprints," *Audubon* 97, no. 5 (May 1995): 75–76.

2. See above, the chapter entitled "The Big Economy and the Great Economy."

3. Frenay, "Biorealism," 74.

4. Cradle-to-grave analysis is increasingly popular in ecosensitive and cost-conscious production circles. By carefully tracking the environmental consequences of production processes from initial acquisition of resources to end use, it results in cleaner production and often lower costs. What we often call "end use" isn't "end use" at all for nature, and product design needs to mimic nature in such a way that this use in turn is part not of a "grave" but of another "cradle."

5. Frenay, "Biorealism," 74.

tainability: understanding creation, or nature, as a genuine community and aligning human configurations to the rest of it. More precisely, the Danes *acted* as though nature *is* a community, whether they understood its detail or not.

On one level this first requisite for sustainability is simple. "Comm-unity" is nature's way. All that exists, coexists. Yet the West, in the grip of a deadly combination generated in recent centuries, now globalized, has not understood this at all. Its confidence in humans as a species apart (some humans far more than others!); its confidence in docetic, "ungrounded," denatured, gnostic reason; and its confidence in earth-oblivious economic messianism as a transforming power for good — these, armed with multiple technologies, have all ripped open the seams of earth, left it bleeding, panting for breath, and exhausted.

This combination has been so powerful as a globalizing culture, a set of institutions, and a way of doing things that we must pause to ask yet again and in more detail what kind of alternative knowledge and socialization are required. The place to begin is the one just proposed — learning from nature.

COMMUNITY

The basic premise for future actions and outlook is the simple sentence above: all that exists, coexists. Community rests at the heart of things. The dance of reality is "a permanent dance of energy and elements" in a "vast communitarian chain" that embraces the entire cosmos.[6] We are consigned to the wonder of a universe whose tapestry is whole.

On one level this is only to acknowledge the shift in science from the mechanistic to the relational as the lead understanding of natural systems. On another level it is to reaffirm the more daring conviction of the doctrine of creation in numerous religious communions: namely, that creation is a community in which the whole and its parts bear an integral dynamism and spirit, both of which are expressions of divine creativity. An interiority inhabits living materiality. By either understanding, scientific or religious, appropriate actions, policies, institutions, and neighborhoods — like Kalundborg — begin with a recognition that comm-unity is nature's way.

A sacramental reservation about messing with natural systems is one response to this, and a proper one. *Sacramentum* is the Latin for the Greek word *mysterion*. *Mysterion* doesn't denote incomprehensibility so much as a profundity that surpasses our words for it. Such reservation, with its sense of a comprehensive spiritual union of earthly elements (the sacramental), is a clue for the "new sort of science and technology governed by a new sort of economics and politics" that Charles Birch called for at the Nairobi assembly of the WCC in 1975.[7] A profound conservatism toward all that makes for life is another way to express

---

6. Leonardo Boff, *Ecology and Liberation* (Maryknoll, N.Y.: Orbis Books, 1995), 40, 36.

7. This excerpt from Birch's address is cited by Dieter T. Hessel, "Where Were/Are the Churches in the Environmental Movement?" *Theology and Public Policy* 7, no. 1 (summer 1995): 21–22.

it. A fearsome respect for creation's integrity, and earth as a slow womb, is yet a third. To tear slow-growing community fiber is a grave infliction — this is the conclusion.

Essential to this is the understanding that no clear line between life and not-life exists. Not when the galactic story, the solar system story, the earth story, the life story, and the human story weave and bind as one integral story.

Try breathing, for example. It only works because of the one-time furnaces of now-dead stars. They were the sources of gases and processes crucial to the eventual formation of oxygen. For that matter *all* the atoms in our bodies, with the exception of primordial hydrogen itself, were produced in supernova explosions. The atoms of the first generation of stars were thrown into space and cycled and recycled into new stars, planets, and eventually living creatures. All of us, all creatures past, present, and future, began in stardust and evolved in the transformations of the universe on its long pilgrimage to date.

Even now, as you read this, the intricate togetherness of things doesn't allow a strong line between life and not-life. If, for example, you were to watch the night side of earth from a satellite's orbit you would see better than one hundred flashes of lightning per second. They are part of the grand nutrient cycles of nature. Lightning annually converts more than three million metric tons of atmospheric nitrogen to nitrogen dioxide. Forest fires set by the lightning release additional nitrogen. Decomposing plant and animal matter contribute more. Rain carries the nitrogen back to water and soil where, after passing through microbial stages, it is taken up again by plants and then animals. On this elaborate journey, nitrogen separates and recombines with hydrogen, oxygen, and carbon as virtually all life-forms absorb and use it and pass it along in altered form. There are innumerable smaller cycles within larger cycles here, but there is no separating life and not-life in the being of it all. All is a part of the other. You, I, and other creatures live because lightning is part of nature's nutrient cycle.[8] If lightning were not "alive," we wouldn't be either.

If we turn again to religion, we note that the perennial wisdom of most religious traditions has typically affirmed that creation is the Great Community, just as they have professed this comprehensive community as the basic referent for our lives and all others. We are kin to all else because we share a common origin in divine creativity. We share an ongoing journey as *creatio continua* as well. And not least, we share a common destiny in the destiny of the universe itself.[9]

This religious and scientific understanding of nature as both the aboriginal and comprehensive community has meaning for our basic disposition as well as a new science and technology governed by a new economics and politics. For starters, it means that if the galaxies are subsystems of the universe, our solar system a subsystem of our galaxy, earth a subsystem of our solar system, life a subsystem of earth, and we a subsystem of earth-life, then we belong here.

---

8. Example taken from Frenay, "Biorealism," 74.

9. See Denis Edwards, *Jesus the Wisdom of God: An Ecological Theology* (Maryknoll, N.Y.: Orbis Books, 1995), 143.

We are at home here. Our lives, their excitement, and their fulfillment are here. We are most ourselves when we are most intimate with the rivers, mountains, forests, meadows, sun, moon, stars, air, soil, rocks, otherkind, and humankind.[10] This and no other is our own primordial community.[11]

This primordial community is a home without an exit. The science, technology, economics, and politics of the industrial revolution never understood this. Most practices to this day still do not or choose to ignore it, probably because it means sure revolutions.

How efficient and realistic is it, for example, to cut down a slow-growing, century-old tree in the Tongass forest of Alaska, sell it for the price of a pizza, ship it off to Japan, with the help of fossil fuels render it there as little bags for snacks, ship these to the United States for sale, then dispose of the bags in a landfill by way of an elaborate, expensive, fossil-fuel burning garbage collection system or, at best, an elaborate, expensive recycling program? If nothing goes away, how realistic and efficient is it to use twenty-six thousand pounds of PVC (polyvinyl chloride, a plastic) annually in Germany for plastic soles and sneakers alone and then add lead as a stabilizer for the PVC? The lead dust from the shoes is carried by rain into the sewers where, with additions from elsewhere, it impedes recycling the sewage sludge for agriculture. The "efficient" and "realistic" linear, industrial solution is to add a treatment plant to remove the lead. Another kind of solution avoids all this with a shoe design that doesn't use PVC and lead, isn't more expensive, and is environmentally benign. But that requires learning how the complex cycles of nature work and designing industrial systems in tune with nature's economy.[12]

How efficient and realistic is it to use petroleum-based ink, which is a hazardous-waste ink, instead of soybean ink, which, though presently slightly more expensive, yields better color and quality, is nontoxic, is not hostage to battles over oil supplies, and derives from a renewable resource?[13] How efficient and realistic is it to continue with what Sim van der Ryn and Stuart Cowan dub "dumb design": that is, design that never asks what the health of ecosystems and human communities require and that results in horrendous waste and injustice? An average house in the United States, for example, uses between 150 and 200 gallons of water per inhabitant per day, and all the water, no matter what its use, exits in water and sewage systems that are coupled in series. Thus we often literally defecate in our water systems in the name of personal hygiene! The average home also produces between 2.5 and 5 tons of garbage per person per year, often with fiber, plastics, paper, wood, glass, and metal tossed into the same

10. Thomas Berry, "The Meadow across the Creek," (unpublished manuscript), 75.

11. Part of the "Desiderata" found in St. Paul's Church, Baltimore, and dated 1692, reads as follows: "Beyond a wholesome discipline be gentle with yourself. You are a child of the universe no less than the trees and stars. You have a right to be here.... Be at peace with God, and whatever your labours and aspirations in the noisy confusion of life, keep peace with your soul. With all its sham and drudgery and broken dreams, it is still a beautiful world."

12. Examples from Frenay, "Biorealism," 70–71, 78.

13. See the discussion in David Wann, *Deep Design: Pathways to a Livable Future* (Washington, D.C.: Island Press, 1996), 63–66.

trash bins. This is design, and these are routine practices, with no thought for their connections to the living systems from which they draw and upon which they depend.[14] It's apartheid design.

Why not find out, to make the point from another tack, what property allows a fly's wing to beat hundreds of times per second without breaking? Tom Eisner and others discovered that fly and dragonfly wing hinges are made of the most perfect rubber known. Or what gives the dragline silk of the golden-silk spider a tensile strength twenty times that of steel and a capacity at the same time to stretch and rebound from 20 percent of its original strength? Biophysicist Lynn Jelinski thinks genetically altered plants might mass produce a spiderlike silk of this kind for any number of uses, from sails to bridges. But it requires understanding spider-silk as well as plants. Michael Braungart notes that the colors of many birds, from green to purple, are produced without blue pigment. The blue of a jay, for example, is created by refractions of light resulting from prismatic structures in the feathers. Since blue fabric dyes are generally highly toxic and toxins are by definition that which nature can only recycle for ill and never good, why not develop fabrics and finishes by using optical properties rather than toxic pigments?[15] Or why not design farming as Braungart and the Hamburg Environmental Institute did on the basis of natural cycles around a series of ponds? The ponds process local sewage and waste by growing aquatic plants and algae. The plants draw off excess nutrients and leave the water a nutritious and safe habitat for fish. The plants are harvested as livestock food or are used with waste from the livestock as fertilizer for nearby fields. Pigs feed on vegetables and snails that grow at the site. Ducks and geese consume algae and provide nutrients for the fish. Water purified by the process flows back into the local watershed. All this arises from knowing local ecosystems and earth as a closed sphere.[16] The Hamburg model is presently being adapted and pursued at sites in Brazil, China, India, Thailand, and Vietnam. The contrast with, say, pig factory-farms in North Carolina or Iowa, cattle ranching in the Amazon, or vegetable farming in the great central valley of California could hardly be more dramatic for earth and community.

As we have seen, industrial ecology can do the same kind of listening to earth as a closed community system. That is the point of Kalundborg. The revolutionary trick is to convert the ingrained linear processes of industrial systems into cyclical and spiral ones that mimic nature as a community.

The same can be done with residential communities. Haymount is now underway as a new community blending traditional and high-technology features appropriate to its natural setting in Caroline County, Virginia. Twenty minutes from Fredericksburg and an hour from Washington, D.C., it will be a walkable, mixed-use town that will retain its woods, wetlands, open meadows, and river. There will be a light-rail commuter stop, an organic farm and farmer's market,

---

14. See the discussion in Sim van der Ryn and Stuart Cowan, *Ecological Design* (Washington, D.C.: Island Press, 1996), 7–15.

15. Examples from Frenay, "Biorealism," 72.

16. Example from ibid., 74.

commercial and light industrial space, parks, churches, a school and a college, as well as four thousand residences, all constructed in accord with the materials of the region and adapted to solar energy. (The life-cycle of materials factors into their selection and use in accord with a general policy of "biotechnical" innovation and adaptation.) The design in fact includes lessons from the environmental history of the region, ancient human settlement included, and fosters systematic restoration of damaged ecosystems. Assuming successful implementation, the area will be more viable ecologically after the town is finished than it was before.[17]

PATHS NOT TAKEN

Being explicit about actions, structures, and paths *not* taken can also clarify which nature is to be mimicked. When the number and cumulative impact of human beings are what they now are, what we do *not* do, or refrain from doing any longer, is as important to sustainability as what we do. The deceptively simple first law of ethics, "Do no harm," is still good advice.

Sustainable development as green globalism is rejected here. It is rejected despite its standing as a sincere attempt at good stewardship of ecumenical earth. Our argument is not with goals or motives. Most sustainable-development goals are, of themselves, proper guides for actions and policies: maintaining the integrity of ecological systems in the process of meeting basic human needs, thus integrating resource development with quality of life advances; observing cultural diversity and promoting local participation and empowerment; pursuing social justice and equity as essential to sustainability itself. Rather, green globalism is rejected because its starting point, framework, and means are fatally askew. Systems and practices largely determine our lives, and the systems and practices of even green globalism are massively faulty. Wrapping the environment around a globalizing economy as the centerpiece of sustainability is the extension of a course with deep roots in earth-destructive modernity, rather than the needed path not yet taken. The roots are institutional (corporations and nation-states) as well as sociopsychological, cultural, and epistemological (humans as a set-apart manager-engineer-entrepreneur-consumer species).

Differently put, spaceship economics and planetary management violate the pluralism of place and the integrity of ecosystems integral to nature's own complex functioning as a community. "Singular man" stands erect vis-à-vis "singular nature,"[18] perpetuating the abstractions of nature/humanity apartheid even when the language mimics that of community empowerment. Thus sustainable development's correct realization that we genuinely belong to the cosmos and that nature is a closed, curved system is arrogantly twisted into a scheme where, finally, some know what is in the interest of the rest. Soon the tilt is toward

---

17. More details about Haymount are available in Wann, *Deep Design*, 121–29.
18. Raymond Williams, *Keywords: A Vocabulary of Culture and Society* (New York: Oxford University Press, 1983), 188–89, 220–21.

yet another master theory, a transcendent technology and a comprehensive set of global institutions and practices. Things are no longer soul-size, with multiple voices attuned to the complexity of things on the ground in places very different from one another. Rather, a loose ensemble of free-trade agreements, planet-spanning information technologies, and the integration of financial markets erases borders and invades communities while uniting the world into a single brutal, lucrative marketplace where all is game and booty. Here, in these global management schemes, the chains of responsibility and accountability are too long and too distant. High levels of participation on the part of all affected do not structure basic decisions, and cannot. Nor is there even much recognition that the size of governing forces, whether political, economic, or cultural, needs to be trimmed back to match the limited talent available in the ordinary flawed human beings of whom even the most impressive social orders are composed. Sustainable development as green globalism on the model of spaceship earth and planetary capitalism is, then, wittingly or not, but the latest version of imperial hubris. It treats Seattle, Boston, Madras, Rio, and Kuala Lumpur as though they were much the same, or should be. It may be very well intended. It is also very wrong, both for different renditions of different *Homo sapiens* in different places and for the rest of wildly variegated nature.

David Korten captures the globalized view nicely by citing a 1993 open letter of Akio Morita, founder and chairman of the Sony Corporation, to heads of state in North America, Europe, and Japan. In it Morita says it is time for all local interests, including local culture, to give way to the larger global good that free-market exchange creates. All economic barriers should be lowered "to begin creating the nucleus of a new world economic order that would include a harmonized world business system with agreed rules and procedures that transcend national boundaries."[19] After showing how such a system is itself more than an economic one, but instead a cultural configuration and an extraordinary concentration of effective governing power, Korten offers his own summary of "the ideal world of the global dreamers":

- The world's money, technology, and markets are controlled and managed by gigantic global corporations;

- A common consumer culture unifies all people in a shared quest for material gratification;

- There is perfect global competition among workers and localities to offer their services to investors at the most advantageous terms;

- Corporations are free to act solely on the basis of profitability without regard to national or local consequences;

---

19. Korten is citing Akio Morita's letter, "Toward a New World Economic Order," *Atlantic Monthly* (June 1993): 88; see David C. Korten, *When Corporations Rule the World* (San Francisco: Berret-Koehler; West Hartford, Conn.: Kumarian Press, 1996), 122.

- Relationships, both individual and corporate, are defined entirely by the market; and

- There are no loyalties to place and community.[20]

Such is the economic centerpiece of a globalized economy, even when "greened" as sustainable development.

Granted, green globalism is better than some other hues, and sustainable development in this scheme is not an unmitigated disaster. It is preferable to the brown globalism of cowboy economics in a freewheeling GATT/WTO version. The political economy of present dominion theology and casino capitalism is even further from effective community than is sustainable development, with the latter's newfound themes of participation, empowerment, and social justice. From the point of view of earth, brown globalism,[21] which is essentially a revived Social Darwinist earth faith and ethic, *is* an unmitigated disaster. Green globalism is a mitigated one.

Take the case of Nauru, that raw, living parable of present dynamics.

Nauru is the world's smallest, most isolated republic, a place "close to nowhere" in the water continent of the western Pacific. Despite its size of eight square miles and location hundreds of miles from the nearest neighbor, the modern world found it. And its seventy-five hundred inhabitants are, as a result of global economic dynamics, rich — at least on paper. The phosphate mines that are the island have brought in tens of millions of dollars each year.

Other than its rich, isolated populace, Nauru's distinction is its devastation. So much of the island has been strip-mined that Naurans are abandoning their depleted home. The name the first Europeans gave it because of its luxuriant

---

20. Korten, *When Corporations Rule the World*, 131.

21. The tenor of brown globalism echoes in the series of advertisements run by Mobil Oil in major newspapers in 1995. In language that matches Schumacher's "forward stampede people" and has no place for "homecomers," the first of the series titled "A Global Vision" begins this way: "As one of the world's largest international oil companies, our vision is simple: To be a *great* global company. And that means seizing opportunities with good economics wherever they occur." The advertisement goes on to say that "with the current world surplus in energy," the difficulty is culling "great" projects from "the basket-load of 'good' projects." Mobil seeks to do this culling. In the process it brings "the synergies of long experience and worldwide operations" to "some areas of the world that are opening up." A series of "partnerships" designed "to assist governments with their economic development plans" is the scheme. Examples are then offered, some "colossal in scale," like the new $20 billion-plus partnership with Qatar for liquefied natural gas. Others have been "enjoyed for years . . . because of a shared vision that host governments have supported with attractive commercial terms."

The only identification of people in this "global vision" is in the words "we" (Mobil), "governments," and "consumers and producers." There is no reference whatsoever to the environment. All the language is of cowboy global economics: getting the hydrocarbons of "resource-rich nations" to "energy-hungry markets elsewhere," "expanding a Mobil presence that dates back in some countries more than a century," or, as noted, newly moving into "some areas of the world that are opening up." Even the complaint is the cowboy complaint of tied hands, too little freedom to venture and "open up the world," and unnecessary imposed limitations: "The pursuit of energy opportunities overseas is . . . needed because potentially rich domestic resources [in the U.S.] — onshore and offshore — remain off limits to the industry." The foregoing is taken from "A Global Vision," an advertisement run on the op-ed page of the *New York Times*, August 10, 1995.

vegetation — Pleasant Island — is now a local joke. Four-fifths of the island is a moonscape of gray limestone pinnacles, some as high as seventy-five feet, while the only inhabitable strip is a coastal fringe of coconut palms and beaches.

Because of the new economy, Naurans no longer farm. So they import everything. And while they are among the world's most affluent peoples, their life expectancy is actually declining. (Presently fifty years for men, fifty-five for women.) It seems that diabetes, high blood pressure, and obesity have come with the diet of fatty, imported food. (A traditional diet of fresh fish, fruit, and vegetables has been replaced by imported canned and frozen foods, much of it attractive "convenience" foods.)

Moreover, the weather has changed. The waves of heat that rise from the mined-out plateau drive away the rain clouds. This leaves the formerly lush island plagued by constant drought. The water taps of these rich entrepreneurs now run dry more than half the time.

Naurans do have property, however. The millions received from mining have purchased office towers, hotels, and golf courses in Australia, Hawaii, and Guam. And somewhere between five hundred million and one billion dollars have been invested in trust funds under Australian and U.S. direction.

There have been losses, of course. Two million dollars were lost in a failed London musical based on the life of Leonardo da Vinci. At least twelve million dollars were lost in an investment scam engineered by eight Australians. Sixty million dollars were used by Australians to buy bogus letters of credit and bank notes (happily, forty-eight million dollars of this has been recovered).

One difficulty is the loss of employable skills. As noted, Naurans stopped farming when they joined the export economy. Now they cannot farm, and thousands of workers from China, India, and the Philippines do the manual labor Naurans either choose not to do or are no longer able to.

Another loss is cultural. Nauran cultural traditions have been basically wiped out in the course of a series of historical events: German occupation in 1888; Australian takeover in World War I and control until independence in 1968 (with the exception of Japanese occupation in World War II); and the further incursion of the modern world via mining, commerce, and satellite media since 1968.

Presently two paths are being contemplated by Naurans remaining in the republic. One plan is to use income from investments to knock down the lifeless limestone coral pinnacles and then lay down enough topsoil to coax mango, breadfruit, and pandanus trees to grow again. Environmentalists say, however, that it is unlikely enough food can be grown to sustain the population. The other plan is to abandon Nauru. The proposal is to use the money to buy a new island home from one of Nauru's Pacific neighbors. "It would be very sad to leave our native island," Minister Aingimea commented to the *New York Times* reporter. "But what else can we do? The land of our ancestors has been destroyed."[22] Earth is no longer *oikos* for Naurans. They have joined the homeless

---

22. Both Aingimea's words and this account as a whole are from "A Pacific Island Nation Is Stripped of Everything," *New York Times*, December 10, 1995, 3.

rich. That may be better than joining the homeless poor, but both groups of homeless share the same aching question. That question was put to a *New York Times* reporter who was covering the movement of Myanmaran refugees. One of the refugees asked: "What I want to know is, does anybody have a plan for people like me, because I want to stay in one place peacefully, growing my garden and living without harm. Is there any place like this?"[23] Better or worse, the lot of neither the Naurans nor these (other) refugees spells sustainability.

In brief, and with this parable and others on the way in mind, what is rejected here as the framework and direction for policy is the spectrum usually identified as *the* range of choice: namely, globalizing capitalism as more ecoqualified or less ecoqualified. What is affirmed is quite another institutional, cultural, and sociopsychological orientation. It turns on community economics, politics, science, and technology, with community understood at various levels from local to regional to transregional and global, but *all* of them within the orbit of nature as *the* primordial, comprehensive, and closed community. This means "day-care" political economy with its margins of safety and room for maneuverability, as well as space for both freedom and surprise. Elasticity and adaptability are prized, within the clear limits of nature's offerings. Community scale here runs to smaller rather than larger, even when there are lots of kids and the overall population is high.

A SKETCH, MORE EXAMPLES, AND A GUIDE

But what are "community scale" and the proposed political economy?

Gregory Bateson enjoyed telling of an exchange at New College, Oxford. The main hall at New College was built in the 1600s using oak beams forty feet long and two feet thick. They eventually suffered from dry rot, and oaks large enough to replace them were not known to the college administrator. It was suggested that the administrator inquire with the college forester whether some of the lands given to Oxford might have trees adequate for the renovation. The forester replied, "We've been wondering when you would ask the question. When the present building was constructed three hundred fifty years ago, the architects specified that a grove of trees be planted and maintained to replace the beams in the ceiling when they suffered from dry rot." Bateson's comment: "That's the way to run a culture."[24]

Like the Kwaakiutl people and their logging practices,[25] many peoples and

---

23. "Exiles Adrift: Nowhere to Run, Nowhere to Hide," *New York Times*, February 23, 1996, A4.

24. As passed along by Frenay, "Biorealism," 106.

25. The Kwaakiutls of the Pacific Northwest practiced a form of logging that took wood from living yew trees. A cut tree was considered "killed," but a standing tree could be "begged from." "Begging" meant notching the trunk of the yew in two places, in accord with the length of the board desired. Wedges were then pounded in from the sides, and a lever was used to split off the "board" from the rest of the tree without killing it. Van der Ryn and Cowan, who report this in *Ecological Design,* 59–60, comment that the difference of "begging" from standing trees that grow for another generation's needs and the present clear-cutting of Pacific Northwest old-growth forests

societies *have* had this sense of responsibility for future generations and a way of life that takes actions such as these. The fabled "seventh-generation" test for actions, policies, buildings, and institutions themselves *is* "the way to run a culture," even if those of the seventh generation should decide the beams in the main hall might not be the best use of the replacement oaks.

In passing, let it be underscored that there is nothing humanly unconstitutional about such future-oriented action. The issue is cultural, not genetic. While it is true that "the present shouts while the future only whispers," and we are often given to shortsightedness, it is also the case that human societies can identify, and have identified, with earth and future generations of both our own kind and otherkind. And we can think cyclically as well as sideways, even when we know that nature's cycles are dynamic rather than screwed tightly. Future thinking is a behavior human beings *have* practiced, and often when it was most needed (under conditions of relative scarcity and the forced requirements of a simplified way of life).

Beyond the seventh-generation test, what else enables a community to last? Wendell Berry once suggested seventeen steps. They're a good start. Among other things, Berry marked out the level of responsibility, investment, and participation we want to emphasize here, together with a strong sense of place. The question What on earth ought to be done? has answers in Berry's "17 Sensible Steps":

1. Always ask of any proposed change or innovation: What will this do to our community? How will this affect our common wealth?

2. Always include local nature — the land, the water, the air, the native creatures — within the membership of the community.

3. Always ask how local needs might be supplied from local sources, including the mutual help of neighbors.

4. Always supply local needs *first*. (And only then think of exporting their products, first to nearby cities, and then to others.)

5. Understand the unsoundness of the industrial doctrine of "labor saving" if that implies poor work, unemployment, or any kind of pollution or contamination.

6. Develop properly scaled value-adding industries for local products to ensure that the community does not become merely a colony of the national or global economy.

7. Develop small-scale industries and businesses to support the local farm and/or forest economy.

8. Strive to produce as much of the community's own energy as possible.

is the difference "between a fundamentally sustainable culture and a fundamentally unsustainable one" (60).

9. Strive to increase earnings (in whatever form) within the community and decrease expenditures outside the community.

10. Make sure that money paid into the local economy circulates within the community for as long as possible before it is paid out.

11. Make the community able to invest in itself by maintaining its properties, keeping itself clean (without dirtying some other place), caring for its old people, teaching its children.

12. See that the old and the young take care of one another. The young must learn from the old, not necessarily and not always in school. There must be no institutionalized "child care" and "homes for the aged." The community knows and remembers itself by the association of old and young.

13. Account for costs now conventionally hidden or "externalized." Whenever possible, these costs must be debited against monetary income.

14. Look into the possible uses of local currency, community-funded loan programs, systems of barter, and the like.

15. Always be aware of the economic value of neighborly acts. In our time the costs of living are greatly increased by the loss of neighborhood, leaving people to face their calamities alone.

16. A rural community should always be acquainted with, and complexly connected with, community-minded people in nearby towns and cities.

17. A sustainable rural community will be dependent on urban consumers loyal to local products. Therefore, we are talking about an economy that will always be more cooperative than competitive.[26]

Berry's own community is rural. The seventeen steps reflect his participation in furthering sustainability at home in Kentucky. Nonetheless, much here is also the gathering wisdom of neighborhood-oriented, community-organizing groups in large cities. The Danes of Kalundborg, both rural and urban, would understand Berry's provincialism, at least in part (their strong dependence on international trade would bend it somewhat). Certainly Common Bread does.

Common Bread is part of the Community Supported Agriculture and Subscription Farming movement. It works at rural/urban intersections. The scheme is cooperation between a grower and a community of citizens in a nearby city who purchase shares in a community farm. The shareholders hire the farmer/gardener (and sometimes the land) and may themselves assist with planting, cultivating, and harvesting. A core group also provides administration and helps with distribution of the produce. Each share entitles the holder to a given volume of fresh, organically grown produce during the growing season and for winter storage.

---

26. Wendell Berry, "Community in 17 Sensible Steps," in *Another Turn of the Crank* (Washington, D.C.: Counterpoint, 1995), 19–21.

The movement, presently a conscious network in the United States, Japan, and Europe, is a response to many things at once: a growing concern for the safety and nutritional value of food purchased supermarket style (we often don't know where our food comes from, how it was grown, how it has been altered, or the human and environmental costs of producing it); the waste and other costs entailed in obtaining food that has often traveled thousands of miles to get from farm to table; the difficulty of farmers, especially small farmers, in making a living without a support system; the ability of the land to meet the needs of future generations (thus the need for conscientious interest and participation in its present welfare on the part of people who hold a stake other than market profit); and the need of urban dwellers to be people of the land and responsible toward soils and farmers upon which and whom they are dependent for life.[27]

This approach has spin-offs and makes connections. It tends to think sideways and cyclically. Common Bread Restaurant and Bakery in South Minneapolis purchases from a community-supported subscription farm thirty miles away. It shares its philosophy. The menu jacket includes this:

At Common Bread we are trying to reduce the distance that food has to travel to reach your plate. We have a commitment to seeking out and purchasing directly from local growers whenever possible. We use foods in season to the extent possible, even going so far as to visit local farms to learn first hand the challenges of growing vegetables, fruits, and poultry in our northern climate.

We live in one of the most abundant agricultural states in this country, yet the average food we eat travels more than 1300 miles to reach our table. For the most part we do not know where our food comes from, how it is grown, and how it has been altered if it has been processed.

We believe that purchasing locally has many benefits. Reducing the distance foods travel guarantees the freshest food possible. Many local varieties are specific to our climate and are consequently better tasting and have a higher nutritional value. Purchasing locally also reduces the use of fossil fuels needed to transport food and we support a diverse and vital local farm economy.

Common Bread is located in one of the poorest, most physically degraded sections of South Minneapolis. It chose that location as part of its commitment. The commitment was not only to wholesome food but to employment and training of local unemployed people and provision of a community-making place for neighborhood groups to meet. (The restaurant's program, which is more than its menu, designed this from the start. Breaking bread together is evidently more than eating, even eating together.)

---

27. This description and rationale are taken from a flyer entitled "Community Supported Agriculture and Subscription Farming," as provided by Daniel Guenther, a farmer-gardener working near Minneapolis.

Here is another instance of a creation patch as a community patch: a rural/ urban farm and an inner-city restaurant, with high local participation and self-conscious concern for sustainability in both countryside and city, and both together. It is an instance of place-based conscious community.

Does Common Bread suffice as an example of sustainability and the meaning of earth community? For that matter, does Kalundborg? No. If we multiplied them ad infinitum, we would still have earth's distress, though considerably less of it. Kalundborg has impressively walked down the Business Council for Sustainable Development's path of eco-efficiency as the second industrial revolution. But like the rest of Denmark, it has not yet dislodged the high levels of consumption of rich nations, or weaned itself from fossil fuels and their contributions to global warming, or disengaged from a system of global trade and finance that promotes affluence in some sectors while generating poverty in others. Until those moves are made and appropriated carrying capacity altered, sustainability will not be Denmark's, Europe's, Latin America's, or the rest of earth's. Kalundborg is only phase 1 in the long transition to sustainability. Like turning an ocean liner around (ships again!), this will take a long time and require lots of room.

For its part, Common Bread does most all the right things. But it will likely not "make it." The support base is small; the enterprise is vulnerable; and most precarious of all, it bucks the system, receiving little support from it. (Its base is a single farm, a modest number of somewhat transient subscribers, a restaurant in a blighted area with little money and unlikely to draw clientele from "outside," and a program that runs on largely volunteer labor; it thus suffers from a lack of systemic supports within an economy that favors agriculture as industry and restaurant chains in well-traveled corridors.)

Yet both Kalundborg and Common Bread, like Berry, know where sustainable community begins: with a cultivated sense of community responsibility, high levels of participation, and an ecological mode of thinking. Continuing this sense of community participation, responsibility, and ecological thinking, what more needs be said?

There is a guide for sustainable-community scale. It is one of the moral norms of sustainability itself: namely, subsidiarity.

Subsidiarity is the means of participation and accountability best tuned to the pluralism of place and the scale most likely to be responsible. We consider it in some detail.

Subsidiarity has a long tradition in Christian ethics. One of the more recent formulations is a papal one: "Just as it is gravely wrong to take from individuals what they can accomplish by their own initiative and industry and give it to the community, so also it is an injustice and at the same time a grave evil and disturbance of right order to assign to a greater and higher association what lesser and subordinate organizations can do."[28] Translated, this means that what can be accomplished on a smaller scale at close range by high participation with

---

28. Pope Pius XI, *Quadragesimo anno* (1931), as cited in *The Westminster Dictionary of Chris-*

available resources should not be given over to, or allowed to be taken over by, larger and more distant organizations. Do not transfer to supposedly higher and larger collectivities what can be provided and performed by (allegedly) "lesser" and "subordinate" ones.

The key is *appropriate* scale and action. For sustainable community, that may in fact mean actions, policies, and institutions that are *more global.* Rectifying ozone damage or reducing greenhouse gases means international treaties with authority ceded from local and regional bodies, for example. Massive waves of refugees and internally displaced persons, resulting from conditions of "overshoot,"[29] also require broader response than local resources can provide. The protection of marine ecosystems and fish populations cannot be done without transnational cooperation. Nor can much control of pollution or any other threats to a shared atmosphere. Oceans, genetic diversity, climate, the ozone layer, and even forests and other great concentrations of green plant matter form a kind of global commons that must be treated as such. In a small and contracting world, global community requires some institutions and policies with genuinely global reach.

But subsidiarity also means massively deconstructing what is now globalized. Food, shelter, livelihood, and other needs that can be met on a community and regional basis, with indigenous resources, talent, and wisdom, should be met there, with firm commitments to Berry's "pluralism of place."[30] Even large cities can better relate to the bioregions in which they and surrounding areas are embedded. Vancouver is not Rio, despite the presence of striking mountains and sea for both.

A movement in Sweden called the Natural Step is giving a national-regional focus to subsidiarity and to deconstructing the global. Thus far, forty-nine local governments, the Swedish Farmers Federation, twenty-two major Swedish corporations, and some ten thousand professionals, business executives, farmers, restaurateurs, students, and government officials in sixteen networks are developing action plans to make Sweden a model of sustainability by achieving nearly 100 percent recycling of metals, eliminating compounds that do not break down naturally in the environment, maintaining biological diversity, and reducing energy levels as close as possible to those of sustainable solar capture.[31]

Most of the other explicit norms for sustainable community — participation, solidarity, sufficiency, material simplicity, spiritual richness, responsibility, and accountability — are also better served when subsidiarity is heeded. If converting linear processes (the industrial paradigm) to cyclical and spiral ones (nature's economy) is one key to sustainability, making "feedback" visible and close to home is another. Local and regional initiative and self-reliance tend to promote these, just as they tend to promote higher degrees of cooperation, mutual sup-

---

tian Ethics, ed. James F. Childress and John Macquarrie (Philadelphia: Westminster, 1986), 608. "Subsidiarity" is from the Latin *subsidium,* meaning "help."

29. See p. 81 above.

30. Recall Berry's "17 Sensible Steps" above.

31. The Sweden example is reported by Korten in *When Corporations Rule the World,* 298.

port, and collaborative problem solving. Not least, subsidiarity tends to preserve resources in the community as the "commons"[32] people depend upon, know best, and care most about.

Stephen Viederman gives particular emphasis to the role of human imagination and collaboration in arriving at sustainable community. That, too, is best served by subsidiarity. Viederman finds the transition to sustainability under present and foreseeable conditions fraught with staggering operational problems. Making the transition from present unsustainability to future sustainability cannot, he is certain, be well addressed either by comprehensive planning or the roulette of the market. "The first task" of sustainability, then, is "to provide the space and time for people to begin to envision the future they desire for their [own] communities, and to ensure access to power that will make it happen."[33] That "first task" is more likely accomplished on the scale for which subsidiarity is the guide.

An example of Viederman's point is the case of the Environmental Protection Agency (EPA) of the U.S. government under the leadership of William Ruckelshaus. Under the Clean Air Act of 1970, the EPA had the authority to decide the fate of a smelting plant that was both a major polluter and a major employer in Tacoma, Washington (the annual payroll was twenty-three million dollars). The issues were thus jobs, the local economy, and health. Instead of making the decision in Washington, D.C., Ruckelshaus took the matter to Tacoma. And there, instead of setting up only the required public hearings, he and other local officials initiated a series of public workshops. These included plant workers, union representatives, various other local citizens, and environmental groups. The same format was used for all meetings. It included education about plant emissions, incidences of disease, local economic implications of possible courses of actions, and so on, as well as time for prepared testimony and open deliberation. What the community eventually decided was not in the minds of Ruckelshaus, local EPA officials, or the citizens themselves when the process began. The collective decision was that Tacoma's economy needed to diversify and there were ways to do that, ways that included retraining present plant workers. The community members together had found a way to recast the keep-the-

---

32. A comment on the famous essay of Garrett Hardin, "The Tragedy of the Commons," is necessary here. Hardin's treatment is deceptive. Commons in preindustrial societies were generally well regulated by the communities to which they belonged and that depended upon them. Hardin's portrayal of the destruction of the commons goes like this. Each farmer discovers that it is in his/her self-interest to add a cow to his/her herd in the pasture (the commons) because he/she accrues all the economic benefits from the additional cow while sharing the costs (the common pasture). When many farmers act on this, the burden on the commons degrades it and may even destroy it as productive pasture. But this logic and practice are not that of the commons at all; rather, this example shows free-standing individuals treating the commons not as a commons but as an open-access system from which each might take as much as he/she can. A calculation of self-interest, with "nature" as essentially free goods and the market as the nexus, is the picture here. This is industrial-system-logic, not commons-logic. Commons-logic means community regulation of shared resources.

33. Stephen Viederman, "The Economics of Sustainability: Challenges," 17 (the paper is available from the Jessie Smith Noyes Foundation, New York City). Used with permission.

plant-open-or-close-it issue and decide what they wanted for their community. Ruckelshaus's later reflection on the process included a line from Thomas Jefferson: "If we think [the people] not enlightened enough to exercise their control with a wholesome discretion, the remedy is not to take it from them, but to inform their discretion." It was a lesson in leadership, community process, and imagination.[34]

Yet even this example of subsidiarity does not mean slavish localism as the standard, since subsidiarity asks not for the most local but for the most appropriate level of organization and response. "Most appropriate" includes the *morally* most appropriate level. Small as beautiful, and the local as the basic unit of the global, may locate the starting blocks, in Tacoma as elsewhere. But in a world of maldistributed resources and power, the local cannot be the only locus of responsible action, just as it is not the only place we meet and live together. Trade and other exchanges of resources, for example, are as necessary in their own way as is transnational cooperation to address global warming and regions suffering "overshoot." The necessary guideline is not "no trade" or "no markets" or even "minimal trade" and "minimal markets." The guideline is to minimize the *appropriation* of carrying capacity *from elsewhere*, thus risking other people's and otherkind's lives in the present and for the future. The guideline is also *restoration* of diminished carrying capacity and the empowerment of peoples whose resources have been diminished as a systemic feature of globalizing dynamics. *Oikos* economics of ecumenical earth uses subsidiarity as a key guide in the pursuit of a just order. What is *not* in view is the plantation and colony economics of "the big house" and a sea of shacks, with its savage inequalities, as the shape of household and habitat.

If there is something that should be added to past accounts of subsidiarity, it is lessons from listening to nature.

Sustainability guided by subsidiarity betters its chances, for example, by incorporating nature's resiliency into social systems. The practices of economy and society ought to be ordered in such a way as to be able to shift and adapt, like nature, to changing conditions. In nature, biodiversity is the mechanism by which adaptation to demanding changes occurs, the means by which nature is resilient in the face of often traumatic change. As such, it is the basic source of all future wealth and well-being.[35]

Paul Hawken's two principles of good social design, drawn from nature, would improve the odds in favor of greater resiliency. They express "day-care" thinking. Good design "changes the least number of elements to achieve the greatest result." In Morocco, a ten kilowatt wind turbine, which would supply

---

34. This example, including the citation from Jefferson, is taken from Ronald Heifetz, *Leadership without Easy Answers* (Cambridge, Mass.: Harvard University Press, 1994), 88–100. Readers may wish to see my chapter, "Shaping Community," which also uses this example, in a volume on essential practices for the twenty-first century as these are instructed by Jewish and Christian traditions. See Dorothy Bass, ed., *Practicing Our Faith: A Way of Life for a Searching People* (San Francisco: Jossey-Bass, 1997).

35. Hawken, *Ecology of Commerce*, 190.

only one American home with electric heat, pumps drinking water for a village of four thousand.[36] And good design "removes stress from a system rather than adding it."[37] Mexico's first wind farm emerged as part of an energy strategy to avoid paying $3.2 million for an electric power line in the Yucatan.[38] In short, easier and simpler is better. Easier and simpler, with room for error, favors sustainability.

The return to how nature works is an apt reminder that sustainability, and thus subsidiarity, necessarily involves the total earth-human process. It encompasses the requirements of sustainable environments, societies, livelihoods, economies, and ways of life. These must function as a phase of earth economics and community.[39] Creation as a community remains the first stipulation, just as the integrity of ecosystems remains the first value.

IN SUM

We do not lack examples of actions appropriate to sustainable community. A running list can easily add to those already cited. In New York City, Bernadette Kosar's Greening of Harlem project targets environmental and social needs in the same action. She gathers Harlem teenagers to reclaim vacant lots and portions of school grounds as neighborhood and school gardens. The project has attracted the participation of the elderly as well as children and is supported by nonprofit groups of all kinds. Kosar sees jobs in this movement as well, just as she sees "growing everything from food to herbs, everything from dyes to potpourris and tomato preserves."[40] Yet her larger point is the evolution of a healthier community on a scale that includes high levels of local participation and pride and often a change of collective character itself. "Nature doesn't make people wild, it doesn't make people uncivilized," she says. "But I think concrete, asphalt, and steel does. It makes people hard, it makes people cold, it makes people inhuman. We have to figure a way to bring nature back to our cities."[41] In Costa Rica, a dozen peasant farmers came together in 1988 to form the San Miguel Association for Conservation and Development. This was their attempt to keep local forests under the control of local communities. They managed to find incentives for local people to harvest and process wood sustainably and develop regional markets for their products. In a country with Central America's highest rate of deforestation at the hands of big companies that have been buying out small local landowners, this was a turnaround. The association has since developed nurseries for native tree stock and edu-

---

36. Wann, *Deep Design*, 196–97.
37. Hawken, *Ecology of Commerce*, 166.
38. Wann, *Deep Design*, 197.
39. See the discussion in the chapter "Conclusions" above.
40. Cited by Wann in *Deep Design*, 140.
41. Ibid.

cated neighboring villagers in sustainable forestry.[42] The Deccan Development Society of Andhra Pradesh, India, organizes *sangams,* communities of women in villages that work toward gender equity, establish credit programs, cultivate and use medicinal herbs, incorporate organic gardening techniques and multiple cropping into local agricultural practices, and plant trees.[43] The Land Resource Management on the Central Plateau in Burkina Faso has found a way for 240 marginal villages to transform ten thousand hectares of previously unproductive drylands in such a manner that the previous average family food deficit of 645 kilograms per year has turned into a 150 kilogram surplus. The key was combining systematic agricultural surveys with a process by which villagers covered the area with miniature dams and embankments and other small-scale irrigation devices they designed themselves, together with manure composting units they designed in order to fertilize previously unfertilized crops.[44] On a different scale and working other areas, Japan's Ministry of International Trade and Industry established the Institute of Innovative Technology with the explicit purpose of "undo[ing] the damage done to the earth over the past two centuries, since the industrial revolution." The institute is exploring alternatives to chlorofluocarbons (CFCs) and trying to develop biodegradable plastics, hydrogen-producing bacteria, carbon-dioxide scrubbers, and genetically engineered algae for higher efficiency photosynthesis. In Germany the Federal Association for Ecologically Conscious Management is developing integrated management systems in accord with six principles: quality throughout the life-cycle of a product; a creative workforce enhanced by a friendly environment that focuses on low noise, healthy food, good air quality, and ecologically oriented architecture and furnishings; employee morale, which improves when corporate goals include both environmental well-being and economics; profitability as increased by adopting cost-reducing ecological innovations and marketing environmentally friendly products; continuity and security understood as avoiding environmental liability risks as well as market risks from decreasing demand for damaging products; and loyalty, built up when employees believe in their company's goals.[45] Germany's "take-back" legislation is still another large-scale example of widespread applicability. Industries are, in effect, forced to use life-cycle analysis in design and production of their goods since products must easily be converted into future raw materials and their components have to be nontoxic to workers. The legislation also tends to end planned obsolescence because increased product life is more cost effective than constant collection and recycling.[46] The Emilia-Romagna district of northern Italy has added important considerations of scale to earth-friendly manufacturing. The results are astounding. Over twenty-five

42. See Aaron Sachs, *Eco-Justice: Linking Human Rights and the Environment* (Washington, D.C.: Worldwatch Institute, 1995), 31–33.
43. Ibid., 32.
44. Ibid., 50.
45. The examples from Japan and Germany are included among others in Wann's *Deep Design,* 43–46.
46. Ibid., 166–67.

years, decentralized "flexible manufacturing networks" have created twenty thousand new jobs, wage rates 175 percent above the Italian average, and a step-up from seventeenth in 1970 to second in 1995 in per capita income among Italy's twenty-one regions. These results were achieved chiefly by changing the size and shape of manufacturing. Now the region has more than 325,000 small firms, 90,000 of them in manufacturing. Ninety percent of these firms employ fewer than twenty-two persons each, with most manufacturing firms employing fewer than five persons. At the same time each tiny company can draw upon a large regional pool of highly trained artisans who have developed impressive technologies and can use the services of trade associations designed to keep the region's companies competitive. David Wann sums up the goals and norms of these varied efforts as "renewability, reversibility, equity, resilience, proximity, and precision."[47] It takes only a moment's reflection to recognize these as traits of sustainable community itself.

We do not lack examples, then. What we lack are the collective mind-set and policies that routinely institutionalize the basic principles of ecological design. These in place, our examples and more would quickly multiply as a critical mass. The general principles, crisply put, are as follows:

- *Solutions grow from place.* What does nature permit in a given locale? and What does nature help us do? are the right questions to pose for this attention to specific site possibilities and limitations. "Without the details nothing can be known, not a lily or a child," writes Diane Ackerman.[48]

- *Ecological accounting informs design.* Just as a full accounting of economic costs is a matter of conventional design now, so must a full accounting of ecological costs be, from resource depletion to pollution and habitat destruction to disposal and reuse.

- *Design should be done with nature.* What we earlier noted as "cradle-to-cradle" design is the going norm.

- *Everyone is a designer.* Good designs grow and evolve organically out of a process of communication that includes widespread and varied participation.

- *Nature should be made visible.* What van der Ryn and Cowan call "flush and forget" technologies do not build a sense of responsibility or nurture earth-oriented mindfulness. Keeping biological and other consequences of our actions in view and experiencing how they act back upon us at shorter rather than longer range foster responsibility and accountability.[49]

---

47. Ibid., 173. The example from Italy is from 172–73.
48. Cited by Elizabeth Hanson in a review of Diane Ackerman's *The Rarest of the Rare* (New York: Random House, 1996), no page given. Hanson's review, "Creature Features," is in the *New York Times Book Review*, February 18, 1996, 11.
49. This is the summary of design principles from van der Ryn and Cowan, *Ecological Design*, 54–56. The rest of their book is an elaboration of these, with multiple examples.

Were we to incorporate these principles into a summary of this whole chapter, the results would align with David Orr's four things always to keep in mind for the transition to sustainable community. They serve nicely as a closing.

First, people are finite and fallible. Our capacity to understand complexity and scale, much less manage it, is limited. Things too big and too complicated are a liability.

Second, a sustainable world can be put together only from the bottom up. The building units are communities, and communities of communities. The social capital of smaller networks enabling larger ones is critical.

Third, the crucial knowledge is knowledge that coevolves from culture and nature together in a given locale (remembering that "locale" in a crowded world full of life and under house arrest has different dimensions, from an immediate neighborhood to the oceans and the atmosphere).

Lastly, the true harvest for ongoing evolution is embedded in nature's own designs. Nature is not first of all a big bank of resources standing at the ready; it is the source and model for the very designs we must draw upon in order to address the problems we face. Our design epistemologies must be compatible with the rest of nature's.[50] Learning to fit into nature's patterns is architectonic of all else we would do.

Is this enough? Maybe. Yet not unless there is an inner world that helps make all this possible.[51]

---

50. The principles are from David W. Orr, *Ecological Literacy: Education and the Transition to a Post-modern World* (Albany: State University of New York Press, 1992), 29–30.

51. In an already lengthy volume, we cannot pursue the details of public policy congruent with the argument and examples of this chapter. Much of that is available, however, in the chapter "Agenda for Change," in Korten's *When Corporations Rule the World*, 307–24.

The Moral Frame

A world within to match the world without is a requirement of sustainability itself. Sustainability will not happen, nor itself be sustained, if no living inner world roughly accords with the outer world we seek. Moral, spiritual, and cultural dimensions are as crucial as technical ones. Indeed, the moral formation of society and world is a critical *action* of its own — or, rather, a million small ones undertaken by everyone.

Broadly speaking, what is untenable for sustainability is a moral universe that circles human creatures only and does not regard other creatures and earth as a whole as imposing moral claims we need worry over. A homocentric moral universe perpetuates apartheid thinking and moral and spiritual autism. Imperviousness and unresponsiveness then reign, an incapacity to be touched in ways we truly feel. The integrity of ecosystem communities and their many citizens isn't recognized.

Mary Wollstonecraft's words of two hundred years ago, from *A Vindication of the Rights of Women*, pertain here: "Those who are able to see pain, unmoved, will soon learn to inflict it."[1] Mary Pellauer, using Wollstonecraft, says that something like this happens to our morality and spirit when our emotional nerve system becomes callused. The point is that like race, gender, and class in many quarters, the standing of nature in modernity is overgrown with moral callousness. Nature's suffering and pain leave us unmoved. So we soon inflict pain, with little understanding; or seeing it, are unmoved. Socialization renders nature a matter of utilitarian interest only or, on our better days, of aesthetic and recreational interest. Then the life of otherkind falters as a realm of binding moral obligation. If nature beyond us is not scenic, edible, or otherwise useful, it does not stir us.

Even worse, there is something like a Western rage against finitude, mortality, and nature itself. At least in its modern guises, much of the West underwrites Loren Eiseley's comment, echoing Pascal, that that which we would destroy we first declare natural.[2]

Like any callused area, we then become morally thick-skinned, hardened, inured, and unsusceptible. Calluses, Pellauer explains, "neither receive nor send messages to the central nervous system."[3] In this case, the moral and emotional

---

1. Mary Wollstonecraft, *A Vindication of the Rights of Women* (New York: Norton, 1967 [1792]), 256.

2. Loren Eiseley, *The Firmament of Time* (New York: Atheneum, 1971), 180.

3. Mary D. Pellauer, "Moral Callousness and Moral Sensitivity: Violence against Women," in *Women's Consciousness, Women's Conscience*, ed. Barbara Hilkert Andolson, Christine E. Gudorf, and Mary D. Pellauer (San Francisco: Harper and Row, 1985), 49.

nervous system, when confronted with nonhuman nature, neither receives nor sends empathic messages. And without those, compassion is stillborn.

Sustainable community requires — and offers — a moral system with more sensitive skin. It does so by according the full sweep of nature inherent moral value. All creatures great and small, and inorganic matter as well, have worth that rests proximately in their membership in the Community of Life. "All that participates in being"[4] has standing. For the religiously inclined, this value ultimately resides in creation as the expression of divine creativity and goodness. Compassion and justice and the mending of community thus envelop more than human members. Suffering and pain are felt and acknowledged, even when they are unavoidable and we can do little about them. They are matters of an aching heart, and while no one wants an aching heart, it is preferable to an empty one.

Within this expanding moral universe, the broadest moral guideline is one suggested by James Gustafson. Human beings, given their power and place in earth's present reality and their nature as self-conscious moral creatures, may inevitably be the measur*e*rs of all things. But the measure itself is that we "relate to all things in a manner appropriate to their relations to God."[5] The "good" all things are is more than their good for us, and our own interests are relative to larger wholes than those of immediate human welfare. Human interests are thus relativized in the interest of the more inclusive life communities of which we are part and upon which we utterly depend. Human beings thereby share with other participants in the Community of Life the need to make those sacrifices required for the welfare and sustainability of this community as a whole. This requires, to recall the previous point, a moral and emotional nervous system that opens out beyond a strict anthropocentric circumference. It requires rubbing away the moral callousness of prominent traditions and ways of living, together with deeply inscribed habits, particularly in the ranks of the powerful and privileged.

For those whose moral universe does not finally rest in the presence and power of the divine, the orientation and socialization suggested here still argue for inherent value and the Community of Life as the human reference point. "The integrity of creation" and biotic rights, or Aldo Leopold's categorical imperative ("A thing is right when it tends to preserve the integrity, stability, and beauty of the biotic community. It is wrong when it tends otherwise"),[6] offer the framework for contemplated actions and policies. So do "justice" and "neighbor" as we have described them,[7] even when shorn of explicitly religious ties.

Not all human communities have lost the emotional sense of a comprehensive moral universe. Many human beings have lived by nonanthropocentric ethical

---

4. H. Richard Niebuhr et al., *The Purpose of the Church and Its Ministry: Reflections on the Aims of Theological Education* (New York: Harper and Row, 1956), 38.

5. James M. Gustafson, *Ethics from a Theocentric Perspective* (Chicago: University of Chicago Press, 1981), 1:113.

6. Aldo Leopold, *A Sand County Almanac* (New York: Ballantine, 1970), 262.

7. See the chapter "ReBeginnings" above.

systems. Some still do. Systems far less desacralizing of nature than modernity's have guided past human communities. But the issue now is how we "reenchant the world"[8] and conduct human life in a manner appropriate to larger wholes and at the same time address earth's distress, billions of human beings' distress included. That is a daunting challenge on a new scale for earth ethics. (Incidentally, Leopold himself argued that all ethics evolve from the single premise that individual persons are members "of a community of interdependent parts" and that his purpose in "The Land Ethic" was no more than to enlarge the boundaries of the community "to include soils, waters, plants, and animals, or collectively: the land" for those whose lives had taken such an odd turn as to forget these.)[9]

The practical reason we need a postmodern ethic of total value has been discussed often enough in this volume. It is the unsustainability of the present course. The addition here is that the discovery of nature as an inclusive community and the realization that we have come to a certain "overlook point"[10] in our evolution also require it. We are all part of a larger organism, our cultures, and our cultures are all part of an even larger organism, the biosphere. Thus "what we do for the earth, we do for ourselves."[11] This requires a holistic pragmatism that marries economics, physics, biology, and ethics, the last-mentioned as the moral frame for the entirety. Such a pragmatism is in contrast to the industrial-information paradigm in which the market's invisible hand works and consequences are neglected until the market has in effect become a fist. This is a paradigm that, in Wann's description, is an "ethically passive, quantity-driven, biologically insensitive way of conducting business."[12] Cultural metabolism here is aligned with evolving technological capabilities, and ethics and biology are little more than an afterthought. Holistic pragmatism and its moral frame of total value are, instead, "actively in search of quality, biologically aware, and socially informed."[13]

This comprehensive communitarian ethic, itself a *decisive earth action* on behalf of sustainability, is not nature romanticism or the simple imitation of nature. It even goes beyond the earlier plea to listen to nature and mimic natural designs. It does so because much of nature is simply *too casual about suffering*. For human morality this is unacceptable, even after acknowledging mortality, un-

---

8. The phrase per se is ours, not Max Weber's. But the reference is to his discussion of processes that "disenchanted" the world. In the early twentieth century, Weber argued that rational effectiveness and efficiency in the organization of economic life in a market-oriented system desacralized nature and removed any sense of magic and mystery about the cosmos of everyday life. See his writings on the ethos of capitalism, especially *The Protestant Ethic and the Spirit of Capitalism* (New York: Charles Scribner's Sons, 1958 [1904]); and *The Sociology of Religion* (Boston: Beacon Press, 1963 [1922]).

9. Leopold, *Sand County Almanac*, 239.

10. The phrase is from David Wann, *Deep Design: Pathways to a Livable Future* (Washington, D.C.: Island Press, 1996), 21.

11. Ibid., 22.

12. Ibid., 172.

13. Ibid.

avoidable suffering, and "death's limited rights."[14] A moral framework inclusive of nature as a subject of high moral standing is not, then, simply a transfer of the rest of nature's behavior to human conduct. In important, even unavoidable, ways, we remain the measurers.

Predation, for example, is an essential part of nature's cycles. It is not a pattern of morality we praise and advocate, however, at least not on our better days. To be sure, "red of tooth and claw" has been justified as conduct redounding to society's good, notably in some versions of fascism, Social Darwinism,[15] and certain phases of revolutions. But it is a course that, like nature in some forms, devours its own children. It is rejected here as a moral paradigm for sustainable community.

In short, while we necessarily draw lines, make distinctions, and intervene in ways that, *like* nature, take life for the sake of life, *unlike* the rest of nature, we also draw lines, make distinctions, and intervene so as to prevent and relieve suffering in ways the rest of nature may not. Goodness is ascribed prima facie to "all that participates in being." But we attempt, for example, to curb, even destroy, viruses and bacteria that bring certain disease, suffering, and premature death. Inherent value is real. But it does not translate as equal value. That said, one of the most difficult and ongoing tasks of earth ethics is to decide where what lines are drawn and where what distinctions are made as humans intervene in the larger panorama of a nature too casual about pain, suffering, and death. Determining what the relevant wholes are and what life requires in the way of a reciprocity of interests, with the concomitant sacrifices, is among the most daunting of moral endeavors. No simple disclosure tells us precisely how we "relate to all things in a manner appropriate to their relations to God."

Differently said: while nature is always a teacher, it is not always a teacher of morality. Human beings, as part of nature, cannot escape their distinctive work as moral creatures. And the stakes are so high because morality is as crucial to sustainability as knowing nature's basic designs. The play of human power on human-to-human terms as well as human-and-the-rest-of-nature terms, the configuration of society in nature as well as nature in society, is determinative of sustainability and unsustainability. And so very much of this turns on the moral character of the social order, together with the treatment of the rest of nature. In the end, the quest is to find the ways of the most comprehensive nonviolence possible within frameworks of shared power. For this a new account of responsibility, as well as different habits and institutions, is needed. It runs well beyond modernity's own and includes both life and not-life.

The proper disposition for this is nicely captured in the centuries-old admonition of the rabbis that we should keep a piece of paper in one pocket reminding

---

14. Dietrich Bonhoeffer, *Ethics,* ed. Eberhard Bethge, 6th ed. (New York: Macmillan, 1965), 79. See the discussion in the chapter "Song of Songs," p. 301, above.

15. Social Darwinism's earth faith conveniently forgot that the "survival of the fittest" principle was supplemented by another that said that creatures who learned to cooperate and adapt stood a better chance of fitting in and surviving.

us that we are dust and ashes, and in the other a slip that says the world is also here for us and its fate in our hands.[16]

In the end, we have come round again to the two issues raised at the outset: sustainability turns on the play of power, both among human beings within society and between human beings cumulatively and the rest of nature; and the key to both is the configuration of culture in nature and as nature. This only underlines yet again the ascendancy of ethics and in fact renders sustainability a virtual synonym for a comprehensive justice. In any event, it is important to highlight the required moral frame, as this chapter has tried to do, and see the striving for its realization as another vital element of what is to be done. The direction overall is a downward distribution of economic and social power and a heightened status for all forms of life, human and other.[17] The direction is also the dissolution of modernity's severe discontinuities of humans from the rest of nature, mind from body, and self from world.

---

16. Michael Lerner, *Jewish Renewal: A Path to Healing and Transformation* (New York: Gross/Putnam, 1994), 107–8.

17. Such a direction entails, as we have argued earlier, race, gender, class, and cultural analysis that gets beyond the deceptive simplisms of "too many people chasing too few fish." See p. 164 above.

By Grace through Faith

Earth community and ethics are humanity's next journey in a world that has become booty and landfill. Compared with the many we are, far too few will have a place worthy of earth creatures. The globalization affecting all things, not least earth's life sources themselves, is not, in fact, truly global.[1] The economic and cultural drive pushing the forward stampede is concentrated in the industrialized world and its extensions in the "developing" world. It is globalization that includes the vast majority of people as measured by impact but excludes most of them as measured by benefit. The ranks "of the window-shoppers and the jobless are growing faster than the ranks of the global army of the employed."[2] This "surplus of gifted, skilled, undervalued, and unwanted human beings is the Achilles heel"[3] of the new world order.[4]

The same globalization continues to silt up harbors from which we should have long since sailed. That some are less silted than a decade ago is hardly reason to continue the dynamics that steadily silt the planet overall. The reality is not only savage inequalities and other social injustice but soil loss exceeding soil formation, species extinction exceeding species evolution, freshwater loss and degradation exceeding freshwater availability and purity, forest and fish loss exceeding forest and fish regeneration and reproduction, carbon emissions exceeding carbon fixation, and population and consumption levels pushing skyward at historically unprecedented rates. Rather than being somewhere *on* the way to sustainability, so many of us are simply *in* the way. The Good Ship Oikoumenē is mired.

At the same time this globalization proceeds, the structures of collective accountability weaken. The sum of separate corporate decisions determines far more of what happens to people's lives and the environment than any public policies arrived at through democratic means and carried out by representative governments. Which is to say that neither businesses nor governments have taken up well the responsibility of fashioning genuine global consciousness and the public morality indispensable to global community. The efforts of the United Nations end up as highly commendable "programmes of action" miserably underpledged — when pledged at all — by those who decide the direction and uses of global wealth and power.[5]

---

1. Richard J. Barnet and John Cavanagh, *Global Dreams: Imperial Corporations and the New World Order* (New York: Simon and Schuster, 1994), 427.
2. Ibid.
3. Ibid., 425.
4. See Jeremy Rifkin, *The End of Work: The Decline of the Global Labor Force and the Dawn of the Post-market Era* (New York: Putnam, 1995).
5. David Hallman reports statistics offered by Morris Miller, a former World Bank executive.

So there is no polite way to say it: the forces of stampeding globalization without global community and responsibility are destroying a beloved world. Theirs is a brown globalism tinged with green at rich world edges. It is not rooted in earth and does not listen to it. It is virile, comprehensive, and alluring as a materialist way of life, but it is tone-deaf to earth's requirements for comprehensive sustainability. Its notion of wealth and value does not include cultural wealth or biological value. It is nature-blind and turns resources into profit without bothering to track the origins of goods and services *in* nature or the consequences of their use *for* nature. Whatever the motives, this amounts to profit by pillage, not least because market mechanisms don't distinguish renewable from nonrenewable sources or worry whether renewables are being used at nonrenewable rates.[6] Such habits tally as earth's distress, ours included. And unfortunately, what distress affluent materialist forces do not manage to inflict is inflicted by the misery of those who remain, precious children of God whose lives and livelihoods are omitted from the positive side of the essential power equations.

There is another way. Millions yearn to walk it and strain to do so. The choice, in Swimme and Berry's somewhat awkward terms, is between "the Technozoic era" of a commercial-industrial-information order with "an even more ordered control of things," and "the Ecozoic era" of mutually enhancing earth-human relationships framed and instructed by creation's integrity.[7] The choice, to recall the Canadian National Round Table on the Environment and the Economy, is between an extension of the cumulative human revolutions in agriculture, industry, and information, all reconfiguring nature for the sake of society (some social strata far more than others), and a fourth revolution that understands nature's economy as basic to the social reconfiguration needed for the sake of both human- and otherkind.

In these pages that choice turns on community and on earth faiths and ethics appropriate to it. All citizens, bar none, are invited and urged to love earth fiercely and vow fidelity to it, to display sacramental sensibilities and covenantal commitments.

As of the early 1990s the turnover in global financial markets was about five hundred billion dollars daily. This capital movement dwarfs the flow of goods and services themselves: for every one dollar of trade in goods and services, twenty-five dollars is traded in financial assets. Moreover, Miller reports, this volume of capital turnover is double what it was as recently as the mid-1980s, and much of it is speculative monies unrelated to the productive aspects of trade and investment. The point above is that most of this happens at the hands of private banks and corporations standing outside the efforts of the United Nations and other institutions seeking global community and fostering responsibility for it (see David Hallman, "Ethics and Sustainable Development," in *Ecotheology: Voices from South and North*, ed. David Hallman [Geneva: WCC Publications; and Maryknoll, N.Y.: Orbis Books, 1994], 273–74).

6. David Wann reports that as recently as 1955, almost three-fourths of the everyday materials U.S. Americans used were from renewable sources, but that by the 1990s that figure was reversed — roughly three-fourths of common materials are made from nonrenewable resources (David Wann, *Deep Design: Pathways to a Livable Future* [Washington, D.C.: Island Press, 1996], 58).

7. Brian Swimme and Thomas Berry, *The Universe Story: From the Primordial Flaring Forth to the Ecozoic Era* (San Francisco: HarperCollins, 1992), 249–50.

Hope is real. There are good grounds for it. Something is clearly afoot, though it is "mostly off-camera."[8] Local citizens' movements and alternative institutions are emerging and are trying to create greater economic self-sufficiency, internalize costs to earth in the price of goods, sustain livelihoods, work out agricultures appropriate to regions, preserve traditions and cultures, revive religious life, maintain human dignity, repair the moral fiber, resist the commodification of all things, be technologically innovative with renewable and nonrenewable resources, revise urban designs and architecture, preserve biological species and protect ecosystems, and cultivate a sense of earth as a sacred good held in trust and in common.

This inspiring mélange has no one configuration. By definition it probably ought not to, given the pluralism of place and peoples and their insistent and varied participation. But it is real and it shares community-focused alternatives to the destructive homogenization of globalization and its monocultures. Richard J. Barnet and John Cavanagh, who judge this disparate movement as presently "the only force we see that can break the global gridlock,"[9] close their study with a judgment about the stakes: "The great question of our age is whether people, acting with the spirit, energy, and urgency our collective crisis requires, can develop a democratic global consciousness rooted in authentic local communities."[10]

Such authentic communities are not *only* local, we have argued. But community from the local to the cosmic *is* the focus and frame for meeting the great adventure hard upon us. That adventure is sustainability as the capacity of earth's communities to survive and thrive together indefinitely.[11]

The issue in the end is not whether rooted and realistic changes *can* be made, radical changes among them. They can. We have listed examples that are multiplying by the day. The issue, rather, is whether we understand the place of faith and the ascendancy of ethics. Do we realize that the key to sustainability of inclusive nature rests in the character of the social order that is ours? Has it hit us full force that sustainability is less a matter of our wits than our collective moralities and cosmologies, that it turns on "justice, peace, and the integrity of creation"?

Beyond this, the issue is sustaining power itself for a long and traumatic journey toward such an order. The issue is the regular renewal of moral-spiritual and sociopsychological energy in a long season of forced society-nature experimentation. The issue is bread for the journey, that is, faith.

While encouraging signs are always helpful, the presence of renewable moral-spiritual energy bears no strict corollary to empirical reality. Life has risen from

---

8. Barnet and Cavanagh, *Global Dreams*, 430.

9. Ibid.

10. Ibid.

11. For an account of what has happened to community in the modern world and what constructive responses need to be made, especially to provide basic moral formation of human character and conscience to address the fraying of social fiber, see Larry L. Rasmussen, *Moral Fragments and Moral Community* (Minneapolis: Fortress, 1993).

the dead in the most degraded places, just as life has often died where all the surface signs bespoke well-being. Something other than steady income and a pension plan generates hope and provokes creative revolt.

It had better. The operational challenges in getting from present headstrong, institutionalized unsustainability to sustainability are daunting. Not the least is that, facing earth's mounting distress, we'd sooner blame the universe on God and the neighbors than change it and ourselves.

Faith is the name of the strong power behind the renewal of moral-spiritual energy. It squarely faces the fact there will never be decisive proof beforehand that life will triumph. Yet it still acts with confidence that the stronger powers in the universe arch in the direction of sustaining life, as they also insist upon justice. World-weariness is combated by a surprising force found *amidst* earth and its distress. Creation carries its own hidden powers. It supports the confidence of the gospel that a steadfast order exists that bends in the direction of life and gives it meaning.

Said differently, the religious consciousness — and dream — that generates hope and a zest and energy *for* life is tapped *in* life itself. The finite bears the infinite, and the transcendent as "the beyond in the midst of life"[12] is as close as the neighbor, soil, air, and sunshine. Life defending itself against the death of life is in fact one of the works of nature itself, in us. God, like the devil and life itself, is in the details. *Spiritus ubique diffussus* — the Spirit permeates all things (to remember the ancient Orthodox teaching). The turn to earth is also a turn to those sources that enable what has not yet come to pass to do so.

Or has it already come to pass, time and again? Dorothee Sölle says she believes in resurrection because it has already happened.[13] So it has, and does. That, not the endless list of woes that tally as earth's distress, is the reason for the epigraph: "The heart, after all, is raised on a mess of stories. Then it writes its own."

We have been raised on a mess of stories. Many no longer lead where we must go. But, with some of those stories and others not yet told, we'll write our own and take our place as tillers, keepers, songsters, and priests of this precious, minuscule patch of creation. As we do so we will learn that faith is the great confronter, uncovering in us a capacity to fight for life in the face of death and venture the risks necessary to be part of a radically changed world. We will also learn, paradoxically, that faith is the great *un*knowing, the experience of an active mystery that surpasses all our words for it and leaves us with either silence or song — a mystery to be experienced without pretension and with an acceptance, excluding despair, of tragic limitations in the human condition. The great unknowing is also the experience of a finitude that crushes any remaining homocentrism and exposes all our efforts to know *the* truth as self-deceptive. Our understandings are provisional and tentative, small children of small times

---

12. Dietrich Bonhoeffer, *Letters and Papers from Prison* (New York: Macmillan, 1972), 282.
13. Dorothee Sölle, "Faith, Theology and Liberation," *Christianity and Crisis* 36, no. 10 (June 7, 1976): 141.

and places we are privileged to share. They are also in glorious degree part of the community of creation that includes and transcends us. Yet precisely here it is not we alone but — to put it in necessarily limited human terms — the *entire* community of creation that "knows" and remembers and learns and evolves and creates, from the cell to the galaxy and beyond. Faith has ears for hearing such stories and warns against mistaking human maps for the territory itself. Mistaking maps for territory is the culprit that turns limited human knowledge into imperial rule and destructive efforts to get a controlling grip on reality. Better to join the solidarity of the shattered and have ears to hear the stories of a compassionate Presence in league with those humble enough to know they are not less than, or more than, frail, beautiful children of the universe.

Such faith stories, such stories of the heart, will be genuine earth stories. Like Alan Paton's opening of *Cry, the Beloved Country,* they will tell of "a lovely road that runs from Ixopo into the hills...lovely beyond any singing of it."[14] Such stories will speak of earth itself as still "lovely beyond any singing of it." They will also reveal Hammerskjold's truth that "God does not die on the day we cease to believe in a personal deity, but we die on the day when our lives cease to be illuminated by the steady radiance, renewed daily, of a wonder, the source of which is beyond all reason."[15] These stories will understand Walker's "Dear God, dear everything" letter and answer Hawken's plea to learn "how to say grace" in the face of the moral question "put sweetly before us." They will also understand with Luther that "faith is not the human notion and dream that some people call faith [but] a divine work in us which changes us and makes us to be born anew of God"; it is "a living, busy, active, mighty thing," "a living, daring confidence in God's grace." Such confidence makes people "glad and bold and happy in dealing with God and with all creatures."[16]

Granted, such confidence is hard-won. It is the grace of Maclean's fly-fishing — "all good things come by grace and grace comes by art and art does not come easily."[17] And it requires eyes to see, just as Baby Suggs told the slaves in the "Clearing": "That the only grace they could have was the grace they could imagine. That if they could not see it, they could not have it."[18]

To close. "The time has come," the Walrus says in Tweedledee's poem for Alice, "to talk of many things: of shoes — and ships — and sealing wax — of cabbages and kings, and why the sea is boiling hot — and whether pigs have wings."[19] We have talked much of things, of "ships" (Progress, Noah's Ark of Life, the Good Ship Oikoumenē), of "shoes" and "sealing wax" (industry, commerce, and the Big Economy), of "cabbages and kings" (the biological expansion

---

14. Alan Paton, *Cry, the Beloved Country* (New York: Scribner's, 1948), 3.

15. Cited in a review of Oscar Hijuelos's *Mr. Ives' Christmas* by Michiko Kakutani, "A Test of Faith for a Father Who Longs for Grace," *New York Times,* November 28, 1995, C17.

16. Martin Luther, *Luther's Works* (Philadelphia: Muhlenberg, 1960), 35:370–71.

17. See the chapter "Creation's Integrity," above, citing Norman Maclean, *A River Runs through It and Other Stories* (Chicago: University of Chicago Press, 1976), 2.

18. See the chapter "Song of Songs," pp. 311–312, above, citing Toni Morrison, *Beloved* (New York: Knopf, 1987), 88.

19. Lewis Carroll, *Through the Looking Glass* (Baltimore: Penguin Books, 1966 [1872]), 240.

of Europe and globalization), of "why the sea is boiling hot" (evolution and earth's slow womb) and "whether pigs have wings" (earth faith and the heart's eccentric stories). The time has come to stop talking, and listen. Just listen. Listen to the unexpected stories of earth, to new stories of the heart, to tales of adventurous faith itself. And listen to God as the great relation of all the relations of the universe. Then, by grace, through faith, we will understand Angelou's poem from the inside out. When we do, it will be "the pulse" of "a new day." And we will have the grace to look up and out

> And into [our] sister's eyes,
> And into [our] brother's face,
> [Our] country, [the planet],
> And say simply
> Very simply
> With hope —
> Good morning.

Index of Persons and Subjects

Scripture Index

HEBREW SCRIPTURES

CHRISTIAN SCRIPTURES

Also in the Ecology and Justice Series

John B. Cobb, Jr., *Sustainability: Economics, Ecology, and Justice*

Charles Pinches and Jay B. McDaniel, editors, *Good News for Animals?*

Frederick Ferré, *Hellfire and Lightning Rods*

Ruben L. F. Habito, *Healing Breath: Zen Spirituality for a Wounded Earth*

Eleanor Rae, *Women, the Earth, the Divine*

Mary Evelyn Tucker and John A. Grim, editors, *Worldviews and Ecology: Religion, Philosophy, and the Environment*

Leonardo Boff, *Ecology and Liberation: A New Paradigm*

Jay B. McDaniel, *With Roots and Wings: Spirituality in an Age of Ecology and Dialogue*

Sean McDonagh, *Passion for the Earth*

Denis Edwards, *Jesus the Wisdom of God: An Ecological Theology*

Rosemary Radford Ruether, editor, *Women Healing Earth: Third-World Women on Ecology, Feminism, and Religion*

Dieter T. Hessel, editor, *Theology for Earth Community: A Field Guide*

Brian Swimme, *The Hidden Heart of the Cosmos: Humanity and the New Story*